PROPERTIES OF MATTER

For B. Sc. Students

PROPERTIES OF MATTER

For B.Sc. Students

R. Murugeshan
M.Sc., M.Phil.
Head of the Department of Physics
Vivekananda College
Tiruvedakam West (Post)
Madurai District
Tamil Nadu

S. CHAND
PUBLISHING

S Chand And Company Limited
(ISO 9001 Certified Company)

S Chand And Company Limited

(ISO 9001 Certified Company)

Head Office: D-92, Sector–2, Noida – 201301, U.P. (India), Ph. 91-120-4682700

Registered Office: A-27, 2nd Floor, Mohan Co-operative Industrial Estate, New Delhi – 110 044, Phone: 011-49731800

www.**schandpublishing.com;** e-mail: **info@schandpublishing.com**

S. CHAND PUBLISHING

Marketing Offices:

Chennai	:	Ph: 23632120; chennai@schandpublishing.com
Guwahati	:	Ph: 2738811, 2735640; guwahati@schandpublishing.com
Hyderabad	:	Ph: 40186018; hyderabad@schandpublishing.com
Jalandhar	:	Ph: 4645630; jalandhar@schandpublishing.com
Kolkata	:	Ph: 23357458, 23353914; kolkata@schandpublishing.com
Lucknow	:	Ph: 4003633; lucknow@schandpublishing.com
Mumbai	:	Ph: 25000297; mumbai@schandpublishing.com
Patna	:	Ph: 4011400; patna@schandpublishing.com

First Edition 1994
Subsequent Edition and Reprints 1995, 98, 99, 2003, 2004, 2005, 2008, 2010
Revised Edition 2012
Reprints 2013, 2014 (Twice), 2016, 2017, 2018, 2019, 2020, 2021, 2022, 2023, 2024 (Twice), 2025

Reprint 2026

ISBN: 978-81-219-0605-0 **Product Code:** H6PTM68PHYS10ENAE12O

PRINTED IN INDIA

By Shri Shyamlal Printing Press Private Limited, Plot 20/4, Site-IV, Industrial Area Sahibabad, Ghaziabad – 201 010 and Published by S Chand And Company Limited, A-27, 2nd Floor, Mohan Co-operative Industrial Estate, New Delhi – 110 044.

Preface to the Fifth Edition

I have great pleasure in presenting the fifth revised and enlarged edition of the book **'Properties of Matter'** to the students. Some important new topics have been included in this edition.

AUTHOR

Preface to the First Edition

This book has been written for the students of B.Sc. Physics of various Indian universities. The book covers widely the syllabi prescribed by Madras, Bharathidhasan, Manonmaniam Sundaranar, Kerala and Mahatma Gandhi Universities.

SI system of units has been used throughout the text. Great care has been taken in dealing with the subject with modern outlook. A large number of questions and problems have been given at the end of each chapter. The students who attempt to tackle them successfully are bound to get a better insight and understanding of the subject.

All available standard books on the subject have been freely consulted for the preparation of this book. I acknowledge gratefully my indebtedness to the authors and the publishers of different books on the subject. I also acknowledge the kind help rendered by my friends and students in preparing this book.

I am extremely thankful to Srimad Nithyanandaji Maharaj, President, Srimad Guhanandaji Maharaj, Secretary, Srimad Sadanandaji Maharaj, Srimad Sasvatanandaji Maharaj and Dr. K. Subrahmanyam, Principal of our college, for their encouragement and help.

Lastly, I would like to thank Management Team & Editorial Staff of S. Chand & Company Ltd. for taking keen interest in bring out the book in the nice form at a short period of time.

R. MURUGESHAN

Lecturer in Physics,
Vivekananda College,
Tiruvedakam West
Madurai District

UNIVERSITY OF MADRAS
B.Sc. DEGREE COURSE IN PHYSICS
SYLLABUS
SEMESTER – I; CORE PAPER – I
P. 101- MECHANICS AND PROPERTIES OF MATTER

Unit 1 : Impulse and Impact

Impulse – impact – Laws of impact – direct impact and oblique impact between two smooth spheres – loss of kinetic energy – motion of two interacting bodies – reduced mass.

Rigid body dynamics

Compound pendulum – theory – equivalent simple pendulum – reversibility of centres of oscillation and suspension – determination of g and k – center of mass – velocity and acceleration of centre of mass – determination of motion of individual particle – system of variable mass.

Unit 2 : Centre of Gravity and Centre of Pressure

Centre of gravity of solid and hollow tetrahedron, solid and hollow hemisphere – Centre of pressure – vertical rectangular lamina – vertical triangular lamina.

Hydrodynamics

Equation of continuity of flow – Venturimeter – Euler's equation of unidirectional flow – Torricelli's theorem – Bernoulli's theorem and its applications.

Unit 3 : Elasticity

Hooke's Law – Stress – Stress – Strain – Elastic constants – Expressions for Poisson's ratio interms of elastic constants – workdone is stretching and twisting a wire – twisting couple on a cylinder – rigidity modulus by static torsion – torsional pendulum – rigidity modulus and moment of inertia.

Unit 4 : Bending of Beams

Cantilever – expression for bending moment – expression for depression – cantilever oscillations – Expression for time period – Experiment to find young's modulus – Non uniform bending – Experiment of determine Young's modulus by Koenig's method – uniform bending – expression for elevation – experiment to determine Young's modulus using microscope.

Unit 5 : Fluid Dynamics

Surface tension-Definition – Excess of pressure over curved surface – Application to spherical and cylindrical drops and bubbles – variation of surface tension with temperature – Jaegar's method.

Vicosity-Definition – Coefficient of viscosity – Rate of flow of liquid in a capillary tube – Poiseuille's formula – variation of viscosity of a liquid with temperature – Application.

SEMESTER – II; CORE PAPER – II

P.102 – THERMAL PHYSICS AND ACOUSTICS

Unit 4 : Waves and oscillations

Simple harmonic motion - combination of two SHMs in a straight line – at right angles – Lissajous's figures – free, damped, forced oscillations and resonance – intensity and loudness of sound – intensity level – decibel – noise pollution.

Unit 5 : Ultrasonics

Ultrasonics – production - piezo electric crystal method – magnetostriction method – applications

Acoustic of buildings – reverberation – Absorption coefficient – Sabine's formula – Acoustics aspects of halls and auditoriums.

BHARATHIAR UNIVERSITY, COIMBATORE

CORE PAPER II - MECHANICS, PROPERTIES OF MATTER AND SOUND

UNIT I (9 hrs)

Conservation Law – Impulse – Direct and oblique impact – Final velocity and loss of kinetic energy – Motion of a particle in a vertical circle – friction – Laws of friction – angle of friction – resultant reaction – cone of friction – Equilibrium of a body on a rough inclined plane to the horizontal and when the inclination is greater than the angle of friction.

UNIT II (9 hrs)

Motion of rigid body

Moment of inertia – Parallel and perpendicular axes theorem – M.I. of rectangular Lamina and triangular lamina – M.I of a solid sphere about an axis through is C.G. – Compound pendulum – torque and angular momentum – Relation – Kinetic rotation – Conservation of angular momentum.

UNIT III (9 hrs)

Gravitation: Kepler's Law of planetary motion – Laws of gravitation – Boy's method for G – Gravitational potential – Gravitational field at a point due to spherical shell – Variation of 'g' with latitude, altitude and depth.

Elasticity: Elastic modulus – Poisson's ratio – relation between them – Expression for bending moment – determination of Young's modulus by uniform and non-uniform bending – I section girders – Static Torsion – Expression for couple per unit twist – Torsional oscillation.

UNIT IV (9 hrs)

Surface Tension: Definition and dimension of surface Tension – Excess of Pressure over a curved surface – Variation of S.T. with temperature – Jaeger's Experiment.

Viscosity: Definition – Rotation viscometer- viscosity of gases, Meyer's modification of Poiseuille's formula – Rankine's method for viscosity of a gas.

UNIT V (9 hrs)

Sound: Simple Harmonic vibration – Progressive waves – properties – Composition of two S.H.M and beats – stationary waves – Properties – Melde's Experiment for the frequency of electrically maintained tuning fork – Transverse and longitudinal modes – Ultrasonics – Properties and application.

PERIYAR UNIVERSITY, SALEM

Semester – I: Mechanics

UNIT : I

Projectile: Range up and down an inclined plane, maximum range - two directions of projections for a given velocity and range.

Impulse-Impact: Laws of impact – coefficient of restitution – impact of a smooth sphere on a fixed smooth plane – Direct impact between two smooth spheres – Losses in kinetic energy in direct impact Oblique impact between two smooth spheres.

UNIT: II

SHM: Composition of two SHM's of same period along a straight line and at the right angles to each other – Lissajous figures.

Dynamics of Rigid Bodies: Compound pendulum – theory – condition for minimum period interchangeability of center of suspension and center of oscillation – g using compound pendulum - Bifilar pendulum – parallel and non-parallel threads.

UNIT: III

Center of gravity: Center of gravity of a solid cone, Solid hemisphere, hollow hemisphere and a tetrahedron.

Friction: Laws of friction – angle of friction – resultant reaction and cone of fiction – equilibrium of a body on an inclined plane under the action of a force.

UNIT: IV

Center of pressure: Definition – center of pressure of a rectangular lamina and triangular lamina.

Hydrodynamics: Equation of continuity of flow – Bernoulli's theorem – venturimeter – Pitot tube.

UNIT: V

Classical Mechanics: Mechanics of system of particles-conservation theorem for angular momentum and energy-constraints – classification – generalized coordinates – transformation equations – principle of virtual work-D'Alembert's principle – derivation of Lagrangian formulation of conservation theorems – generalized momentum, energy and linear momentum.

Semester – III, Properties of Matter and Sound

UNIT I : ELASTICITY

Three types of elastic moduli - Poisson's ratio - Bending of beams - Expression for bending moment - Depression of the loaded end of a Cantilever - uniform - non uniform bending - theory - experiment, pin and microscope method - work done in uniform bending - Koenig method - non uniform bending - theory - expression for couple per unit twist - determination of rigidity modulus - Static torsion method with scale and telescope - Torsional oscillation of a body, Rigidity modulus by torsion pendulum - with mass.

UNIT II : VISCOSITY

Coefficient of critical velocity - Poiseuilles formula coefficient viscosity - Correction for the formula - determination of coefficient of viscosity by capillary flow method - comparison of viscosities Oswald's viscometer - viscosity of a highly viscous liquid - Stoke's method for the coefficient of a highly viscous liquid - variation of viscosity with temperature and pressure - viscosity of gases – Mayer's formula for the rate of flow of a gas through a capillary tube - Rankine's method for the determination of viscosity of a gas.

UNIT III : SURFACE TENSION AND OSMOSIS

Surface energy - angle of contact and its determination - excess of pressure inside curved surface - formation of drops - Experimental study of variation of Surface tension with temperature - drop weight method of determining surface tension and interfacial surface tension - angle of contact of mercury - Quinck's method - surface tension and vapour pressure - osmosis - experimental determination of osmosis pressure - Laws of osmosis pressure - osmosis and vapour pressure of a solution.

UNIT IV : SOUND

Theory of free, damped and forced vibration - resonance - Sharpness of resonance - Fourier's theorem - application for Saw-tooth wave and square wave - Sonometer - determination of a.c. frequency using sonometer - determination of frequency using Melde's apparatus.

UNIT V: ULTRASONICS AND ACOUSTICS

Ultrasonics - Production - Piezo electric method - magnetostriction method - detection - properties - application.

Acoustics of buildings - reverberation time - derivation of Sabine's formula - determination of absorption Coefficient.

Contents

Chapters	Page No.
1. Elasticity	1–33
2. Viscosity	34–52
3. Surface Tension	53–72
4. Hydrodynamics	73–83
5. Osmosis	84–86
6. Gravitation	87–104
7. Moment of Inertia	105–113
8. Impact of Elastic Bodies	114–122
9. Projectile on Inclined Plane, Motion of Two Interacting Bodies and Bifilar Pendulum	123–129
10. Mechanics of a Rigid Body	130–137
11. Waves and Oscillations	138–168
12. Composition of Simple Harmonic Motions	169–173
13. Beats	174–176
14. Intensity of Sound	177–180
15. Progressive Waves	181–182
16. Stationary Waves	183–185
17. Determination of Frequency	186–189
18. Motion in a Vertical Circle	190–191
19. Systems of Particles	192–197
20. Centre of Gravity	198–200
21. Centre of Pressure	201–204
22. Friction	205–207
23. Classical Mechanics	208–220

1 Elasticity

CHAPTER

1.1. INTRODUCTION

A body can be deformed (*i.e.*, changed in shape or size) by the suitable application of external forces on it. A body is said to be perfectly elastic, if it regains its original shape or size, when the applied forces are removed. This property of a body to regain its original state or condition on removal of the applied forces is called *elasticity*. A body which does not tend to regain its original shape or size, even when the applied forces are removed, is called a perfectly plastic body. No body, in nature, is either perfectly elastic or perfectly plastic. Quartz fibre is the nearest approach to a perfectly elastic body.

Stress : When an external force is applied on a body, there will be relative displacement of the particles and due to the property of elasticity, the particles tend to regain their original positions. *Stress is defined as the restoring force per unit area.* If a force F is applied normally to the area of cross-section A of a wire, then stress = F/A. Its dimensions are $ML^{-1} T^{-2}$.

Thermal Stresses : Suppose the ends of a rod are rigidly fixed, so as to prevent expansion or contraction. If the temperature of the rod is changed, tensile or compressive stresses, called thermal stresses, will be set up in the rod. If these stresses are very large, the rod may be stressed even beyond its breaking strength. The stress is *tensile* when there is an increase in length. The stress is *compressive* when there is a decrease in length. A tangential stress tries to slide each layer of the body over the layer immediately below it.

Strain : When a deforming force is applied, there is a change in length, shape or volume of the body. The ratio of the change in any dimension to its original value is called strain. It is of three types :-

(1) The ratio of change in length (l) to original length (L) is called *longitudinal strain* (l/L).

(2) Let $ABCD$ be a body with the side CD fixed (Fig. 1.1). Suppose a tangential force F is applied on the upper face AB. The shape of the body is changed to $A'B'CD$. The body is sheared by an angle ϕ. This angle ϕ measured in radians is called the *shearing strain* (ϕ).

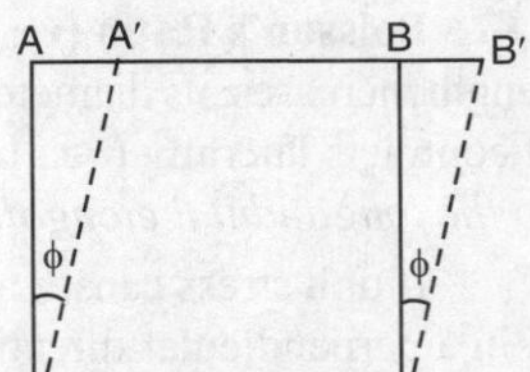

Fig. 1.1

(3) *Volume strain* (*Bulk strain*) : The ratio of change in volume (v) to original volume (V) is called *volume strain* (v/V).

Hooke's Law : *Within elastic limit, the stress is directly proportional to strain.* Stress $\propto$ strain or stress/strain = E . E is a constant called modulus of elasticity.

The dimensional formula of modulus of elasticity is $ML^{-1} T^{-2}$. Its units are Nm^{-2}.

1.2. DIFFERENT MODULI OF ELASTICITY

(1) **Young's modulus (E) :** *It is defined as the ratio of longitudinal stress to longitudinal strain within elastic limits.* Let a wire of length L and area of cross-section A undergo an increase in length l when a stretching force F is applied in the direction of its length.

Then, longitudinal stress = F/A and longitudinal strain = l/L.

$$\therefore \qquad E = \frac{F/A}{l/L} = \frac{FL}{Al}.$$

(2) **Rigidity modulus (G) :** *It is defined as the ratio of tangential stress to shearing strain.*

Consider a solid cube *ABCDEFGH* (Fig. 1.2). The lower face *CDGH* is fixed and a tangential force *F* is applied over the upper face *ABEF*. The result is that each horizontal layer of the cube is displaced, the displacement being proportional to its distance from the fixed plane. Point *A* is shifted to *A*′, *B* to *B*′, *E* to *E*′ and *F* to *F*′ through an angle ϕ, where $AA' = EE' = l$.

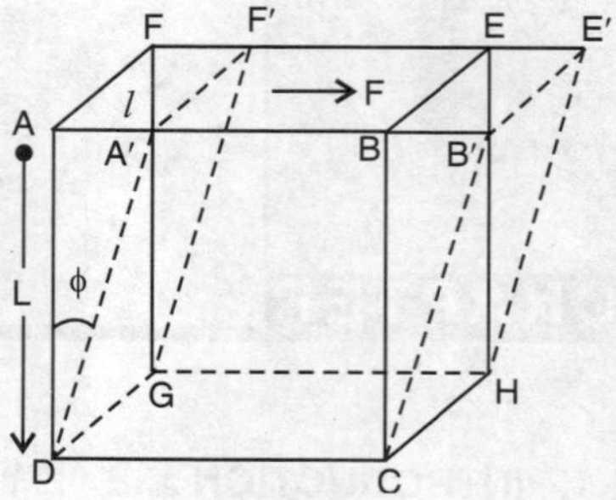

Fig. 1.2

Clearly $\phi = l/L$ where l is the relative displacement of the upper face of the cube with respect to the lower fixed face, distant *L* from it.

This angle ϕ through which a line originally perpendicular to the fixed face is turned, is a measure of the shearing strain.

$$\text{Now, Rigidity modulus } (G) = \frac{\text{Tangential stress}}{\text{Shearing strain}} = \frac{F/A}{\phi}$$

Here, $A = L^2$ = Area of face *ABEF*.

$\therefore \qquad G = T/\phi$ where T = Tangential stress.

(3) **Bulk Modulus (K):** *It is defined as the ratio of volume stress (Bulk Stress) to the volume strain.*

When three equal stresses (F/A) act on a body in mutually perpendicular directions, such that there is a change of volume v in its original volume *V*, we have, Stress = pressure $P = F/A$. Volume strain $= -v/V$. The negative sign indicates that if pressure increases, volume decreases.

$$\therefore \qquad K = \frac{\text{Bulk strees}}{\text{Volume strain}} = \frac{F/A}{-v/V} = \frac{P}{-v/V}$$

Poisson's Ratio (ν): When a wire is stretched, it becomes longer but thinner, *i.e.*, although its length increases, its diameter decreases. When a wire elongates freely in the direction of a tensile stress, it contracts laterally (*i.e.*, in a direction perpendicular to the force). *The ratio of lateral contraction to the longitudinal elongation is called Poisson's ratio*. It is denoted by the letter ν.

If unit stress causes an extensional strain λ in its own direction and lateral contractional strain μ in a perpendicular direction, $\nu = \mu/\lambda$.

1.3. RELATION BETWEEN ANGLE OF SHEAR AND LINEAR STRAIN

Consider a cube *ABCD* having each side equal to *L* with its face *DC* fixed (Fig. 1.3). Let a shearing force acting along the face *AB* deform the cube into the rhomboid *A'B'CD*. The angle through which the face *AD* or *BC* has been turned is evidently the shearing strain. The diagonal *DB* has elongated to *DB*′ while the other diagonal *AC* has shortened to *A′C*. Draw *BK* perpendicular to *DB*′. Since ϕ is very small, $\Delta BKB'$ may be assumed to be a right angled isosceles triangle and $\angle BB'K = 45°$. Further, $DB \approx DK$.

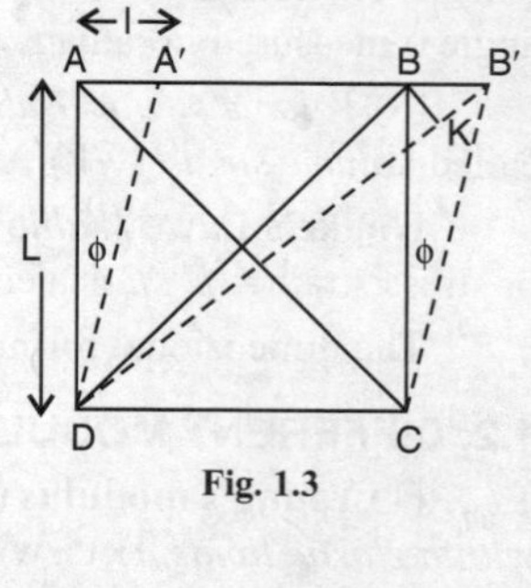

Fig. 1.3

$$\left.\begin{array}{l}\text{Tensile strain}\\ \text{along diagonal } DB\end{array}\right\} = e = \frac{DB' - DB}{DB} = \frac{DB' - DK}{DB} = \frac{KB'}{DB}$$

But in $\Delta BKB'$, $KB'/BB' = \cos 45°$

or $KB' = BB' \cos 45° = BB'/\sqrt{2}$

We have, $DB = \sqrt{L^2 + L^2} = L\sqrt{2}$

$$\therefore \quad e = \frac{KB'}{DB} = \frac{BB'/\sqrt{2}}{L\sqrt{2}} = \frac{BB'}{2L} = \frac{\phi}{2} \quad [\because BB'/L = \phi]$$

Similarly, compressive strain along diagonal $AC = \phi/2$. Hence, a shearing strain ϕ is equivalent to a tensile strain and a compressive strain at right angles to each other, each of value $\phi/2$.

Similarly, it can be proved that a shearing stress is equivalent to a linear tensile stress and an equal compressive stress at right angles to each other.

1.4. RELATION BETWEEN VOLUME STRAIN AND LINEAR STRAIN

Consider a unit cube. Suppose it is subjected to three equal stresses, all tending to expand the cube, in three mutually perpendicular directions. Each side becomes $(1 + e)$ where e is the linear strain. Hence the new volume of the cube is $(1 + e)^3 \approx 1 + 3e$.

Increase in volume = $3e$.

$$\therefore \quad \text{Volume strain} = \frac{\text{increase in volume}}{\text{original volume}} = \frac{3e}{1} = 3e$$

1.5. WORK DONE IN A STRAIN

When a body is strained, work has to be done to deform the body. This work done is stored up in the body as potential energy. It can be shown that the work done per unit volume in any kind of strain (linear, shear or bulk) is equal to $[\frac{1}{2} \times (\text{stress}) \times (\text{strain})]$.

***(a)* Linear Strain:** Let a force F act on a wire of length L and area of cross-section A such that the increase in length is l.

Young's modulus $= E = \dfrac{FL}{Al}$ or $F = \dfrac{EAl}{L}$.

Work done in producing a stretching $dl = F \,.\, dl = \dfrac{EAl}{L} dl$

$$\therefore \quad \left.\begin{array}{l}\text{Total work done to produce a}\\ \text{stretching of the wire from 0 to } l\end{array}\right\} = W = \int_0^l F.\,dl$$

$$= \int_0^l \frac{EAl}{L} dl = \frac{EA}{L}\left[\frac{l^2}{2}\right]_0^l = \frac{1}{2}\frac{EAl^2}{L} = \frac{1}{2}.\frac{EAl}{L} l = \frac{1}{2}.F.l$$

$$= \frac{1}{2} \times \text{Stretching force} \times \text{Elongation produced.}$$

Now, volume of the wire = $A.L$.

Hence, work done per unit volume of the wire

$$= \frac{\frac{1}{2}F.l}{AL} = \frac{1}{2}\frac{F}{A}\cdot\frac{l}{L} = \frac{1}{2}\text{Stress} \times \text{Strain}.$$

(b) Shearing Strain: Let a tangential force F, acting in the direction AB of cube $ABCD$ shear it through an angle ϕ. In Fig. 1.3, $AD = L$ and $AA' = l$.

Stress $= F/L^2$ and shearing strain $= \phi = l/L$

$$\therefore \qquad G = \frac{F/L^2}{l/L} = \frac{F}{L.l} \text{ or } F = GLl.$$

Work done to displace the layer AB by $dl = F\,dl = GLl\,dl$.

$$\therefore \qquad \left.\begin{array}{l}\text{Total work done during the}\\ \text{whole displacement from 0 to } l\end{array}\right\} = \int_0^l F\,dl = \int_0^l GLl\,dl$$

$$= \frac{1}{2} GL \cdot l^2 = \frac{1}{2} GLl \cdot l = \frac{1}{2} F.l$$

$$= \frac{1}{2} \text{ Tangential force} \times \text{displacement.}$$

Volume of the cube $= L^3$.

$$\text{Hence, work done per unit volume} = \frac{1}{2}\frac{Fl}{L^3} = \frac{1}{2}\frac{F}{L^2}\frac{l}{L}$$

$$= \frac{1}{2}(F/A).\,\phi$$

$$= \frac{1}{2} \text{ Stress} \times \text{Strain.}$$

(c) Volume Strain: Let a stress or pressure P be applied uniformly all over a body of volume V such that its volume decreases by v.

Stress $= P$ and strain $= v/V$.

$$\text{Hence} \qquad K = \frac{P}{v/V} \text{ or } P = \frac{Kv}{V}$$

Work done to produce a small decrease in volume $dv = P\,dv$

$$\therefore \qquad \left.\begin{array}{l}\text{Total work done for the whole}\\ \text{decrease in volume from 0 to } v\end{array}\right\} = \int_0^v P.\,dv$$

$$= \int_0^v \frac{Kv}{V}\,dv = \frac{1}{2}\frac{Kv^2}{V} = \frac{1}{2}\cdot\frac{Kv}{V}\,v = \frac{1}{2}P.v$$

$$= \frac{1}{2} \text{ stress} \times \text{change in volume.}$$

$$\text{Therefore, work done per unit volume} = \frac{1}{2}\frac{Pv}{V} = \frac{1}{2}P.\frac{v}{V}$$

$$= \frac{1}{2} \text{ Stress} \times \text{Strain}$$

Example 1: *Find the energy stored in a wire 5 metres long and 10^{-3} metre in diameter when it is stretched through 3×10^{-3} metre by a load. Young's modulus of material is 2×10^{11} Nm^{-2}.*

The energy stored in the wire is equal to the work done in stretching it.

$$W = \frac{1}{2} \text{ Stretching force} \times \text{Elongation produced.}$$

Since $E = \dfrac{F/A}{l/L}$, we have, $F = \dfrac{E.A.l}{L}$

$\therefore$ Work done $= \dfrac{1}{2}\dfrac{EAl}{L}\,l = \dfrac{1}{2}\dfrac{EAl^2}{L}$

Here $E = 2 \times 10^{11}\ \text{Nm}^{-2};\ r = 0.5 \times 10^{-3}$ m

$$A = \pi r^2 = \pi\,(0.5 \times 10^{-3})^2 = \pi \times 0.25 \times 10^{-6}\ \text{m}^2$$

$$l = 3 \times 10^{-3}\ \text{m};\ L = 5\ \text{m}$$

$$\therefore \quad W = \frac{1}{2}\frac{2\times10^{11}\ (\pi\times 0.25\times10^{-6})\,(3\times10^{-3})^2}{5} = 0.1414\ \text{J}.$$

Example 2: *Calculate the elastic energy stored up in a wire originally 5 metres long and 10^{-3} m in diameter which has been stretched by 3×10^{-4} m due to a load of 10 kg.*

$$W = \frac{1}{2}\ \text{Stretching force} \times \text{Elongation produced} = \frac{1}{2}F \times l$$

Stretching force $= F = 10 \times 9.8$ N $= 98$ N

Elongation produced $= l = 3 \times 10^{-4}$ m

$$\therefore \quad W = \frac{1}{2} \times 98 \times 3 \times 10^{-4} = 1.47 \times 10^{-2}\ \text{J}.$$

Example 3: *A steel wire 2×10^{-3} m in diameter is just stretched between two points at a temperature of 20°C. Determine its tension when the temperature falls to 10°C. Linear expansivity of steel = $1.1 \times 10^{-5}\ K^{-1}$. Young's modulus for steel = $2.1 \times 10^{11}\ Nm^{-2}$.*

The total contraction of the wire when the temperature falls by 10° C $\Big\}$ $= L\,\alpha\,t$

i.e., $l = L \times 1.1 \times 10^{-5} \times 10 = L \times 1.1 \times 10^{-4}$

$$\therefore \quad F = \frac{EAl}{L} \qquad \left(\because\ E = \frac{F/A}{l/L}\right)$$

$$= \frac{2.1\times10^{11}\times\pi\,(10^{-3})^2\ L\times1.1\times10^{-4}}{L} = 72.56\ \text{N}.$$

1.6. BEHAVIOUR OF A WIRE UNDER PROGRESSIVE TENSION

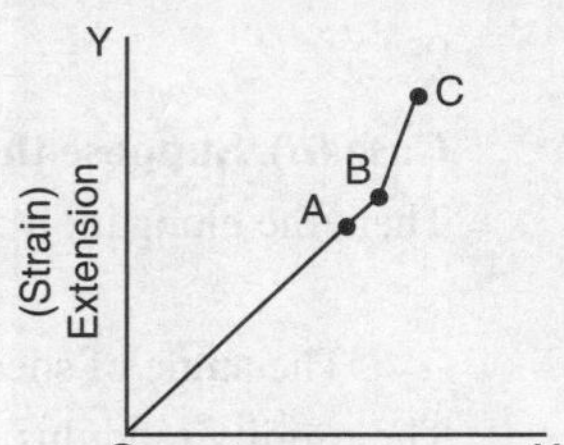

Fig. 1.4

If we subject a wire to gradually increasing load and plot a graph between load and extension, we obtain a curve of the form shown in Fig. 1.4. It is called the *stress–strain* diagram. In the part *OA*, which is straight, the extension is proportional to the load and the wire obeys Hooke's law. In this range, the wire regains its original length when unloaded and so it is called the range of *perfect elasticity*. At *A*, the wire reaches the elastic limit.

If the wire is loaded beyond *OA*, the extension is no longer proportional to the load and Hooke's law is not obeyed. If the load is now removed, the wire will not regain its original length, but a permanent elongation will be produced in it. At the point *B*, even the addition of a very small load causes enormous elongation. This point is called the *yield point*. After the yield point, the extension increases very rapidly and depends on the time for which the load acts. The extension of the wire goes on increasing and the area of cross-section decreases until the breaking point is reached and finally the wire will break. The load at which the wire breaks is known as the *breaking weight*. The maximum load to which the wire can be subjected, divided by its original cross-sectional area, is called the *breaking stress*.

Elastic after-effect. It is found that even within elastic limit, practically all substances take some time to reach their original length after the deforming force is removed. This delay in recovering back the original condition on removal of deforming forces is called *elastic after-effect*. Quartz, phosphor bronze, silver and gold have very little elastic after-effect and hence they are used as suspension wires in Boys' experiment, quadrant electrometer, moving coil galvanometer, etc.

Elastic Fatigue. If a body is continuously subjected to stress and strain, it gets fatigued. Consider two torsional pendulums A and B having similar wires. A is set into vibration for a fairly long time continuously, while B is at rest. Now if A and B are set into vibration, with the same amplitude to start with, A comes to rest earlier than B. This is due to *elastic fatigue* of the suspension wire. Elastic fatigue can be removed by giving sufficient rest to the wire.

1.7. RELATION BETWEEN THE ELASTIC MODULI

Suppose three stresses P, Q and R are acting perpendicular to the three faces $ABCD$, $ADHE$ and $ABFE$ of a unit cube of an isotropic material (Fig. 1.5). Each one of these stresses will produce an extension in its own direction and a compression along the other two perpendicular directions. If λ is the extension per unit stress, the elongation along the direction of P will be λP. If μ is the contraction per unit length per unit stress, then the contraction along the direction of P due to the other two stresses will be μQ and μR.

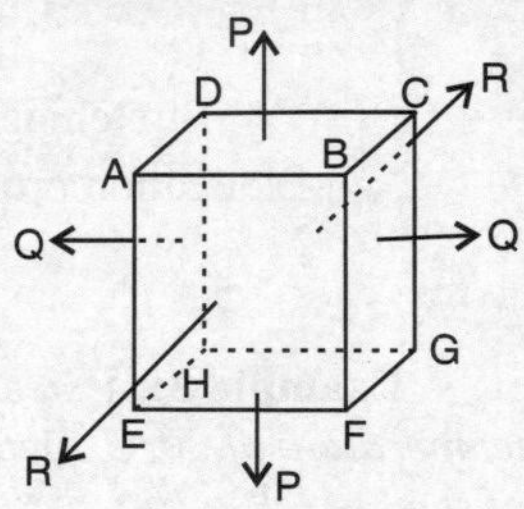

Fig. 1.5

Let all the three stresses act simultaneously on the cube.

Net elongation along the direction of $P = e = \lambda P - \mu Q - \mu R$.

Net elongation along the direction of $Q = f = \lambda Q - \mu P - \mu R$;

Net elongation along the direction of $R = g = \lambda R - \mu P - \mu Q$.

We can express the three elastic constants E, G and K in terms of λ and μ.

Case (*i*). Suppose only the stress P acts and $Q = R = 0$.

We have then the case of a simple longitudinal stress.

The linear strain $= e = \lambda . P$

$$\therefore \quad \text{The Young's modulus} = E = \frac{\text{Stress}}{\text{Linear strain}} = \frac{P}{\lambda . P} = \frac{1}{\lambda}$$

or
$$\lambda = \frac{1}{E} \qquad \text{...(1)}$$

Case (*ii*). Suppose the stress $R = 0$ and $P = -Q$.

Then, the elongation along the direction of P is

$$e = \lambda . P - \mu(-P) = (\lambda + \mu) P$$

The angle of shear $\phi = 2e = 2(\lambda + \mu) P$

The rigidity modulus G is given by

$$G = \frac{\text{Stress}}{\text{Angle of shear}} = \frac{P}{\phi} = \frac{P}{2(\lambda + \mu) P} = \frac{1}{2(\lambda + \mu)}$$

or
$$2(\lambda + \mu) = \frac{1}{G} \qquad \text{...(2)}$$

Case (*iii*). Let $P = Q = R$. Since the body is now subjected to uniform stress in all directions, the increase in volume is

$$3e = 3(\lambda - 2\mu) P \qquad [\because e = \lambda P - \mu P - \mu P = (\lambda - 2\mu) P]$$

∴ The bulk strain = 3(λ – 2μ) *P*.

$$\text{The bulk modulus} = K = \frac{\text{Stress}}{\text{Bulk strain}} = \frac{P}{3(\lambda - 2\mu)P} = \frac{1}{3(\lambda - 2\mu)}$$

or $$(\lambda - 2\mu) = \frac{1}{3K} \qquad ...(3)$$

1. Relation between E, G and K

We have, from (2), $2(\lambda + \mu) = \dfrac{1}{G}$

or $$2\lambda + 2\mu = \frac{1}{G} \qquad ...(4)$$

From (3), $$\lambda - 2\mu = \frac{1}{3K} \qquad ...(5)$$

Adding (4) and (5), $$3\lambda = \frac{1}{G} + \frac{1}{3K} = \frac{3K + G}{3GK}$$

or $$\lambda = \frac{3K + G}{9\,GK} \qquad ...(6)$$

From (1), $$\lambda = 1/E$$

∴ $$E = \frac{9\,GK}{3K + G} \qquad ...(7)$$

2. Relation between G, K and *v*

By definition, Poisson's ratio v is given by $v = \mu/\lambda$.

From (2), $$\lambda + \mu = \frac{1}{2G}$$

From (3), $$\lambda - 2\mu = \frac{1}{3K}$$

Subtracting (3) from (2),

$$3\mu = \frac{1}{2G} - \frac{1}{3K}$$

∴ $$\mu = \frac{3K - 2G}{18\,GK} \qquad ...(8)$$

Hence $$v = \frac{\mu}{\lambda} = \frac{(3K - 2G)}{18\,GK} \times \frac{9\,GK}{(3K + G)} \qquad \text{(Using Eq. 6)}$$

or $$v = \frac{(3K - 2G)}{6K + 2G} \qquad ...(9)$$

3. Relation between E, G and *v*

We have, from (1), $\lambda = 1/E$

and from (2), $\lambda + \mu = 1/(2G)$

Dividing (2) by (1), $\dfrac{\lambda + \mu}{\lambda} = \dfrac{E}{2G}$

or $$1 + \frac{\mu}{\lambda} = \frac{E}{2G} \text{ or } 1 + v = \frac{E}{2G}$$

or $$v = \frac{E}{2G} - 1 \qquad ...(10)$$

4. Limits to the value of *v*.

We have from (9), $$v = \frac{3K - 2G}{6K + 2G}$$

or $$3K(1 - 2v) = 2G\,(1 + v) \qquad ...(11)$$

Now both K and G must be positive quantities. Hence, if v is positive, $(1 + v)$ is positive and R.H.S. is positive. Hence L.H.S. must be positive or $(1 - 2v) > 0$ or $v < \frac{1}{2}$. If v is negative, $(1 - 2v)$ will be positive and hence $(1 + v)$ must be positive, or $v > -1$. Hence the theoretical limits to the value of v are $\frac{1}{2}$ and -1. In actual practice, v is always a positive quantity and lies between 0 and 0.5.

Example 4: *Calculate G and v for silver, given E and K for silver = 7.25 × 10*^{10} *Nm*$^{-2}$ *and 11 × 10*^{10} *Nm*$^{-2}$.

Here, $E = 7.25 \times 10^{10}\ \text{Nm}^{-2}$; $K = 11 \times 10^{10}\ \text{Nm}^{-2}$; $G = ?$

Now, $$E = \frac{9\,GK}{3K + G}$$

$$\therefore \quad G = \frac{3\,KE}{9\,K - E} = \frac{3(11 \times 10^{10})\,(7.25 \times 10^{10})}{9\,(11 \times 10^{10}) - (7.25 \times 10^{10})}$$

$$= 2.607 \times 10^{10}\ \text{Nm}^{-2}.$$

Here, $E = 7.25 \times 10^{10}\ \text{Nm}^{-2}$; $G = 2.607 \times 10^{10}\ \text{Nm}^{-2}$; $v = ?$

$$v = \frac{E}{2G} - 1 = \frac{7.25 \times 10^{10}}{2 \times (2.607 \times 10^{10})} - 1 = 1.391 - 1 = 0.391.$$

1.8. DETERMINATION OF POISSON'S RATIO v FOR RUBBER

A rubber tube R (like a cycle tyre) about 1 m in length and 2 cm in diameter is taken. It is suspended in a vertical position as shown in Fig. 1.6. The ends A and B are tightly closed with rubber corks. A graduated glass capillary tube (G) open at both ends is inserted inside the rubber tube through the upper end A. The rubber tube is completely filled with water till water rises in the tube G. A pointer P is fixed to the lower end B of the rubber tube. The lower end B carries a pan with weights (W).

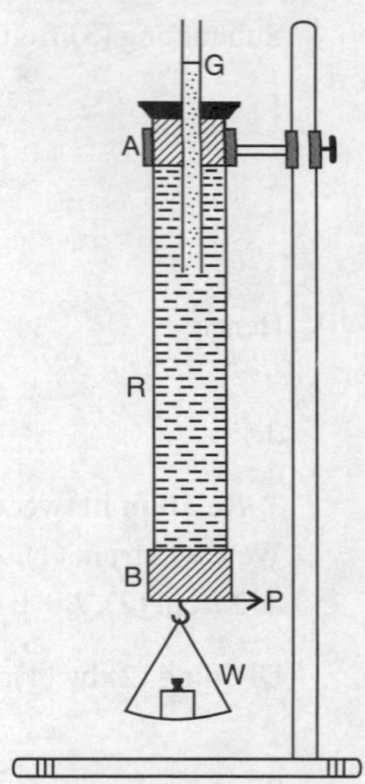

Fig. 1.6

When conditions become steady, the positions of the water meniscus in G and the pointer P are noted with the help of two separate travelling microscopes. When a suitable weight (W) is placed in the scale pan, the length of rubber tube increases and its area of cross-section decreases. The internal volume of the tube also increases. Consequently, the water level in G falls. The increase in the length of the tube is determined by noting the position of the pointer P in the microscope. The increase in volume is determined by noting the water level in the tube G. v can be calculated by using the formula proved below.

Relation. Let L, D, A and V be the respective initial values of length, diameter, area of cross-section and internal volume of the rubber tube.

Then, $$V = A \times L.$$

When the load is applied, let dV be the increase in volume, dL the increase in length and dA the decrease in area of cross-section.

New volume

$$= V + dV = (A - dA)(L + dL)$$

$$= AL - L\,dA + A\,dL \qquad \text{(Neglecting the term } dA \times dL\text{)}$$

$$\therefore \qquad dV = A\,dL - L\,dA \qquad ...(1)$$

Also we know that $\quad A = \dfrac{\pi D^2}{4} \quad$ or $\quad dA = \dfrac{\pi D\,dD}{2}$

$$\text{or} \qquad dA = \left(\frac{2A}{D}\right) dD \qquad ...(2)$$

Substituting this value of dA in Eq. (1)

$$dV = A\,dL - \left(\frac{2AL}{D}\right) dD \quad \text{or} \quad \frac{1}{A}\frac{dV}{dL} = 1 - \frac{2L}{D}\frac{dD}{dL}$$

$$\text{or} \qquad \frac{1}{A}\frac{dV}{dL} = 1 - 2\,\frac{dD/D}{dL/L}$$

$$\text{But,} \qquad \frac{dD/D}{dL/L} = \frac{\text{Lateral strain}}{\text{Longitudinal strain}} = \nu$$

$$\therefore \qquad \frac{1}{A}\frac{dV}{dL} = 1 - 2\nu$$

$$\text{or} \qquad \nu = \frac{1}{2}\left[1 - \frac{1}{A}\frac{dV}{dL}\right] \qquad ...(3)$$

TORSION

1.9. TORSION OF A BODY

When a body is fixed at one end and twisted about its axis by means of a torque at the other end, the body is said to be under torsion. Torsion involves shearing strain and so the modulus involved is the rigidity modulus.

Torsion of a cylinder–Expression for torque per unit Twist

Consider a cylindrical wire of length L and radius a fixed at its upper end and twisted through an angle θ by applying a torque at the lower end. Consider the cylinder to consist of an infinite number of hollow co-axial cylinders. Consider one such cylinder of radius x and thickness dx [Fig. 1.7(*i*)].

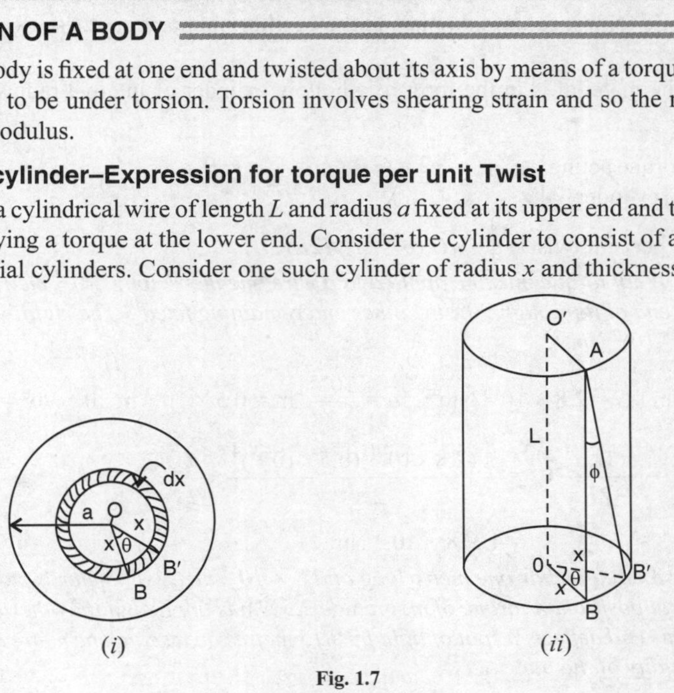

Fig. 1.7

A line such as AB initially parallel to the axis OO' of the cylinder is displaced to the position AB' through an angle ϕ due to the twisting torque [Fig. 1.7(*ii*)]. The result of twisting the cylinder is a shear strain. The angle of shear = $\angle BAB' = \phi$.

Now $\quad BB' = x.\theta = L\phi$ or $\phi = x.\theta/L$

We have, rigidity modulus = $G = \dfrac{\text{Shearing stress}}{\text{Angle of shear } (\phi)}$

$\therefore$ shearing stress $= G \,.\, \phi = Gx\,\theta/L$

But, shearing stress $= \dfrac{\text{Shearing force}}{\text{Area on which the force acts}}$

$\therefore$ Shearing force = Shearing stress × Area on which the force acts.

The area over which the shearing force acts = $2\pi\, x\, dx$

Hence, the shearing force $= F = \dfrac{G\,x\,\theta}{L} \times 2\pi x\, dx$

$$\left.\begin{array}{l}\text{The moment of this}\\ \text{force about the axis}\\ OO' \text{ of the cylinder}\end{array}\right\} = \frac{G\,x\,\theta}{L} 2\pi\, x\, dx . x = \frac{2\pi G\theta}{L} x^3\, dx$$

$$\therefore \left.\begin{array}{l}\text{Twisting torque on}\\ \text{the whole cylinder}\end{array}\right\} = C = \int_0^a \frac{2\,\pi\,G\,\theta}{L} x^3\, dx$$

or

$$C = \frac{\pi\,G\,a^4\,\theta}{2L}$$

$$\left.\begin{array}{l}\text{The torque per unit twist (}i.e.,\\ \text{the torque when } \theta = 1 \text{ radian)}\end{array}\right\} = c = \frac{\pi\,G\,a^4}{2L}$$

Note 1: When an external torque is applied on the cylinder to twist it, at once an internal torque, due to elastic forces, comes into play. In the equilibrium position, these two torques will be equal and opposite.

Note 2: If the material is in the form of a hollow cylinder of internal radius a and external radius b, then,

$$\left.\begin{array}{l}\text{The torque acting}\\ \text{on the cylinder}\end{array}\right\} = C = \int_a^b \frac{2\,\pi\,G\,\theta}{L} x^3\, dx = \frac{\pi\,G\,\theta}{2L}(b^4 - a^4)$$

Torque per unit twist = $c = \pi G\,(b^4 - a^4)/(2L)$

Example 5: *What torque must be applied to a wire one metre long, 10^{-3} metre in diameter in order to twist one end of it through 90°, the other end remaining fixed ? The rigidity of the material of the wire is 2.8×10^{10} Nm^{-2}.*

Here, $L = 1$ m ; $G = 2.8 \times 10^{10}$ Nm^{-2} ; $a = \dfrac{10^{-3}}{2}$ m $= 0.5 \times 10^{-3}$ m ; $\theta = 90° = \pi/2$ radians ;

$$\therefore \quad C = \frac{\pi\,G\,a^4}{2L}\theta = \frac{\pi\,(2.8\times10^{10})(0.5\times10^{-3})^4}{2\times1}\times\frac{\pi}{2}$$

$$= 4.318 \times 10^{-3} \text{ Nm}$$

Example 6: *A circular bar one metre long and 8×10^{-3} metre diameter is rigidly clamped at one end in a vertical position. A torque of magnitude 2.5 Nm is applied at the other end. As a result, a mirror fixed at this end deflects a spot of light by 0.15 metres on a scale one metre away. Calculate the modulus of rigidity of the bar.*

For a twist θ, the mirror turns through θ, and the reflected beam through 2θ. If the deflection is d on a scale D away, $2\theta D = d$ or $\theta = \dfrac{d}{2D} = \dfrac{0.15}{2 \times 1} = 0.075$ radians;

Here, $C = 2.5$ Nm ; $a = 4 \times 10^{-3}$ m ; $\theta = 0.075$ radians; $L = 1$ m ; $G = ?$

Hence, $C = \pi Ga^4\theta/2L$ or $G = C\,.\,2L/\pi a^4\theta$

i.e.,
$$G = \frac{2.5 \times 2 \times 1}{\pi (4 \times 10^{-3})^4 (0.075)} = 8.290 \times 10^{10} \text{ Nm}^{-2}$$

Example 7: *A steel wire of diameter* 3.6×10^{-4} *m and length 4 m extends by* 1.8×10^{-3} *m under a load of 1 kg and twists by 1.2 radians when subjected to a total torsional torque of* 4×10^{-5} *Nm at one end. Find the values of E, G and v for steel.*

We have,
$$E = \frac{F/A}{l/L} = \frac{FL}{A \cdot l}$$

Here, $F = mg = 1 \times 9.8 = 9.8$ N : $L = 4$ m ;

$A = \pi a^2 = \pi(1.8 \times 10^{-4})^2$ m² and $l = 1.8 \times 10^{-3}$ m.

∴
$$E = \frac{9.8 \times 4}{\pi(1.8 \times 10^{-4})^2 \times 1.8 \times 10^{-3}} = 2.139 \times 10^{11} \text{ Nm}^{-2}$$

$$\left.\begin{array}{l}\text{Torque which must be applied}\\ \text{to twist one end of the wire}\\ \text{through an angle } \theta \text{ radians}\end{array}\right\} = C = \frac{\pi Ga^4 \theta}{2L}$$

or $G = C\,2L/\pi a^4\theta$

Here, $C = 4 \times 10^{-5}$ Nm ; $L = 4$ m ; $a = 1.8 \times 10^{-4}$ m ;

$\theta = 1.2$ radians

∴
$$G = \frac{C\,2\,L}{\pi\, a^4 \theta} = \frac{(4 \times 10^{-5}) \times 2 \times 4}{\pi\,(1.8 \times 10^{-4})^4 \times 1.2} = 0.8083 \times 10^{11} \text{ Nm}^{-2}$$

$$v = \frac{E}{2G} - 1 = \frac{2.139 \times 10^{11}}{2 \times 0.8083 \times 10^{11}} - 1 = 1.323 - 1 = 0.323.$$

Example 8: *Explain why a hollow rod is a better shaft than a solid one of the same mass, length and material.*

Consider a solid cylinder of length L, radius r and shear modulus G.

$$\left.\begin{array}{l}\text{The torque required to twist the}\\ \text{solid cylinder through an angle } \theta\end{array}\right\} = C_1 = \frac{\pi\, Gr^4\theta}{2\,L} \qquad ...(1)$$

Let r_1 and r_2 be the inner and outer radii of the hollow cylinder of the same length, mass and material.

Then the torque required to twist it through the same angle θ is

$$C_2 = \frac{\pi\, G(r_2^{\,4} - r_1^{\,4})\theta}{2L} \qquad ...(2)$$

Hence,
$$\frac{C_2}{C_1} = \frac{r_2^{\,4} - r_1^{\,4}}{r^4} = \frac{(r_2^{\,2} + r_1^{\,2})\,(r_2^{\,2} - r_1^{\,2})}{r^4}$$

Since the two cylinders have the same mass,

$$\pi\,(r_2^{\,2} - r_1^{\,2})\, l\rho = \pi\, r^2 l.\rho$$

(where ρ is the density of the material of the cylinders).

or $\qquad r_2^2 - r_1^2 = r^2$ ...(3)

Adding $2r_1^2$ to both sides, $\quad r_2^2 + r_1^2 = r^2 + 2\,r_1^2$ ...(4)

Hence,
$$\frac{C_2}{C_1} = \frac{(r^2 + 2r_1^2)\,r^2}{r^4} = \frac{r^2 + 2r_1^2}{r^2}$$

or
$$\frac{C_2}{C_1} = 1 + \frac{2\,r_1^2}{r^2}$$

$\therefore$ $C_2 > C_1$ *i.e.*, the twisting torque for a hollow cylinder is greater than that for a solid cylinder of the same mass, length and material. Hence a hollow cylinder is stronger and a better shaft than a solid one of the same mass, length and material.

1.10. DETERMINATION OF RIGIDITY MODULUS—STATIC TORSION METHOD

Searle's apparatus : The experimental rod is rigidly fixed at one end *A* and fitted into the axle of a wheel *W* at the other end *B* (Fig. 1.8). The wheel is provided with a grooved edge over which passes a tape. The tape carries a weight hanger at its free end. The rod can be twisted by adding weights to the hanger. The angle of twist can be measured by means of two pointers fixed at *Q* and *R* which move over circular scales S_1 and S_2. The scales are marked in degrees with centre zero.

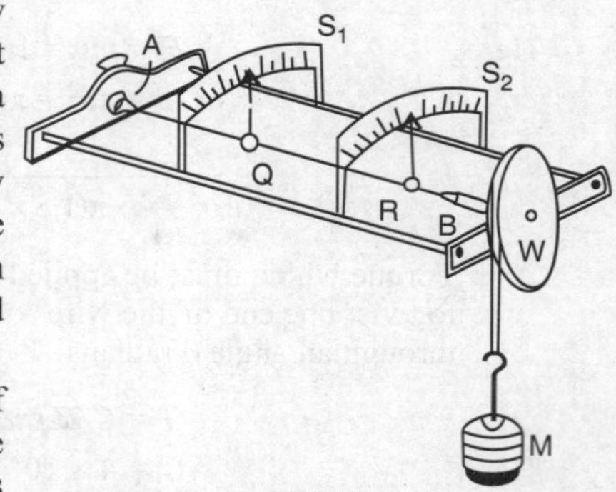

Fig. 1.8

With no weights on the hanger, the initial readings of the pointers on the scales are adjusted to be zero. Loads are added in steps of *m* kg (conveniently 0.2 kg). The readings on the two scales are noted for every load, both while loading and unloading. The experiment is repeated after reversing the twisting torque by winding the tape over the wheel in the opposite way. The observations are tabulated.

The readings in the last column give the twist for a load of *M* kg for the length *QR* (= *L*) of the rod. The radius *a* of the rod and the radius *R* of the wheel are measured.

If a load of *M* kg is suspended from the free end of the tape, the twisting torque = *MgR*.

The angle of twist = θ degrees = θ. π/180 radians.

$$\therefore \quad \text{The restoring torque} = \frac{\pi G a^4}{2L} \cdot \frac{\theta \pi}{180}.$$

For equilibrium,
$$MgR = \frac{\pi G a^4}{2L} \cdot \frac{\theta . \pi}{180} \quad \text{or} \quad G = \frac{360\, M\, g\, R\, L}{\pi^2 a^4 \theta}$$

Since *a* occurs in the fourth power in the relation used, it should be measured very accurately.

Notes: (1) We eliminate the error due to the eccentricity of the wheel by applying the torque in both clockwise and anticlockwise directions.

(2) We eliminate errors due to any slipping at the clamped end by observing readings at two points on the rod.

1.11. DETERMINATION OF RIGIDITY MODULUS—STATIC TORSION METHOD. (SEARLE'S APPARATUS—SCALE AND TELESCOPE)

A plane mirror strip is fixed to the rod at a distance *L* from the fixed end of the rod [Fig. 1.9]. A vertical scale (*S*) and telescope (*T*) are arranged in front of the mirror. The telescope is focussed

Load	Reading on S_1					Reading on S_2					$\theta = \theta_2 - \theta_1$	θ for M kg
	Torque Clockwise		Torque Anticlockwise		Mean θ_1	Torque Clockwise		Torque Anticlockwise		Mean θ_2		
	Load increasing	Load decreasing	Load increasing	Load decreasing		Load increasing	Load decreasing	Load increasing	Load decreasing			

on the mirror and adjusted so that the reflected image of the scale in the mirror is seen through the telescope. With some dead load W on the weight-hanger, the reading of the scale division coinciding with the horizontal cross-wire is taken. Weights are added in steps of m kg and the corresponding scale readings are taken. Weights are then decreased continuously in steps of m kg and the readings taken again. The torque is reversed now, by passing the tape anticlockwise on the wheel. The readings are taken as before. From these readings, the shift in scale reading (s) for a load m kg is found.

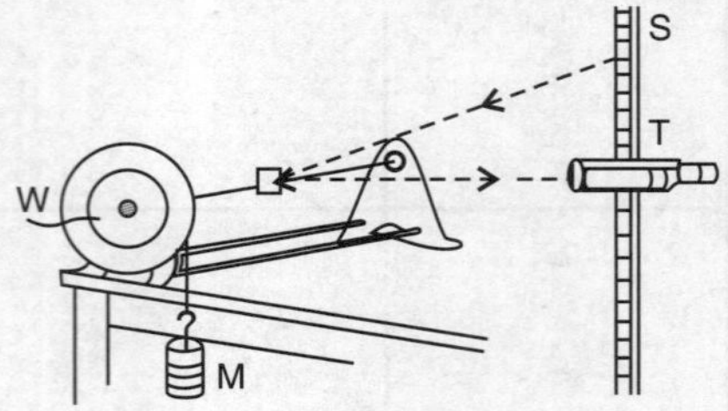

Fig. 1.9

The length L of the rod from the fixed end to the mirror is measured. The mean radius a of the rod is accurately measured with a screw gauge. The radius (R) of the wheel is found by measuring its circumference with a thread. The distance (D) between the scale and the mirror is measured with a metre scale.

G is calculated using the formula $G = \dfrac{4\,mg\,R\,L\,D}{\pi\, a^4 s}$

Load in kg	*Telescope Reading*						$\frac{X \sim Y}{2}$	*Shift in scale reading for 4m kg*
	Torque clockwise			*Torque anticlockwise*				
	Loading	*Unloading*	*Mean (X)*	*Loading*	*Unloading*	*Mean (Y)*		
W								
W + m								
W+ 2 m								
W + 3 m								
W + 4 m								
W + 5 m								
W + 6 m								
W + 7 m								

1.12. WORK DONE IN TWISTING A WIRE

Consider a cylindrical wire of length L and radius a fixed at its upper end and twisted through an angle θ by applying a torque at the lower end.

If c is the torque per unit angular twist of the wire, then the torque required to produce a twist θ in the wire is

$$C = c\,\theta.$$

The work done in twisting the wire through a small angle $d\theta$ is

$$C\,d\theta = c\theta\,d\theta.$$

$\therefore$ The total work done in twisting the wire through an angle θ $\Big\}$ $= W = \int_0^\theta c.\theta\,d\theta$

or
$$W = \frac{1}{2}c.\theta^2$$

The work done in twisting the wire is stored up in the wire as potential energy.

Example 9: *Find the amount of work done in twisting a steel wire of radius 10^{-3} m and of length 0.25 m through an angle of 45°. Given G for steel = 8×10^{10} Nm^{-2}.*

We have, $W = \dfrac{1}{2}c.\theta^2 = \dfrac{1}{2}\dfrac{\pi G a^4}{2L}\theta^2 \qquad \left(\because c = \dfrac{\pi G a^4}{2L}\right)$

Here, $G = 8 \times 10^{10}\ \text{Nm}^{-2}$; $a = 10^{-3}$ m ; $\theta = 45° = \pi/4$ rad ;

$L = 0.25$ m

$$\therefore \quad W = \frac{1}{2}\,\frac{\pi\,(8\times10^{10})(10^{-3})^4\,(\pi/4)^2}{2\times0.25} = 0.1550\ \text{J}.$$

1.13. TORSIONAL OSCILLATIONS OF A BODY

Suppose a wire is clamped vertically at one end and the other end carries a body (*i.e.*, a disc, bar or a cylinder) of moment of inertia I about the wire as the axis. Let the length, radius and rigidity modulus of the wire be respectively l, a and G. When the body is given a slight rotation by applying a torque, say by the hand, the wire is twisted. If the body is released, the body oscillates in the horizontal plane about the wire as axis. These oscillations are called *Torsional oscillations* and the arrangement is known as a *Torsion pendulum.*

Let us consider the energy of the vibrating system when the angle of twist is θ. Let ω be the angular velocity of the body.

The potential energy of the wire due to the twist $= \frac{1}{2}c.\theta^2$.

$$\left.\begin{array}{l}\text{The kinetic energy of the}\\ \text{body due to its rotation}\end{array}\right\} = \frac{1}{2}I\omega^2 = \frac{1}{2}I\left(\frac{d\theta}{dt}\right)^2$$

$$\left.\begin{array}{l}\text{The total energy}\\ \text{of the system}\end{array}\right\} = \frac{1}{2}I\left(\frac{d\theta}{dt}\right)^2 + \frac{1}{2}c\theta^2 = \text{constant}$$

Differentiating this with respect to t,

$$\frac{1}{2}I\,.\,2\,\frac{d\theta}{dt}\cdot\frac{d^2\theta}{dt^2} + \frac{1}{2}c\,2\theta\,\frac{d\theta}{dt} = 0.$$

or
$$I\,\frac{d^2\theta}{dt^2} + c\theta = 0 \text{ or } \frac{d^2\theta}{dt^2} + \frac{c}{I}\theta = 0$$

The body has simple harmonic motion and its period is given by

$$T = 2\pi\sqrt{\frac{I}{c}}$$

Rigidity modulus by Torsion pendulum (Dynamic torsion method) :

The torsion pendulum consists of a wire with one end fixed in a split chuck and the other end to the centre of a circular disc as in Fig. 1.10.

Two equal symmetrical masses (each equal to m) are placed along a diameter of the disc at equal distances d_1 on either side of the centre of the disc. The disc is rotated through an angle and is then released. The system executes torsional oscillations about the axis of the wire. The period of oscillations T_1 is determined.

Then
$$T_1 = 2\pi\sqrt{\frac{I_1}{c}}$$

or
$$T_1^{\,2} = \frac{4\pi^2}{c}I_1.$$

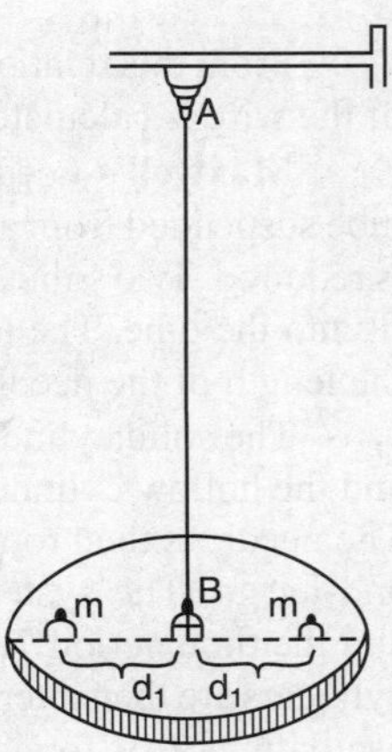

Fig. 1.10

Here, I_1 = Moment of inertia of the whole system about the axis of the wire and
c = torque per unit twist.

Let I_0 = M.I. of the disc alone about the axis of the wire.
i = M.I. of each mass about a parallel axis passing through its centre of gravity.

Then by the parallel axes theorem,

$$I_1 = I_0 + 2i + 2md_1^2$$

$$\therefore \quad T_1^2 = \frac{4\pi^2}{c}[I_0 + 2i + 2m \cdot d_1^2]. \qquad ...(1)$$

The two masses are now kept at equal distances d_2 from the centre of the disc and the corresponding period T_2 is determined. Then,

$$T_2^2 = \frac{4\pi^2}{c}[I_0 + 2i + 2md_2^2] \qquad ...(2)$$

$$\therefore \quad T_2^2 - T_1^2 = \frac{4\pi^2}{c} \cdot 2m \cdot (d_2^2 - d_1^2) \qquad ...(3)$$

But $$c = \pi Ga^4/2L$$

Hence $$T_2^2 - T_1^2 = \frac{4\pi^2\, 2m\,(d_2^2 - d_1^2)\, 2L}{\pi Ga^4}$$

or $$G = \frac{16\pi Lm\,(d_2^2 - d_1^2)}{a^4\,(T_2^2 - T_1^2)}$$

Using this relation, G is determined.

M.I. of the disc by torsional oscillations. The two equal masses are removed and the period T_0 is found when the disc alone is vibrating. Then,

$$T_0^2 = \frac{4\pi^2}{c} I_0 \text{ or } I_0 = \frac{cT_0^2}{4\pi^2} \qquad ...(4)$$

From (3), $$c = \frac{4\pi^2\, 2m\,(d_2^2 - d_1^2)}{T_2^2 - T_1^2}$$

Hence $$I_0 = \frac{4\pi^2\, 2m\,(d_2^2 - d_1^2)}{T_2^2 - T_1^2} \cdot \frac{T_0^2}{4\pi^2} = \frac{2m\,(d_2^2 - d_1^2)\,T_0^2}{T_2^2 - T_1^2}$$

From this relation, the moment of inertia of the disc about the axis of the wire is calculated.

Maxwell's needle: Maxwell's needle consists of a hollow metal tube suspended from a torsion head T by a wire whose rigidity modulus is required. Two solid and two hollow cylinders of equal lengths exactly fit into the tube. The length of each cylinder is equal to one quarter of the length of the needle.

The solid cylinders (S and S) are first placed in the inner position and the hollow cylinders (H and H) at the end as shown in Fig. 1.11(*i*). The needle is then rotated through a small angle about the wire as axis and let go. The system oscillates torsionally about the wire as axis and the time period T_1 is found. The positions of the solid and hollow cylinders are then interchanged [Fig. 1.11(*ii*)] and the period of torsional oscillations T_2 is found. The length of the wire (L) and the radius of the wire (a) are also measured.

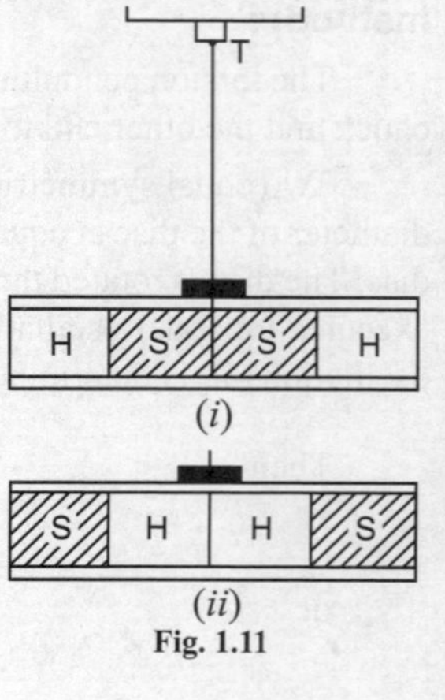

Fig. 1.11

Let I_0 = M.I. of the hollow tube about the wire as axis,

i_1 = M.I. of the solid cylinder about an axis through its C.G. parallel to the axis of the wire.

i_2 = M.I. of the hollow cylinder about an axis through its C.G. parallel to the axis of the wire,

m_1 = mass of the solid cylinder,

m_2 = mass of the hollow cylinder,

d = the length of each cylinder.

Then, the M.I. I_1 of the system about the axis of the wire in the first case is given by,

$$I_1 = I_0 + 2i_1 + 2i_2 + 2m_1\,(d/2)^2 + 2m_2\,(3d/2)^2 \qquad ...(1)$$

Further, $$T_1 = 2\pi\sqrt{I_1/c} \text{ or } T_1^2 = 4\pi^2 I_1/c \qquad ...(2)$$

Similarly, the M.I. I_2 of the system about the axis of the wire in the second case is given by,

$$I_2 = I_0 + 2i_1 + 2i_2 + 2m_2\,(d/2)^2 + 2m_1\,(3d/2)^2 \qquad ...(3)$$

and $$T_2^2 = 4\pi^2\, I_2/c \qquad ...(4)$$

Subtracting (2) from (4), $T_2^2 - T_1^2 = \dfrac{4\pi^2}{c}(I_2 - I_1)$

Now, subtracting (1) from (3)

$$I_2 - I_1 = 4d^2\,(m_1 - m_2) \text{ and } c = \pi\, Ga^4/2L$$

Hence, $$T_2^2 - T_1^2 = \frac{4\,\pi^2 \times 2L}{\pi\, G\, a^4}\,4d^2\,(m_1 - m_2)$$

or $$G = \frac{32\,\pi\, Ld^2\,(m_1 - m_2)}{(T_2^2 - T_1^2)\,a^4} \qquad ...(5)$$

Example 10: *A metal disc of 0.1 m radius and mass 1 kg is suspended in a horizontal plane by a vertical wire attached to its centre. If the diameter of the wire is 10^{-3} m, its length 1 m, and the period of torsional vibrations is 4 seconds, find the rigidity modulus of the wire.*

The metal disc executes torsional oscillations about the axis of the wire. Hence

$$T = 2\,\pi\sqrt{I/c} \text{ or } T^2 = 4\pi^2\, I/c,$$

where I = moment of inertia of the disc about the axis of the wire and

c = torque per unit twist = $\pi\, Ga^4/2\, L$.

$$\therefore \quad T^2 = 4\,\pi^2 \frac{I}{(\pi\, Ga^2/2\,L)} \text{ or } G = \frac{8\,\pi\, IL}{T^2 a^4}$$

Here, $I = MR^2/2$ [where M = mass of the disc = 1 kg and R = radius of the disc = 0.1 m].

$$\therefore \quad I = 1 \times (0.1)^2/2 = 0.005 \text{ kg m}^2.$$

L = Length of the wire = 1 m ; $a = 0.5 \times 10^{-3}$ m ; $T = 4$ s.

$$\therefore \quad G = \frac{8\,\pi\, IL}{T^2 a^4} = \frac{8\,\pi\,(0.005)\times 1}{4^2(0.5\times 10^{-3})^4} = 1.256\times 10^{11}\ \text{Nm}^{-2}$$

Example 11: *A steel bar is suspended in a horizontal position by a vertical wire attached to its centre. A horizontal torque of moment 5 Nm twists the bar horizontally through an angle of 12°. When the bar is released, it oscillates as a torsion pendulum with a period of $\frac{1}{2}$ s. Determine the moment of inertia.*

A torque of moment 5 Nm produces a twist of 12° or $12 \times \pi/180$ rad = 0.2090 rad.

Torque per unit twist = c = 5/0.2090 = 23.88 Nm.

Period of oscillation = $T = 2\pi\sqrt{I/c}$ or $T^2 = 4\pi^2 I/c$

∴ Moment of inertia = $I = cT^2/4\pi^2$.

Here, c = 23.88 Nm, T = 0.5 s

∴ $I = (23.88)(0.5)^2/4\pi^2 = 0.1513$ kg m^2.

Example 12: *A body, suspended symmetrically from the lower end of a wire, 1 m long, 1.22×10^{-3} m in diameter, oscillates about the wire as axis with a period of 1.25 seconds. If the modulus of rigidity of the material of the wire is 8×10^{10} Nm^{-2}, calculate the moment of inertia of the body about the axis of rotation.*

The body executes torsional oscillations about the axis of the wire with a time-period $T = 2\pi\sqrt{I/c}$

where I = M.I. of the body about the axis of the wire or the axis of rotation and

c = torque per unit twist = $\pi G \cdot a^4/2L$.

Hence $T^2 = 4\pi^2 I/c$.

or
$$I = T^2 \frac{c}{4\pi^2} = T^2 \frac{\pi G a^4}{2L\, 4\pi^2}$$

∴
$$I = \frac{T^2 G a^4}{8\pi L}$$

Here, T = 1.25 s ; $G = 8 \times 10^{10}$ Nm^{-2} ; $a = 0.61 \times 10^{-3}$ m ; L = 1 m.

Hence
$$I = \frac{(1.25)^2 (8\times10^{10})(0.61\times10^{-3})^4}{8\pi\times1}$$

$= 6.885 \times 10^{-4}$ kg m^2.

BENDING OF BEAMS

1.14. DEFINITIONS

Beam : *A beam is defined as a rod or bar of uniform cross-section (circular or rectangular) whose length is very much greater than its thickness.*

Bending Couple : If a beam is fixed at one end and loaded at the other end, it bends. The load acting vertically downwards at its free end and the reaction at the support acting vertically upwards constitute the bending couple. This couple tends to bend the beam clockwise. Since there is no rotation of the beam, the external bending couple must be balanced by another equal and opposite couple which comes into play inside the body due to the elastic nature of the body. The moment of this elastic couple is called the internal bending moment. When the beam is in equilibrium,

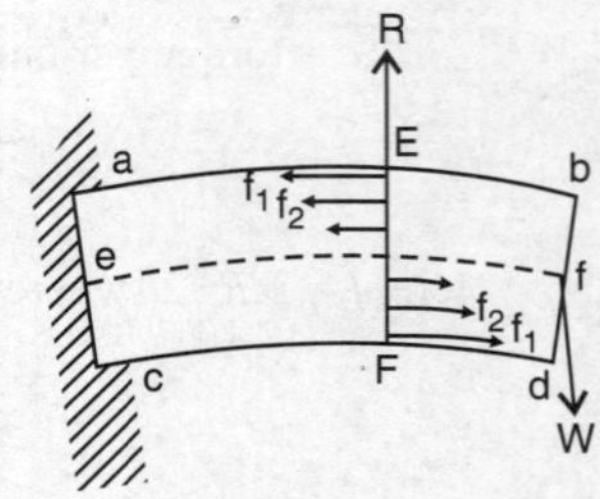

Fig. 1.12

the external bending moment = the internal bending moment.

Plane of Bending : The plane of bending is the plane in which the bending takes place and the bending couple acts in this plane. In Fig. 1.12, the plane of paper is the plane of bending.

Neutral Axis : When a beam is bent as in Fig. 1.12, filaments like *ab* in the upper part of the beam are elongated and filaments like *cd* in the lower part are compressed. Therefore, there must be a filament like *ef* in between, which is neither elongated nor compressed. Such a filament is known as the neutral filament and the axis of the beam lying on the neutral filament is the neutral axis. The change in length of any filament is proportional to the distance of the filament from the neutral axis.

1.15. EXPRESSION FOR THE BENDING MOMENT

Consider a portion of the beam to be bent into a circular arc, as shown in Fig. 1.13. *ef* is the neutral axis. Let R be the radius of curvature of the neutral axis and θ the angle subtended by it at its centre of curvature C.

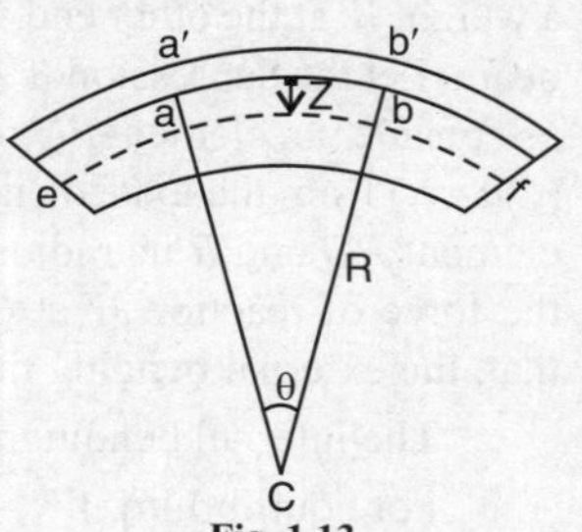

Fig. 1.13

Filaments above *ef* are elongated while filaments below *ef* are compressed. The filament *ef* remains unchanged in length.

Let $a'b'$ be a filament at a distance z from the neutral axis. The length of this filament $a'b'$ before bending is equal to that of the corresponding filament on the neutral axis ab.

We have, original length $= ab = R\theta$.

Its extended length $= a'b' = (R + z)\,\theta$

Increase in its length $= a'b' - ab = (R + z)\,\theta - R\theta = z\,.\,\theta$.

$$\therefore \qquad \text{Linear strain} = \frac{\text{increase in length}}{\text{original length}} = \frac{z\,.\,\theta}{R\,.\,\theta} = \frac{z}{R}$$

If E is the Young's modulus of the material,

$$E = \text{Stress/Linear strain}$$

i.e., $$\text{Stress} = E \times \text{Linear strain} = E\,(z/R)$$

If δA is the area of cross-section of the filament,

the tensile force on the area δA = stress × area $= \dfrac{E\,.\,z}{R}\,\delta A$.

Moment of this force about the neutral axis *ef*

$$= \frac{E\,.\,z}{R}\,\delta A\,.\,z = \frac{E}{R}\,\delta A\,.\,z^2.$$

$$\left.\begin{array}{l}\text{The sum of the moments of}\\ \text{forces acting on all the filaments}\end{array}\right\} = \Sigma\,\frac{E}{R}\,\delta A\,.\,z^2$$

$$= \frac{E}{R}\,\Sigma\,\delta A\,.\,z^2$$

$\Sigma\delta A\,.\,z^2$ is called the geometrical moment of inertia of the cross-section of the beam about an axis through its centre perpendicular to the plane of bending. It is written as equal to Ak^2. *i.e.*, $\Sigma\delta A\,.\,z^2 = Ak^2$. ($A$ = Area of cross-section and k = radius of gyration).

But the sum of moments of forces acting on all the filaments is the internal bending moment which comes into play due to elasticity.

Thus, bending moment of a beam $= EAk^2/R$.

Notes : (*i*) For a rectangular beam of breadth b, and depth (thickness) d, $A = bd$ and $k^2 = d^2/12$.

$$\therefore \qquad Ak^2 = bd^3/12.$$

(*ii*) For a beam of circular cross-section of radius r, $A = \pi\,r^2$ and $k^2 = r^2/4$.

$$\therefore \qquad Ak^2 = \pi r^4/4.$$

(*iii*) EAk^2 is called the flexural rigidity of the beam.

1.16. DEPRESSION OF THE LOADED END OF A CANTILEVER

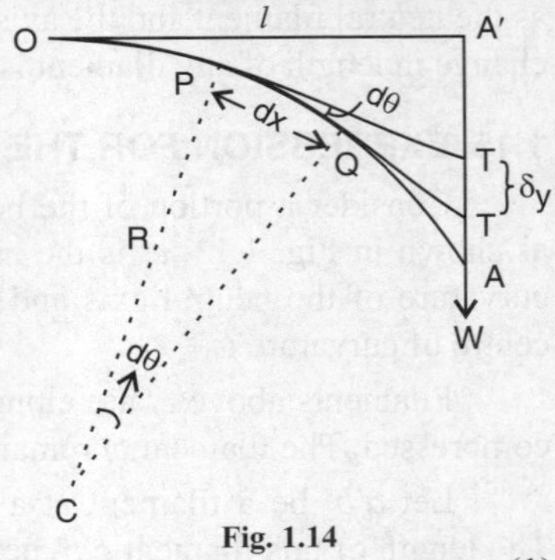

Fig. 1.14

Cantilever : A cantilever is a beam fixed horizontally at one end and loaded at the other end.

Let OA be a cantilever of length l fixed at O and loaded with a weight W at the other end. OA' is the unstrained position of the beam. Let the depression $A'A$ of the free end be y (Fig. 1.14). Let us consider an element PQ of the beam of length dx at a distance $(QA = x)$ from the loaded end. C is the centre of curvature of the element PQ and R its radius of curvature. The load W at A and the force of reaction W at Q constitute the external couple, so that, the external bending moment $= W.x$.

The internal bending moment $= EAk^2/R$.

For equilibrium, $Wx = EA\,k^2/R$ or $R = EA\,k^2/Wx$ ...(1)

Draw tangents at P and Q meeting the vertical line at T and S respectively. Let $TS = dy$ and $d\theta$ = Angle between the tangents. Then, $\angle PCQ$ also $= d\theta$.

Now, $$PQ = dx = R\,d\theta \text{ or } d\theta = \frac{dx}{R} = dx \cdot \frac{Wx}{E\,A\,k^2} \quad \text{(From Eq. 1)}$$

We have, $$dy = x\,d\theta = x\,.\,\frac{Wx\,dx}{EAk^2} = \frac{Wx^2\,dx}{E\,A\,k^2} \quad ...(2)$$

$$\therefore \left.\begin{array}{l}\text{the total depression of}\\ \text{the end of the cantilever}\end{array}\right\} = y = \int_0^l \frac{Wx^2}{E\,A\,k^2}\,dx = \frac{W\,l^3}{3E\,A\,k^2}$$

Angle between the tangents at the ends of a cantilever :

Since the beam is fixed horizontally at O, the tangent at O is horizontal. If a tangent is drawn at A (the free end of the bent bar), it makes an angle θ with the horizontal.

$$\left.\begin{array}{l}\text{Angle between the}\\ \text{tangents at } P \text{ and } Q\end{array}\right\} = d\theta = \frac{Wx}{EA\,k^2}\,dx.$$

$$\left.\begin{array}{l}\text{Angle between the}\\ \text{tangents at } O \text{ and } A\end{array}\right\} = \theta = \int_0^l \frac{Wx}{E\,A\,k^2}\,dx$$

$$\therefore \qquad \theta = \frac{W\,l^2}{2E\,A\,k^2}.$$

Work done in uniform bending. Consider a beam bent uniformly by an external couple. Let A be the area of cross-section of the beam. Consider a filament of area of cross-section δA at a distance z from the neutral axis (Fig. 1.13). Then,

the tensile force on the area $\delta A = \dfrac{Ez}{R}\,\delta A.$

The linear strain of this filament $= z/R$. If l is the length of the filament, then,

the extension of the filament $= zl/R$.

$$\left.\begin{array}{l}\text{The work done in}\\ \text{bending the filament}\end{array}\right\} = \frac{1}{2}\text{ force} \times \text{extension}$$

$$= \frac{1}{2}\frac{Ez}{R}\,\delta A \times \frac{zl}{R} = \frac{1}{2}\frac{El}{R^2} \times z^2\,.\,\delta A.$$

For uniform bending R is constant. Hence, the work done in bending the whole beam is

$$W = \frac{1}{2}\frac{El}{R^2}\Sigma z^2\, \delta A = \frac{1}{2}\frac{El}{R^2} \times Ak^2 = \frac{1}{2}\frac{EAk^2}{R} \times \frac{l}{R}$$

Here, EAk^2/R = the bending moment and l/R = the angle subtended by the bent beam at its centre of curvature.

$\therefore$ The work done in uniform bending = $\frac{1}{2}$ (bending moment) $\times$ (Angle subtended by the bent beam at its centre of curvature).

Example 13 : *Obtain an expression for the depression at the free end of a heavy beam clamped horizontally at one end and loaded at the other end.*

Consider an element PQ of the beam of length dx at a distance x from the fixed end O (Fig. 1.14). Now, in addition to the load W acting at A, a weight equal to that of the portion $(l - x)$ of the beam also acts at its mid-point. Let W_1 be the weight of the beam. Then, the weight per unit length of the beam = W_1/l. Now, we have an additional weight $W_1\,(l - x)/l$ acting at a distance $(l - x)/2$ from Q. Therefore,

$$\left.\begin{matrix}\text{total moment of the}\\ \text{external couple applied}\end{matrix}\right\} = W\,(l - x) + \frac{W_1}{l}(l - x)\frac{(l - x)}{2}$$

$$= W\,(l - x) + \frac{W_1}{2l}(l - x)^2$$

The beam being in equilibrium, this must be balanced by the bending moment EAk^2/R. Therefore,

$$W\,(l - x) + \frac{W_1}{2l}(l^2 - 2l\,.\,x + x^2) = \frac{EAk^2}{R} = EAk^2\left(\frac{d^2y}{dx^2}\right)$$

Integrating,

$$W\left(lx - \frac{x^2}{2}\right) + \frac{W_1}{2l}\left(l^2\,.\,x - l\,.\,x^2 + \frac{x^3}{3}\right) = EAk^2\,\frac{dy}{dx} + C$$

where C is a constant of integration.

Since at $x = 0$, $dy/dx = 0$, we have $C = 0$.

Integrating once again,

$$EAk^2\int_0^y dy = W\int_0^l (lx - x^2/2)\,dx + \frac{W_1}{2l}\int_0^1 (l^2x - lx^2 + x^3/3)\,dx$$

or $$EAk^2y = W\left(\frac{l^3}{3}\right) + \frac{W_1}{2l}\left(\frac{l^4}{4}\right)$$

or $$EAk^2y = \frac{Wl^3}{3} + \frac{W_1l^3}{8},$$

or $$y = \left(W + \frac{3}{8}W_1\right)\frac{l^3}{3E\,A\,k^2}$$

1.17. MEASUREMENT OF E

(1) Cantilever depression : The given beam is clamped rigidly at one end (Fig. 1.15). A weight-hanger (H) is suspended at the free end of the beam. A pin (P) is fixed vertically by some wax at the free end of the beam. A travelling microscope (M) is focussed on the pin. The microscope is adjusted so that the horizontal cross-wire coincides with the tip of the pin and the reading on the vertical scale of the microscope is noted. Then weights m, $2\,m$, $3\,m$, $4\,m$, etc., are added to the weight-hanger. The microscope is adjusted each time to make the horizontal cross-wire coincide with the tip of the pin and the reading on the vertical scale of the microscope is noted in each case. Observations are made for decreasing loads also. The results are tabulated as follows:

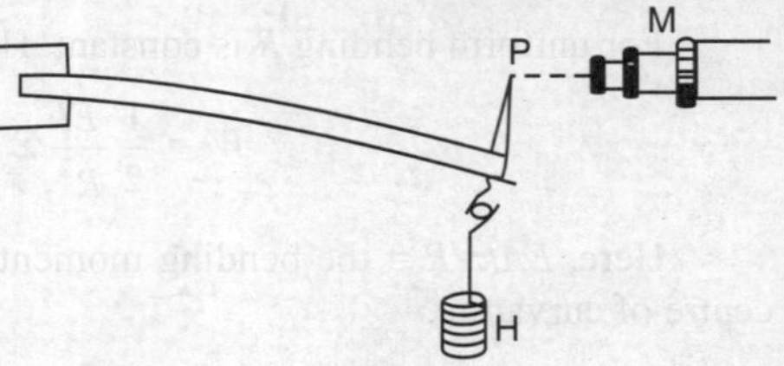

Fig. 1.15

Load in kg	*Microscope Reading*			*Depression for M (= 4 m)*	*Mean depression for a load of M kg*
	Load increasing	*Load decreasing*	*Mean*		
W			x		
$W+m$			x_1		
$W+2m$			x_2		
$W+3m$			x_3		
$W+4m$			x_4	x_4-x	
$W+5m$			x_5	x_5-x_1	
$W+6m$			x_6	x_6-x_2	
$W+7m$			x_7	x_7-x_3	

The mean depression (y) for a load M kg is found.

The length of the beam (l) between the clamped end and the loaded end is measured. The mean breadth (b) of the beam and its mean thickness (d) are determined.

If y is the depression produced for a load of Mg, then,

$$y=\frac{Mg\,l^3}{3EAk^2} \text{ or } E=\frac{Mg\,l^3}{3Ak^2y}$$

Now, $Ak^2 = bd^3/12$ for a rectangular beam.

Hence,
$$E=\frac{Mg\,l^3}{3\,(b\,d^3/12)y}=\frac{4Mg\,l^3}{b\,d^3\,y}.$$

The Young's modulus of the material of the beam is calculated using this relation.

(2) E – by measuring the tilt in a loaded cantilever.

The given rectangular beam is rigidly clamped at one end and a small plane mirror M is fixed at the free end [Fig. 1.16]. A weight-hanger (H) is attached at the free end of the beam. A vertical scale (S) and telescope (T) are arranged in front of the mirror. The telescope is focussed so that the image of the vertical scale due to reflection in the mirror is obtained in the telescope. The reading on the scale which coincides with the horizontal cross-wire is noted. Then weights m, $2\,m$, $3\,m$, $4\,m$ etc., are added to the weight-hanger and the readings of the scale as observed in the telescope are noted in each case. Observations are made for decreasing loads also. The results are tabulated as follows :

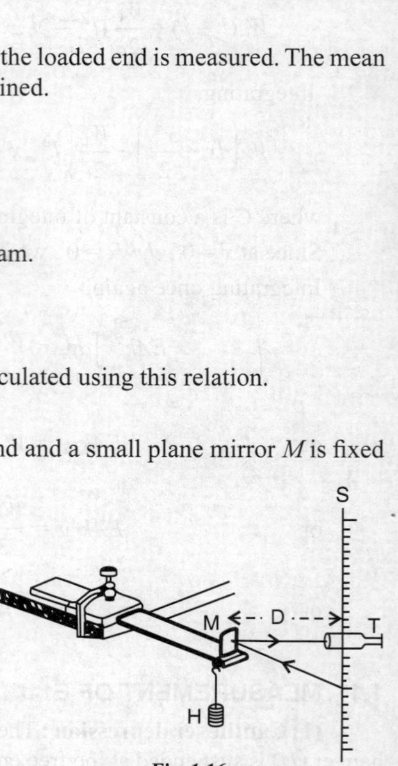

Fig. 1.16

Load in kg	Readings on the scale			Change in scale reading for M kg
	Load increasing	Load decreasing	Mean	

The average of the readings in the last column gives the mean change in scale reading s for a load of M kg. The distance D between the mirror and the scale is found. The breadth (b) and the thickness (d) of the beam are accurately measured.

$$\left.\begin{array}{l}\text{The angle between the two ends}\\ \text{of the centilever for a load of } M \text{ kg}\end{array}\right\} = \theta = \frac{s}{2D} \qquad ...(1)$$

But $$\theta = \frac{Mg\,.\,l^2}{2\,EA\,k^2} = \frac{Mg\,.\,l^2}{2\,E\,.\,bd^3\,/12} \qquad \left(\text{Since } A\,k^2 = \frac{b\,d^3}{12}\right)$$

$$\therefore \qquad \theta = \frac{6\,Mgl^2}{bd^3 E} \qquad ...(2)$$

From (1) and (2), $\dfrac{s}{2D} = \dfrac{6\,Mg\,.\,l^2}{bd^3\,E}$ or $E = \dfrac{12\,Mg\,.\,l^2 D}{bd^3 s}$

1.18. OSCILLATIONS OF A CANTILEVER

Let OA be a cantilever of length l, of negligible mass fixed at O. Let a mass M be attached at the other end A (Fig. 1.17). If the mass is slightly depressed and then released, the cantilever will execute simple harmonic motion about its original depressed position.

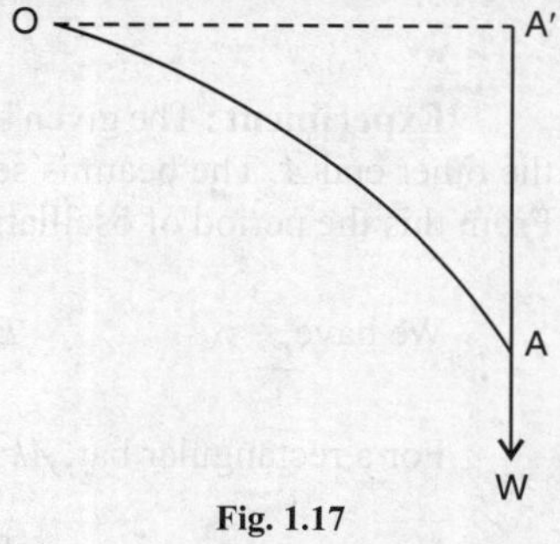

Fig. 1.17

The depression of the loaded end of the cantilever is

$$y = \frac{W\,l^3}{3\,E\,A k^2}$$

or $$W = \frac{3\,E\,A k^2}{l^3}\,y.$$

This must be equal to the elastic reaction of the cantilever balancing it and hence directed opposite to it.

If M is the mass of the weight W and d^2y/dt^2, the acceleration (upwards), we have,

$$\text{elastic reaction} = M\,\frac{d^2 y}{dt^2}$$

$$\therefore \qquad -M\,\frac{d^2 y}{dt^2} = \frac{3\,E\,A\,k^2}{l^3}\,y$$

or $$\frac{d^2 y}{dt^2} = \frac{-3\,E\,A\,k^2}{Ml^3}\,y$$

But, $$\frac{3\,E\,A\,k^2}{M\,l^3} = \text{A constant}$$

The acceleration of mass M or the free end of the cantilever is thus proportional to its displacement and is directed opposite to it.

It, therefore, executes a *S.H.M.* of time period T, given by

$$T = 2\pi\sqrt{\frac{\text{Displacement}}{\text{Acceleration}}} = 2\pi\sqrt{\frac{y}{\left(\dfrac{3EAk^2 y}{M.l^3}\right)}} = 2\pi\sqrt{\frac{Ml^3}{3EAk^2}}$$

If the mass of the cantilever is not negligible, it can be shown that,

$$T = 2\pi\sqrt{\frac{\left(M+\frac{1}{3}m\right)l^3}{3EAk^2}} \quad \text{where } m = \text{mass of the cantilever.}$$

The mass of the cantilever can be eliminated by finding the periods T_1 and T_2 for two different masses M_1 and M_2 attached to the cantilever at the same length. Then,

$$T_1^2 = 4\pi^2\frac{\left(M_1+\frac{1}{3}m\right)l^3}{3EAk^2} \text{ and } T_2^2 = 4\pi^2\frac{\left(M_2+\frac{1}{3}m\right)l^3}{3EAk^2}$$

or $$T_2^2 - T_1^2 = \frac{4\pi^2(M_2-M_1)l^3}{3EAk^2}$$

$\therefore$ $$E = \frac{4\pi^2(M_2-M_1)l^3}{3Ak^2(T_2^2-T_1^2)}$$

Experiment : The given beam is rigidly clamped at O. A certain load of M_1 kg is suspended from the other end A. The beam is set in transverse oscillations and the time for 25 oscillations is found. From this the period of oscillation T_1 is calculated. Similarly, the period T_2 with a load M_2 is found.

We have, $$E = \frac{4\pi^2(M_2-M_1)l^3}{3Ak^2(T_2^2-T_1^2)}$$

For a rectangular bar, $Ak^2 = bd^3/12$.

Hence, $$E = \frac{16\pi^2 l^3(M_2-M_1)}{bd^3(T_2^2-T_1^2)}$$

The length of the cantilever l, the breadth b and depth d are measured. E is calculated using the above formula.

Example 14 : *A steel bar, 0.3 m long, 2 ×10⁻² m broad and 2 × 10⁻³ m thick is clamped at one end and loaded at the other with a mass of 0.01 kg. Calculate the period of vibration of the bar, neglecting the effect of weight of the bar. (E for steel = 20 × 10¹⁰ Nm⁻²).*

For a cantilever of length l, loaded with a mass M, period of vibration $\Big\}$ $= T = 2\pi\sqrt{\dfrac{Ml^3}{3EAk^2}}$

Here, $l = 0.3$ m , $M = 0.01$ kg ; $E = 20 \times 10^{10}$ Nm^{-2}

$$Ak^2 = bd^3/12 = (2\times10^{-2})(2\times10^{-3})^3/12 = \frac{4}{3}\times10^{-11}$$

$$\therefore \quad T = 2\pi \sqrt{\frac{(0.01) \times (0.3)^3}{3 \times (20 \times 10^{10}) \times \left(\frac{4}{3} \times 10^{-11}\right)}} = 0.0365 \text{ second}$$

1.19. DEPRESSION AT THE MID-POINT OF A BEAM LOADED AT THE MIDDLE

Let AB represent a beam of length l, supported on two knife-edges at A and B and loaded with a weight W at the centre C. The reaction at each knife-edge is $W/2$ acting vertically upwards. The beam bends as shown in Fig. 1.18, the depression being maximum at the centre. The bending is non-uniform. Let $CD = y$.

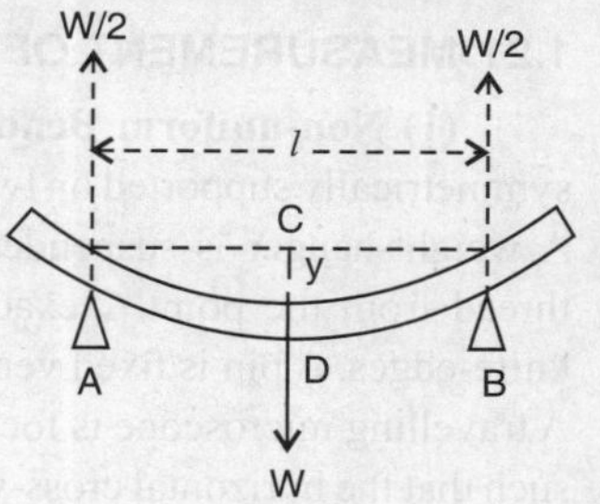

Fig 1.18

The portion DA of the beam may be considered as a cantilever of length $l/2$, fixed at D and bending upwards under a load $W/2$. Hence the elevation of A above D or,

$$\text{the depression of } D \text{ below } A = y = \frac{(W/2)\,(l/2)^3}{3EAk^2} = \frac{Wl^3}{48\,EAk^2}$$

Note : The inclination of the tangent at the points A and B is given by

$$\tan\theta = \frac{dy}{dx} = \frac{W\,l^2}{16\,EAk^2}$$

Since θ is small, $\quad \tan\theta = \theta$.

$$\therefore \quad \theta = \frac{W\,l^2}{16\,EA\,k^2}$$

1.20. UNIFORM BENDING OF A BEAM

Consider a beam of negligible mass supported symmetrically on two knife-edges A and B in a horizontal level (Fig. 1.19). Let $AB = l$.

Let equal weights W, W be added to the beam at its ends C and D. Let $AC = BD = a$. Then the beam is bent into an arc of a circle. The reactions on the knife-edges will then be W and W, acting vertically upwards. Consider the cross-section of the beam at any point P. The only forces acting on the part PC of the beam are the forces W at C and the reaction W at A.

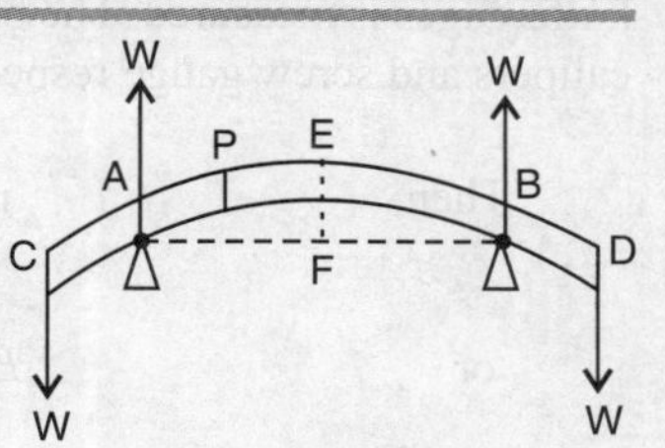

Fig. 1.19

The external bending moment with respect to P

$$= W\,.\,CP - W\,.\,AP = W\,(CP - AP) = W\,.\,AC = Wa.$$

This must be balanced by the internal bending moment EAk^2/R.

Hence, $\quad Wa = EAk^2/R \quad$ ...(1)

Since for a given load W, E, a and Ak^2 are constant, R is a constant. The bending is then said to be uniform. If y is the elevation of the mid-point of AB above its normal position (Fig. 1.20),

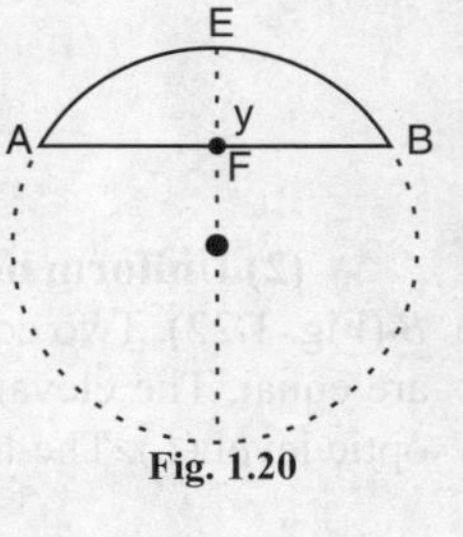

Fig. 1.20

$$EF\,(2R - EF) = AF^2$$

$$y\,(2R - y) = (l/2)^2$$

$$y\,.\,2R = l^2/4 \qquad (\because y^2 \text{ is negligible})$$

$$y = l^2/8R$$

From (1), $$\frac{1}{R} = \frac{Wa}{E\,A\,k^2}$$

$$\therefore \quad y = \frac{Wal^2}{8\,E\,A\,k^2}$$

1.21. MEASUREMENT OF YOUNG'S MODULUS—BY BENDING OF A BEAM

(1) Non-uniform Bending : The given beam is symmetrically supported on two knife-edges (Fig. 1.21). A weight-hanger is suspended by means of a loop of thread from the point C exactly midway between the knife-edges. A pin is fixed vertically at C by some wax. A travelling microscope is focussed on the tip of the pin such that the horizontal cross-wire coincides with the tip of the pin. The reading in the vertical traverse scale of microscope is noted. Weights are added in equal steps of m kg and the corresponding readings are noted. Similarly, readings are noted while unloading. The results are tabulated as follows :

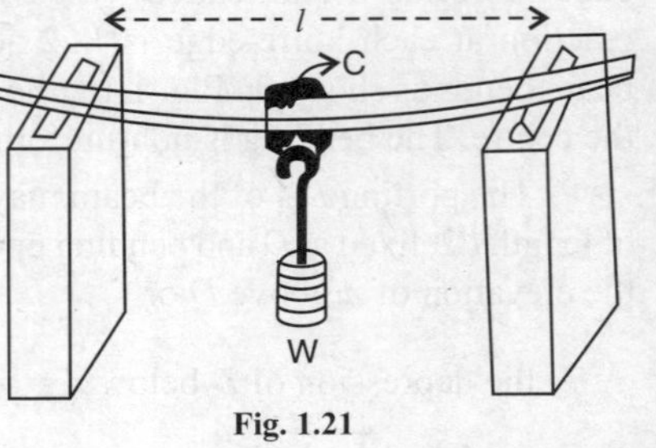

Fig. 1.21

Load in kg	*Readings of the microscope*			*y for M kg*
	Load increasing	*Load decreasing*	*Mean*	

The mean depression y is found for a load of M kg. The length of the beam (l) between the knife-edges is measured. The breadth b and the thickness d of the beam are measured with a vernier calipers and screw gauge respectively.

Then, $$y = \frac{Wl^3}{48\,EA\,k^2} \text{ or } E = \frac{Wl^3}{48\,Ak^2\,y}$$

or $$E = \frac{Mg\,l^3}{48 \times (bd^3/12) \times y} \qquad (\because W = Mg \text{ and } Ak^2 = bd^3/12)$$

$$\therefore \quad E = \frac{Mgl^3}{4\,bd^3\,y}$$

Example 15 : *In an experiment a rod of diameter 0.0126 m was supported on two knife-edges, placed 0.7 metre apart. On applying a load of 0.9 kg exactly midway between the knife-edges, the depression on the middle point was observed to be 0.00025 m. Calculate the Young's modulus of the substance.*

$$E = \frac{Mgl^3}{12 y \pi r^4} = \frac{(0.9)(9.8)\,(0.7)^3}{12\,(0.00025)\,\pi\,(0.0063)^4}$$

$$\therefore \quad E = 2.039 \times 10^{11} \text{ Nm}^{-2}$$

(2) Uniform bending : The given beam is supported symmetrically on two knife-edges A and B (Fig. 1.22). Two equal weight-hangers are suspended, so that their distances from the kinfe-edges are equal. The elevations of the centre of the beam may be measured accurately by using a single optic level (L). The front leg of the single optic lever rests on the centre of the loaded beam and the hind legs are supported on a separate stand. A vertical scale (S) and telescope (T) are arranged in front of the mirror. The telescope is focussed on the mirror and adjusted so that the reflected image of the scale in the mirror is seen through the telescope. The load on each hanger is increased in equal steps of m kg and the corresponding readings on the scale are noted. Similarly, readings are noted while unloading. The results are tabulated as follows :

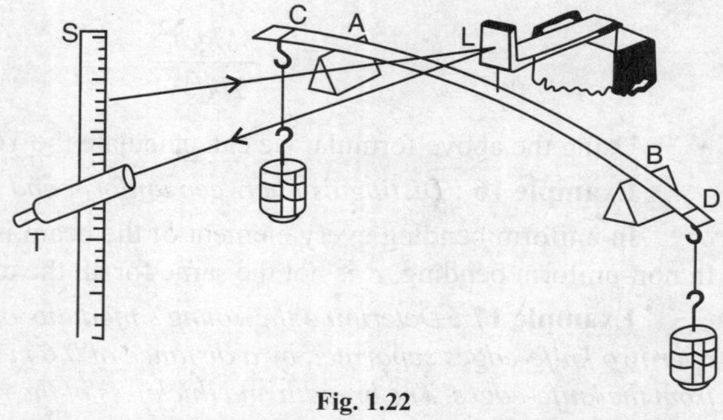

Fig. 1.22

Load in kg	Readings of the scale as seen in the telescope			Shift in reading for M kg
	Load increasing	Load decreasing	Mean	

The shift in scale reading for M kg is found from the table. Let it be S. If

D = The distance between the scale and the mirror,

x = the distance between the front leg and the plane containing the two hind legs of the optic lever, then

$$y = Sx/2D.$$

The length of the beam l between the knife-edges, and a, the distance between the point of suspension of the load and the nearer knife-edge ($AC = BD = a$) are measured. The breadth b and the thickness d of the beam are also measured.

Then,
$$y = \frac{Wal^2}{8\,E\,A\,k^2} \text{ or } \frac{Sx}{2D} = \frac{Mgal^2}{8\,E\,(b\,d^3/12)}$$

[Since $W = Mg$ and $Ak^2 = b\,d^3/12$]

$$\therefore \quad E = \frac{3\,Mgal^2D}{Sxbd^3}$$

Pin and Microscope Method : The given beam is supported symmetrically on two knife-edges A and B. Two equal weight-hangers are suspended so that their distances from the knife-edges are equal. A pin is placed vertically at the centre of the beam. The tip of the pin is viewed by a microscope. The load on each hanger is increased in equal steps of m kg and the corresponding microscope readings are noted. Similarly, readings are noted while unloading. The results are tabulated as follows :

Load in kg	Readings of the microscope			y for M kg
	Load increasing	Load decreasing	Mean	

The mean elevation (y) of the centre for M kg is found. The length of the beam l between the knife-edges and a, the distance between the point of suspension of the load and the nearer knife-edge ($AC = BD = a$) are measured. The breadth b and the thickness d of the beam are also measured.

$$y = \frac{Wal^2}{8EAk^2} = \frac{Mgal^2}{8E(bd^3/12)} \quad \left(\because\ W = Mg \text{ and } Ak^2 = \frac{bd^3}{12}\right)$$

$$\therefore \qquad E = \frac{3Mgal^2}{2bd^3 y}$$

Using the above formula, we can calculate the Young's modulus of the material of the beam.

Example 16 : *Distinguish between uniform and non-uniform bending.*

In uniform bending every element of the beam is bent with the same radius of curvature (R). In non-uniform bending, R is not the same for all the elements in the beam.

Example 17 : *Determine the Young's modulus of the material of a rod, if it is bent uniformly over two knife-edges separated by a distance of 0.6 m and loads of 2.5 kg are hung at 0.18 m away from the knife-edges. The breadth and thickness of the rod are 0.025 m and 0.005 m respectively. The elevation at the middle of the rod is 0.007 m.*

$$E = \frac{3Mgal^2}{2bd^3 y} = \frac{3 \times 2.5 \times 9.8 \times 0.18 \times (0.6)^2}{2 \times 0.025 \times (0.005)^3 \times 0.007}$$

$$= 1.088 \times 10^{11} \text{ Nm}^{-2}.$$

I Section of girders : A girder supported at its two ends as on the opposite walls of a room, bends under its own weight and/or, under the load placed above it. The middle portion gets depressed. The depression (y) at the mid-point of a rectangular beam is proportional to $Wl^3/Eb\,d^3$. For the depression (y) to be small for a given load (W), the length of the girder (l) should be small and its breadth (b), depth (d) and Young's modulus for its material (E) must be large.

Due to depression, the upper parts of the beam above the neutral surface contract, while those below the neutral surface extend. Hence the stresses have a maximum value at the top and bottom and progressively decrease to zero as we approach the neutral surface from either face. Therefore, the upper and the lower surfaces of the beam must be stronger than the intervening part. That is why the two surfaces of a girder or iron rails (for railway tracks etc.) are made much broader than the rest of it, thus giving its cross-section the shape of the letter *I*. In this manner, material will be saved, without appreciably impairing its strength.

1.22. SEARLE'S METHOD FOR DETERMINING E, G AND ν

To determine E : Two identical cylindrical bars AB and CD are suspended from a rigid support by two parallel torsionless strings attached to hooks at the mid-points of the bars. The centres of the bars are joined by the short wire of length l whose elastic constants are required. When the ends A and C of the bars are pulled closer to each other symmetrically, the wire is bent into a circular arc [Fig. 1.23(i)]. When the bars are released, the bars oscillate in the horizontal plane with supporting threads as axes.

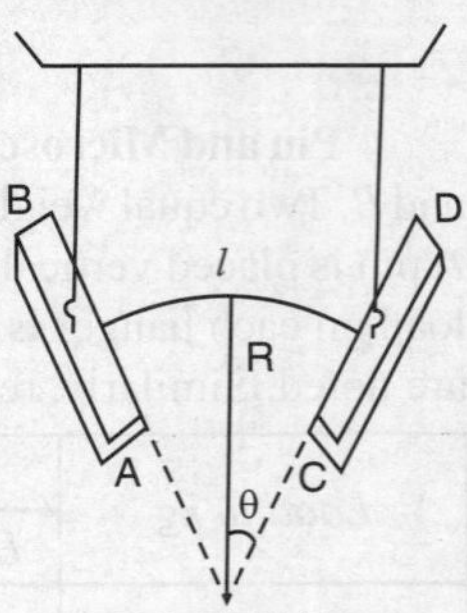

Fig. 1.23 (*i*)

Let θ be the angular displacement of each bar at any instant during the oscillations. The radius of curvature of the wire at this instant is given by $R = l/2\,\theta$ [from Fig. 1.23 (i)].

Hence the bending moment of the wire is equal to the couple exerted by it on each bar and is given by

$$\frac{-EAk^2}{R} = -EAk^2\,\frac{2\theta}{l} \qquad \left(\because \frac{1}{R} = \frac{2\theta}{l}\right)$$

The negative sign indicates that the restoring couple is directed towards the undisturbed position.

Here E = Young's modulus of the material of the wire and

Ak^2 = geometrical moment of inertia of the wire of radius a about the neutral surface.

Now, if I is the M.I. of each bar about the suspension thread (*i.e.*, about an axis passing through its C.G. and perpendicular to its length) and $d^2\theta/dt^2$ is the instantaneous angular acceleration,

the couple acting on each bar $= I\dfrac{d^2\theta}{dt^2}$

$$\therefore \qquad I\frac{d^2\theta}{dt^2} = -EAk^2\frac{2\theta}{l} \text{ or } \frac{d^2\theta}{dt^2} + \frac{2EAk^2}{Il}\theta = 0.$$

This shows that the motion of each bar is simple harmonic and its time period

$$T_1 = 2\pi\sqrt{\frac{Il}{2\,EAk^2}}$$

$$\text{or} \qquad T_1^{\,2} = 4\pi^2\frac{Il}{2\,EAk^2} \text{ or } E = \frac{4\pi^2 I\, l}{2Ak^2\,T_1^2}$$

For a wire of circular cross-section of radius a,

$$Ak^2 = \pi a^4/4.$$

$$\therefore \qquad E = \frac{8\pi I l}{T_1^2\,a^4}$$

This period of oscillation T_1 is obtained by counting the time for 50 oscillations. The radius of the wire a is measured by a screw gauge. Moment of inertia can be known from the mass and dimensions of the bars. If the bars are cylindrical, of mass M, length L and radius R, then

$$I = M\left[\frac{L^2}{12} + \frac{R^2}{4}\right]$$

To determine G : One of the bars, say AB, is clamped horizontally in a stand and the other bar CD hangs vertically below it at the end of the wire [Fig. 1.23 (*ii*)]. The bar CD is given a slight rotation and released; it executes torsional oscillations about the wire. The period T_2 is found.

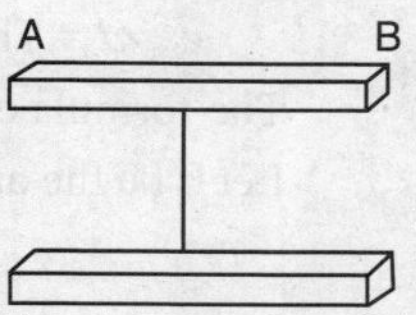

Fig. 1.23 (*ii*)

Then $T_2 = 2\pi\sqrt{I/c}$ where c = couple per unit twist in the wire $= \pi G a^4/2l$, a being the radius of the wire and G the rigidity modulus of the material of the wire. Substituting for c,

$$T_2 = 2\pi\sqrt{\frac{I2l}{\pi Ga^4}} \text{ or } T_2^2 = 4\pi^2\frac{I\,.\,2l}{\pi\,Ga^4}$$

$$\therefore \qquad G = \frac{8\pi I l}{T_2^2\,a^4}.$$

Thus G is calculated.

To determine ν : We know that,

$$\nu = \frac{E}{2G} - 1 = \frac{(8\pi I l/T_1^2\,a^4)}{2\,(8\pi I l/T_2^2\,a^4)} - 1 = \frac{T_2^2}{2T_1^2} - 1.$$

Substituting the values of T_1 and T_2, the value of ν is obtained.

The advantages of this method are : (*i*) Only a short length of the wire is required for the experiment.

(*ii*) This experiment gives the value of ν in terms of two accurately measurable quantities T_1 and T_2. This eliminates altogether the chief source of error, namely, the measurement of the radius (a) of the wire.

1.23. KÖNIG'S METHOD

The beam is supported on two knife-edges K_1 and K_2 separated by a distance l. Two plane mirrors m_1 and m_2 are fixed near the two ends of the beam at equal distances beyond the knife-edges. [Fig. 1.24 (*a*)]. The two plane mirrors face each other and they are inclined slightly outwards from the vertical.

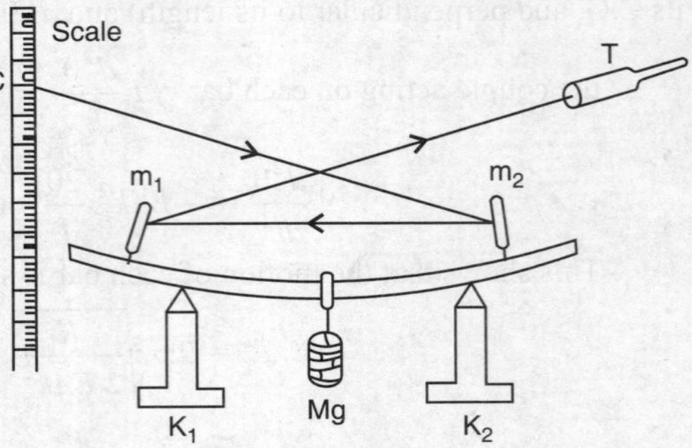

Fig. 1.24 (*a*)

An illuminated translucent scale and a telescope (T) are arranged as shown. The reading of a point C on the scale as reflected first by m_2 and then by m_1 is viewed in the telescope. Let the load suspended at the mid-point of the beam be M. The beam is then bent and the bending is non-uniform. The mirrors at the ends are turned towards each other [Fig. 1.24 (*b*)]. Let the shift in the scale reading be s. The Young's modulus of the material of the beam is then calculated from the relation

$$E = \frac{3Mgl^2\ (2D + L)}{2bd^3\, s}$$

where l = Distance between the knife-edges

D = Distance between the scale and the remote mirror, m_2

L = Distance between the two mirrors.

s = Shift in scale reading for a load of M kg

b = Breadth of the beam

d = Thickness of the beam

The formula can be deduced as explained below.

Let θ be the angle through which each end of the beam has been turned due to loading. Then,

$$\theta = \frac{Wl^2}{16\, EAk^2}$$

The mirrors m_1 and m_2 also turn through the same angle θ due to loading. In Fig. 1.24(*b*), m_1 and m_2 represent the initial and m_1' and m_2' the displaced positions of the mirrors. Originally, the image of the scale division at C coincides with the cross-wire and finally when the load is applied, H is seen to be in coincidence with the cross-wire. For convenience in evaluating θ, consider the rays of light to be reversed in their path.

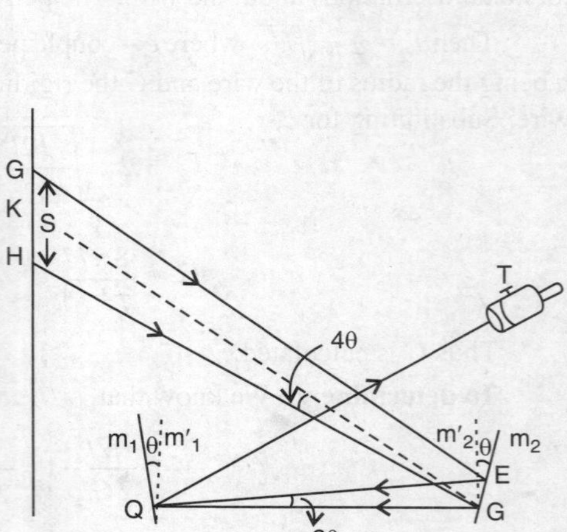

Fig. 1.24 (b)

$TQEC$ will be the original path. When m_1 is turned through an angle θ to the position m_1', QE is turned through 2θ and strikes m_2 at G. Then $EG = L2\theta$. The ray GH is turned through an angle 4θ, since, in addition to QE having moved through 2θ, m_2 itself has turned through θ. Draw GK parallel to EC. Then, $\angle KGH = 4\theta$ and $CK = EG$. $KH = D\,4\theta$

$\therefore$ The total shift in scale reading $= s = CK + KH$

$$= EG + KH \qquad (\because \; CK = EG)$$
$$= L2\theta + D\,4\theta$$
$$= (L + 2D)\,2\,\theta$$

But $\theta = \dfrac{Wl^2}{16\,EAk^2}$

Hence, $s = (L + 2D) \times 2 \times \dfrac{Wl^2}{16\,EAk^2}$.

$$\therefore \quad E = \frac{Wl^2 (L + 2D)}{8\,Ak^2 s}$$

Now $Ak^2 = bd^3/12$ for a beam of rectangular cross-section and

$$W = Mg.$$

$$\therefore \quad E = \frac{Mgl^2\;(L + 2D)}{8\,(bd^3/12)s} = \frac{3Mgl^2\;(2D + L)}{2bd^3 s}$$

EXERCISE – 1

SECTION A

1. What are the units and dimensions of stress, strain, Poisson's ratio and modulus of elasticity ?
2. Define E, G, K and v.
3. Show that a shear is equivalent to an elongation strain and a compression strain at right angles to each other.
4. Find the formula for the work done in stretching a wire. Hence deduce an expression for the energy per unit volume of the wire.
5. Two identical springs, one of steel, the other of copper, are stretched by an identical amount. On which operation must more work be expended?

   [**Ans :** $W = \dfrac{1}{2}\dfrac{EAl}{L}\,l$. For both the springs A, L and l are the same. E is greater for steel than for copper. So more work must be expended on the steel spring.]
6. Show that the theoretical limiting values of Poisson's ratio are – 1 and 0.5.
7. What are the theoretical limits to the value of v ? For a material $v = 0.7$. Comment.
8. Calculate the work done in twisting a wire.
9. Derive an expression for the period of oscillation of a torsion pendulum.
10. What is meant by a beam ? Explain the terms neutral surface, neutral axis, plane of bending, and bending moment of a beam.
11. Explain why steel girders and rails are made in the form of I-section.
12. Explain why a hollow shaft is stronger than a solid one of the same mass, length and material.
13. Give the reasons of the following :
    (*i*) In a girder of rectangular cross-section, the longer side is used as depth.
    (*ii*) A cantilever of uniform section is more likely to break near its fixed end
    (*iii*) Hollow rods are used in constructing cycle frames.

SECTION B

14. Define E, G, K, and v. Obtain the relations connecting these quantities.
15. Show that (*i*) $E = 2G\,(1 + v)$; (*ii*) $E = \dfrac{9\,GK}{3K + G}$

16. Derive an expression for the moment of the couple required to twist one end of a cylinder when the other is fixed.

17. Describe, with necessary theory, how the rigidity modulus of the material of a rod is determined by the static torsion method.

18. Describe, with necessary theory, how you would determine the rigidity modulus of a wire experimentally by using the torsion pendulum.

19. Derive an expression for the couple required to bend a uniform straight metallic strip into an arc of a circle of small curvature.

20. What is a cantilever ? Obtain an expression for the depression at the free end of a thin light beam clamped horizontally at one end and loaded at the other.

21. Derive an expression for the bending of a bar supported at the two ends and loaded in the middle. Describe an experiment to determine E by bending.

22. Describe, with relevant theory, an experiment to determine the Young's modulus of the material of a bar by uniform bending.

23. Describe, with necessary theory, the oscillation method to determine E for the material of a cantilever.

24. Describe Searle's method for determining the elastic constants of a material given in the form of a thin wire. Derive the necessary formula.

25. Describe König's method for the determination of Young's modulus of a beam.

26. Describe how the dimensions of a beam are altered when the beam is bent.

SECTION C

27. What will be the density of lead under a pressure of 2×10^8 Nm^{-2} ? Density of lead =11450 kg/m^3 and bulk modulus of lead = 0.8×10^{10} N/m^2. **[Ans.** 11685 kgm^{-3}].

28. A wire 3 m long and 3×10^{-4} m in diameter elongates 1.32×10^{-3} m when stretched by a force of 0.600 kg weight. Find (*a*) Young's modulus and (*b*) the energy stored up in the wire.

[Ans. (*a*) 1.9×10^{11} Nm^{-2}. (*b*) 0.813 Jm^{-3}]

29. Calculate the work done in producing an extension of 2×10^{-2} m in a steel wire of length 2 m and diameter 10^{-3} m. E of steel =2×10^{11} N/m^2. **[Ans.** 0.157 J]

30. The volume of oil contained in a certain hydraulic press is 0.2 m^3. Find the decrease in volume of the oil when subjected to a pressure of 2.04×10^7 Nm^{-2}. The compressibility of the oil is 20×10^{-6} atm^{-1}. Take 1 atm = 1.02×10^5 Nm^{-2}

Hint. Compressibility = $\dfrac{-v/V}{P} = \dfrac{-v}{V.P}$. Here, compressibility = 20×10^{-6} atm^{-1}, $V = 0.2$ m^3,

$P = 2.04 \times 10^7$ Nm^{-2} = $\dfrac{2.04 \times 10^7}{1.02 \times 10^5}$ = 200 atmospheres. v = ? **[Ans:** -8×10^{-4} m^3]

31. A load of 2 kgf is applied to the ends of a wire 4m long, and produces an extension of 2.4×10^{-4} m. If the diameter of the wire is 2×10^{-3} m, calculate the stress on the wire, its strain, and the value of Young's modulus. **[Ans:** 6.24×10^6 Nm^{-2}, 6×10^{-5} and 1.04×10^{11} Nm^{-2}.]

32. Find the greatest length of a wire that can be suspended vertically without breaking. Its breaking stress and density are equal to 7.2×10^8 Nm^{-2} and 7.8×10^3 kg m^{-3} respectively. **[Ans.** 9.4×10^3 m]

33. When a rubber cord is stretched, the change in volume is almost negligible compared with the change in shape. Obtain the value of Poisson's ratio for rubber. **[Ans.** $\nu = 0.5$]

34. A copper wire 3m long and 10^{-6} m^2 in cross-section is fixed at one end and a weight of 1 kg is attached at the other end. If $E = 12.5 \times 10^{10}$ Nm^{-2} and $\nu = 0.25$, what lateral compression is produced in it ?

[Ans. 2.497×10^{-9} m.]

35. Calculate the load that must be suspended from a steel wire of 1 mm in diameter to produce an elongation of 0.02% of its original length. $E = 20 \times 10^{10}$ Nm^{-2}. **[Ans.** 3.207 kg]

36. A wire of length 1m and diameter 10^{-3} m is fixed at one end and twisted at the other end through an angle of 70° by applying a couple of value 0.01 Nm. Evaluate the rigidity modulus of the wire.

[Ans. 8.281×10^{10} Nm^{-2}].

37. A circular disc of mass 0.250 kg and radius 2.5×10^{-2} m is fixed perpendicularly to the end of a wire of length 0.3 m and diameter 10^{-3} m. Find the rigidity modulus of the wire if the period of torsional oscillations is 1.05 s. [**Ans.** 3.418×10^{10} Nm^{-2}]

38. A sphere of mass 0.8 kg and radius 0.03 m is suspended from a wire of length 1m and radius 5×10^{-4} m. If the period of torsional oscillations is 1.23 s, calculate the rigidity modulus of the material of the wire. [**Ans.** 7.652×10^{10} Nm^{-2}]

39. A gold wire, 3.2×10^{-4} m in diameter, elongates by 10^{-3} m when stretched by a force of 0.330 kg weight and twists through 1 radian when equal and opposite torques of 1.45×10^{-5} Nm are applied at its ends. Find the value of Poisson's ratio for gold. [**Ans.** 0.429]

40. A steel transmission shaft 2 cm in diameter and 5 cm long, rotates at 1500 rpm. The torsion in the shaft is 0.02 radian. Taking modulus of rigidity of steel to be 8×10^{10} Nm^{-2}, find the number of kilowatts of power transmitted by the shaft.

[**Hint.** Twisting torque = 5.029 Nm. Angular velocity = 50 π rad s^{-1}. Power transmitted = twisting torque × Angular velocity = 0.7898 kW].

41. A shaft of diameter 0.08 m and length 5m is transmitting power of 8 kW at 300 revolutions per minute. If the coefficient of rigidity of the material of the shaft be 8×10^{10} Nm^{-2}, find the relative shift between the ends of the shaft. [**Ans.** 0.01981 m].

42. Two solid cylinders of the same material having lengths l, $2l$ and radii r, $2r$ respectively are joined coaxially. Under a couple applied between the free ends, the shorter cylinder shows a twist of 30°. Calculate the twist of the larger cylinder.

[**Hint :** $C \propto r^4 \theta / l$ $\quad \therefore \quad \dfrac{(2r)^4 \theta}{2l} = \dfrac{r^4 \, 30°}{l}$ or $\theta = 3.75°$

43. A uniform bar of length 1.5 m is clamped horizontally at one end and a load of 0.1 kg attached at its free end. Calculate the depression at the loaded end if $E = 9.78 \times 10^{10}$ Nm^{-2} and the bar is of breadth 0.024 m and thickness 0.005 m. Find also the period of vibration if the bar is set in transverse vibrations (neglecting the effect of the mass of the bar). [**Ans.** 0.0451 m and 0.426 s].

44. For the same cross-sectional areas, show that the beam of a square cross-section is stiffer than one of circular section of the same material. Show also that for a given load the depressions are in the ratio 3:π.

45. A rectangular bar, 0.02 m in breadth and 0.01 m in depth and 1m in length, is supported at the ends and a load of 2 kg is applied at its middle. Calculate the depression, if the Young's modulus of the material of the bar is 2×10^{11} Nm^{-2}.

[**Hint :** $y = \dfrac{Mgl^3}{4\,bd^3E} = \dfrac{2 \times 9.8 \times 1^3}{4 \times 0.02 \times (0.01)^3 \times (2 \times 10^{11})} = 0.001225$ m]

46. A brass bar of 1 cm square in cross-section is supported on two knife-edges 100 cm apart. A load of 1 kg at the centre of the bar depresses that point by 2.51 mm. What is the Young's modulus of brass ? [**Ans.** 9.76×10^{10} Nm^{-2}]

47. A uniform rectangular bar 1m long, 0.02 m broad and 0.003 m thick is supported on its flat surface symmetrically on two knife-edges 0.7 m apart. When the loads of 0.2 kg are hung from the ends, the elevation of the bar above its normal position is found to be 0.0022 m. Find the Young's modulus of the material of the bar.

[**Hint :** $E = \dfrac{3Mgal^2}{2\,bd^3y} = \dfrac{3 \times 0.2 \times 9.8 \times (0.15)\,(0.7)^2}{2 \times 0.02 \times (0.003)^3\,(0.0022)}$

$= 1.819 \times 10^{11}$ Nm^{-2}

48. On two knife-edges separated by a distance of 0.6 m, a cylindrical beam of length 1 m is placed horizontally and symmetrically. When a load of 7.3 kg is suspended from each end, the elevation of the centre is 4.75×10^{-3} m. Find the Young's modulus of the beam. The radius of the beam = 7×10^{-3} m. [**Ans.** 7.23×10^{10} Nm^{-2}]

CHAPTER 2

Viscosity

2.1 INTRODUCTION

When two parallel layers of a liquid are moving with different velocities, they experience tangential forces which tend to retard the faster layer and accelerate the slower layer. These forces are called forces of viscosity. Consider two layers of liquid separated by a distance dz (Fig. 2.1). Let v and $v + dv$ be the velocities of two layers. So the velocity gradient is dv/dz. Let A be the surface area of the layer. The viscous force is directly proportional to the surface area A and velocity gradient dv/dz

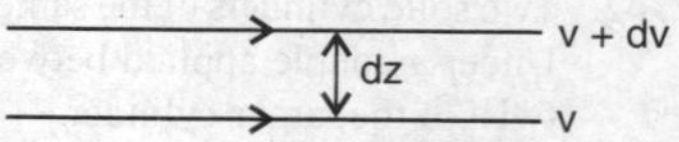

Fig. 2.1

$$i.e., \quad F \propto A\frac{dv}{dz} \quad \text{or} \quad F = \eta\, A\frac{dv}{dz} \qquad ...(1)$$

where η is a constant for the liquid and called coefficient of viscosity. If $A = 1$ and $dv/dz = 1$, we have $F = \eta$.

The coefficient of viscosity is defined as the tangential force per unit area required to maintain a unit velocity gradient.

Unit of η is Nsm^{-2}. It is called the pascal second.

$$\text{Dimensions of } [\eta] = \frac{[F]}{[A]\,[(dv/dz)]} = \frac{MLT^{-2}}{L^2\,(LT^{-1}/L)} = ML^{-1}T^{-1}$$

2.2 STREAMLINE FLOW AND TURBULENT FLOW

Consider a liquid flowing in a pipe. Let the velocity of flow be v_1 at A, v_2 at B and v_3 at C (Fig. 2.2). If as time passes, the velocities at A, B, and C are constant in magnitude and direction, then the flow is said to be steady. In a steady flow, each particle follows exactly the same path and has exactly the same velocity as its predecessor. In such a case, the liquid is said to have an orderly or *streamline flow*.

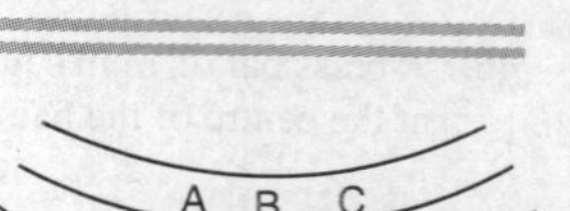

Fig. 2.2

The line ABC is called a streamline, which is the path followed by an orderly procession of particles. The tangent to the streamline at any point gives the velocity of the liquid at that point.

The flow is steady or streamlined only as long as the velocity of the liquid does not exceed a limiting value, called the *critical velocity*. When the external pressure causing the flow of the liquid is excessive, the motion of the liquid takes place with a velocity greater than the critical velocity and the motion becomes unsteady or *turbulent*. This causes eddies and whirlpools in the motion of the liquid. This turbulent motion is also known as *vortex motion*.

The distinction between streamline flow and turbulent flow can be demonstrated by injecting a jet of ink axially in a wider tube in which water is made to flow axially. When the velocity of the liquid is small, the ink will move in a straight line. As the speed of flow is increased beyond the critical velocity, the ink will spread out, showing that the motion has become turbulent.

Definition of critical velocity. *Critical velocity of a liquid is the velocity below which the motion of the liquid is orderly and above which the motion of the liquid becomes turbulent.*

Expression for the critical velocity. *The critical velocity of a liquid may depend upon* (*i*) the coefficient of viscosity of the liquid (η), (*ii*) the density of the liquid (ρ) and (*iii*) the radius r of the tube through which the liquid is flowing. We may write

$$v_c = k\,\eta^a \rho^b r^c$$

where k is constant called *Reynold's number.*

Writing the dimensions of these quantities,

$$[LT^{-1}] = [ML^{-1}\,T^{-1}]^a\,[ML^{-3}]^b\,[L]^c$$

$$[LT^{-1}] = [M^{a+b}\,L^{-a-3b+c}\,T^{-a}]$$

$$\therefore \quad a + b = 0\,;\ -a - 3b + c = 1 \text{ and } -a = -1$$

From these equations we have, $a = 1$, $b = -1$ and $c = -1$.

$$\therefore \quad v_c = \frac{k\,.\,\eta}{\rho\, r}$$

Significance of Reynold's number :

$$k = \frac{v_c \rho\, r}{\eta}$$

The significance of the Reynold's number k is that its value determines the nature of flow of a liquid through a tube. In the case of apparatus, geometrically similar, whatever their actual dimensions, turbulence sets in at the same constant value of Reynold's number in all cases of liquid flow. The flow will be steady and streamline in each individual case, until this number is not exceeded. After exceeding this number, the flow becomes turbulent. Even though the values of r, ρ and η may all vary from each other, but so long as k remains the same, the liquid flow will be similar in all the cases.

2.3. POISEUILLE'S FORMULA FOR THE FLOW OF A LIQUID THROUGH A CAPILLARY TUBE

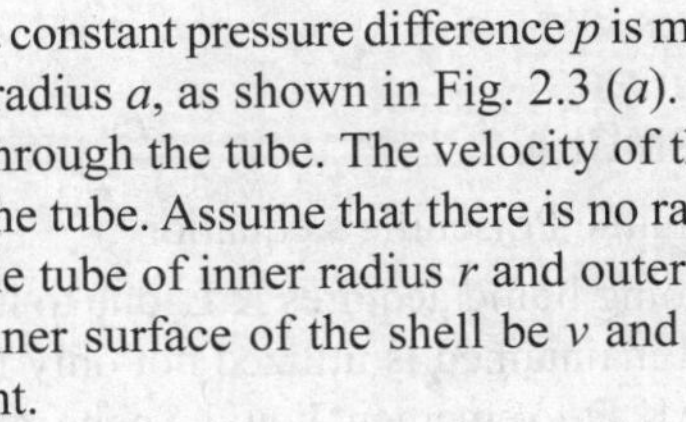

Fig. 2.3(*a*)

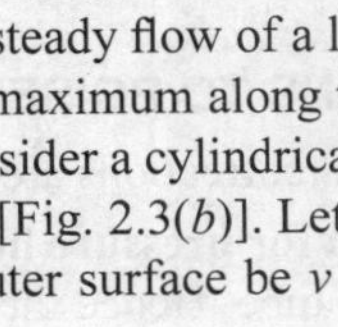

Fig. 2.3(*b*)

Suppose a constant pressure difference p is maintained between the two ends of the capillary tube of length l and radius a, as shown in Fig. 2.3 (*a*). Consider the steady flow of a liquid of coefficient of viscosity η through the tube. The velocity of the liquid is a maximum along the axis and is zero at the walls of the tube. Assume that there is no radial flow. Consider a cylindrical shell of the liquid co-axial with the tube of inner radius r and outer radius $r + dr$ [Fig. 2.3(*b*)]. Let the velocity of the liquid on the inner surface of the shell be v and that on the outer surface be $v - dv$. (dv/dr) is the velocity gradient.

The surface area of the shell $= A = 2\,\pi\, r\, l$.

According to Newton's law of viscous flow, the backward dragging tangential force exerted by the outer layer on the inner layer, opposite to the direction of motion

$$F_1 = -\eta A \frac{dv}{dr} = -\eta\, 2\,\pi\, r\, l \frac{dv}{dr}$$

The driving force on the liquid shell, accelerating it forward

$$F_2 = p\pi r^2$$

where, p = pressure difference across the two ends of the tube and

πr^2 = Area of cross-section of the inner cylinder.

When the motion is steady,

backward dragging force (F_1) = The driving force (F_2)

$$-\eta\, 2\,\pi\, r\, l \frac{dv}{dr} = p\,\pi r^2 \text{ or } dv = \frac{-p}{2\,\eta\, l} r\, dr.$$

Integrating, $$v = \frac{-p}{2\eta\, l}\frac{r^2}{2} + C.$$

where C is a constant of integration.

When $r = a$, $v = 0$. Hence, $0 = \dfrac{-p}{2\eta\, l}\dfrac{a^2}{2} + C$ or $C = \dfrac{pa^2}{4\,\eta\, l}$

$$\therefore \qquad v = \frac{p}{4\,\eta l}(a^2 - r^2)$$

This gives us the average velocity of the liquid flowing through the cylindrical shell.

Hence the volume of the liquid that flows out per second through this shell

$$dV = \begin{pmatrix}\text{Area of cross-section of the shell}\\ \text{of radius } r \text{ and thickness } dr\end{pmatrix} \times \text{Velocity of flow}$$

$$= 2\pi\, r\, dr \frac{p}{4\,\eta\, l}(a^2 - r^2) = \frac{\pi\, p}{2\,\eta\, l}(a^2 r - r^3)\, dr$$

The volume of the liquid that flows out per second is obtained by integrating the expression for dV between the limits $r = 0$ to $r = a$.

$$V = \int_0^a \frac{\pi\, p}{2\eta\, l}(a^2 r - r^3)\, dr = \frac{\pi\, p}{2\,\eta\, l}\left[a^2 \frac{r^2}{2} - \frac{r^4}{4}\right]_0^a$$

$$= \frac{\pi\, p}{2\,\eta\, l}\frac{a^4}{4}$$

or $$V = \frac{\pi\, p\, a^4}{8\,\eta\, l}$$

2.4. CORRECTIONS TO POISEUILLE'S FORMULA

Two important corrections are to be applied in the Poiseuille's equation :

(*i*) **Correction for pressure head :** The outgoing liquid acquires K.E. due to its velocity after passing through the tube. Hence the pressure-head maintained is utilized not only for overcoming viscous resistance but also in imparting considerable K.E. to emergent liquid. So the effective pressure is less and is given by

$$p_1 = p - \frac{V^2 \rho}{\pi^2 a^4}.$$

This can be deduced as follows :

The K.E. given to the liquid of density ρ per second

$$E' = \int_0^a \frac{1}{2}(2\pi r\, dr v\rho)\, v^2 = \pi\rho \int_0^a r v^3\, dr$$

But
$$v = \frac{p}{4\eta l}(a^2 - r^2)$$

∴
$$E' = \pi\rho \int_0^a r\left(\frac{p}{4\eta l}\right)^3 (a^2 - r^2)^3\, dr$$

$$= \pi\rho\left(\frac{p}{4\eta l}\right)^3 \frac{a^8}{8} = \left(\frac{\pi p a^4}{8\eta l}\right)^3 \frac{\rho}{\pi^2 a^4}$$

$$= \frac{V^3 \rho}{\pi^2 a^4}$$

The work done in overcoming viscosity is $p_1 V$ whereas total work done per unit volume is pV. Here p_1 is the effective pressure.

∴
$$pV = p_1 V + \frac{V^3 \rho}{\pi^2 a^4}$$

or
$$p_1 = p - \frac{V^2 \rho}{\pi^2 a^4}$$

∴
$$p_1 = g\rho\left(h - \frac{V^2}{\pi^2 a^4 g}\right)$$

Thus $[V^2/(\pi^2 a^4 g)]$ is the correction factor to the pressure head for gain of kinetic energy by the emergent liquid.

(*ii*) **Correction for length of tube:** At the inlet end of the tube, the flow of the liquid is not streamline for some distance. Consequently the liquid is accelerated. The effective length of the tube is thus increased from l to $l + 1.64\, a$. Thus, the corrected relation for η becomes

$$\eta = \frac{\pi a^4}{8V(l + 1.64\, a)}\left(h - \frac{V^2}{\pi^2 a^4 g}\right) g\rho$$

2.5. POISEUILLE'S METHOD FOR DETERMINING COEFFICIENT OF VISCOSITY OF A LIQUID

The liquid is taken in the constant level tank up to a height h (Fig. 2.4). A capillary tube AB is fixed to the bottom of the tank. A weighed beaker is placed below the free end B of the capillary tube. The mass m of the liquid collected in it in time t is found out.

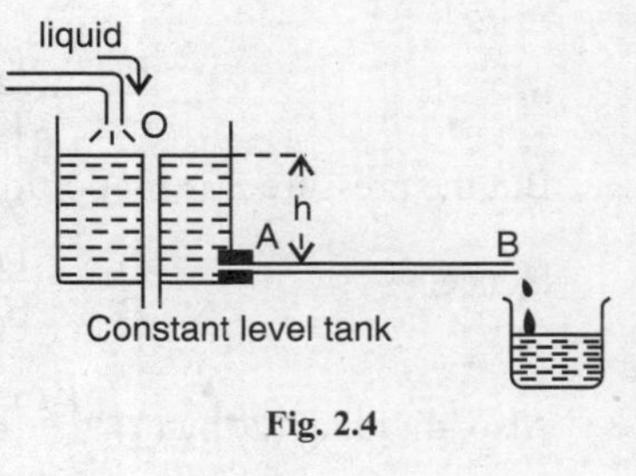

Fig. 2.4

Volume of liquid flowing per second = $V = m/(\rho\, .\, t)$ where ρ is the density of the liquid. The length l of the capillary tube is measured by a metre rod. The radius a of the capillary tube is determined very accurately, using the travelling microscope. Then from the relation

$$\eta = \frac{\pi\, pa^4}{8\, V\, l} \qquad \text{(where } p = h\,\rho g\text{)}$$

the value of η for the liquid can be easily calculated.

Comparison of viscosities : The liquid whose viscosity is η_1 is first used in the constant level tank and the volume of liquid flowing per second $= V_1 = m_1/\rho_1 \,.\, t$ is determined with a capillary tube. The tube is then taken out and cleaned well. The experiment is repeated for the other liquid whose viscosity is η_2 and the volume of liquid flowing per second $= V_2 = m_2/\rho_2 .\, t$ is determined for the same pressure head and with the same capillary tube. If l is the length of the tube, a its radius and ρ_1 and ρ_2 the densities of the two liquids,

$$\eta_1 = \frac{\pi\, h\, \rho_1 g\, a^4}{8\, V_1\, l} \text{ and } \eta_2 = \frac{\pi\, h\, \rho_2 g\, a^4}{8 V_2\, l} \quad \therefore \quad \frac{\eta_1}{\eta_2} = \frac{\rho_1 V_2}{\rho_2 V_1}$$

ρ_1/ρ_2 can be determined with a Hare's apparatus. Thus the viscosities of two liquids can be compared.

2.6 OSTWALD'S VISCOMETER

This instrument is used to compare the viscosities of two liquids. It is also used to study the variation of viscosity of a liquid with temperature.

The apparatus consists of two glass bulbs A and B joined by a capillary tube DE bent into a U-form (Fig. 2.5). The bulb A is connected to a funnel F. The bulb B is connected to an exhaust pump through a stop-cock S. K, L, and M are fixed marks, as shown in the figure. The whole apparatus is placed inside a constant temperature bath.

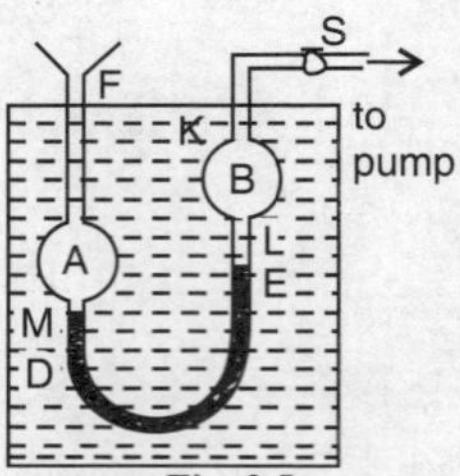

Fig. 2.5

The liquid is then introduced into the apparatus through the funnel and its volume is adjusted, so that the liquid occupies the portion between the marks K and M, when the stop-cock is closed. The stop-cock is now opened and with the help of the exhaust pump the liquid is sucked up above the mark K. The stop-cock is closed and the exhaust pump is removed. The stop-cock is again opened. The liquid is allowed to flow through the capillary tube.

The time (t_1) taken by the liquid to fall from the mark K to the mark L is noted. The experiment is then repeated with the second liquid and the time (t_2) taken by it to fall from K to L is noted.

Theory : Let η_1 and η_2 be the coefficients of viscosity and ρ_1 and ρ_2 the densities of the two liquids respectively. Let the volume of liquid between K and L be V. Then,

the rate of flow of the first liquid $= V_1 = V/t_1$ ...(1)

and the rate of flow of the second liquid $= V_2 = V/t_2$ ...(2)

Now, $\eta_1 = \dfrac{\pi \,.\, P_1 \,.\, a^4}{8\, V_1 \,.\, l}$ and $\eta_2 = \dfrac{\pi \,.\, P_2 \,.\, a^4}{8\, V_2 \,.\, l}$

or $$\frac{\eta_1}{\eta_2} = \frac{V_2}{V_1} \times \frac{P_1}{P_2} \qquad ...(3)$$

But the pressure P is proportional to the density of the liquid used ($P = h\,\rho g$)

Hence, $$\frac{P_1}{P_2} = \frac{\rho_1}{\rho_2} \qquad ...(4)$$

Also, dividing (2) by (1), $$\frac{V_2}{V_1} = \frac{t_1}{t_2} \qquad ...(5)$$

Hence, $$\frac{\eta_1}{\eta_2} = \frac{t_1 . \rho_1}{t_2 . \rho_2} \qquad ...(6)$$

From equation (6), η_1/η_2 can be calculated.

2.7. POISEUILLE'S METHOD FOR DETERMINING COEFFICIENT OF VISCOSITY OF A LIQUID [VARIABLE PRESSURE HEAD]

The given liquid is poured into a graduated burette. The capillary tube is fixed as shown in figure (Fig. 2.6). The clip is opened fully. The liquid is allowed to flow slowly through the capillary tube. When the liquid level in the burette crosses the zero marking, a stop-clock is started. The readings of the stop clock are noted when the liquid level crosses the 10cc, 20cc, 30cc. etc., markings. The vertical height h between the capillary tube and midpoints of the range 0–10cc, 10–20cc, 20–30cc etc., are measured.

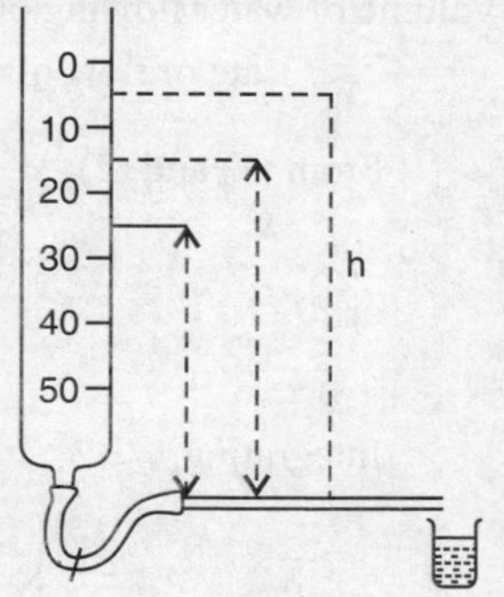

Fig. 2.6

The length of the capillary tube (l) is measured. The radius a of the capillary tube is measured using mercury thread or microscope. The density of the liquid ρ is determined using Hare's apparatus. The readings are tabulated as follows :

Burette Reading cc	*Stop-clock reading seconds*	*Burette reading range*	*Volume of liquid flowing m^3*	*Mean pressure head h m*	*Time of flow t seconds*	$\frac{h.t}{V}$
0						
10		0 to 10 cc	10×10^{-6}			
20		10 to 20 cc	"			
30		20 to 30 cc	"			

Mean $h . t/V$ =

The coefficient of viscosity is calculated using the formula,

$$\eta = \frac{\pi \rho g a^4}{8l}\left(\frac{h.t}{V}\right)$$

Example 1 : *Water flows through a horizontal tube of length 0.2 metres and internal radius 8.1×10^{-4} metre under a constant head of the liquid 0.2 metres high. In 12 minutes 8.64×10^{-4} m^3 of liquid issues from the tube. Calculate the coefficient of viscosity of water. (The density of water = 1000 kg m^{-3} and g = 9.81 ms^{-2}).*

From Poiseuille's formula, $V = \frac{\pi p . a^4}{8 \eta . l}$

Here, $p = h . \rho . g = 0.2 \times 1000 \times 9.8$ Nm^{-2} ; $a = 8.1 \times 10^{-4}$ m

$$l = 0.2\ m,\ V = \frac{8.64 \times 10^{-4}}{12 \times 60} = 1.2 \times 10^{-6}\ m^3 ;\ \eta = ?$$

Now, $$\eta = \frac{\pi p . a^4}{8 V . l} = \frac{(3.14)(0.2 \times 1000 \times 9.8)(8.1 \times 10^{-4})^4}{8(1.2 \times 10^{-6})\,0.2}$$

$$= 1.38 \times 10^{-3}\ Nsm^{-2}$$

Example 2 : *A vessel of cross-section 0.002m^2 has at the bottom a horizontal capillary tube of length 0.1m and internal radius 0.0005m. It is initially filled with water up to a height of 0.2m above the capillary tube. Find the time taken by the vessel to empty one-half of its contents, given the viscosity of water = 10^{-3} Nsm^{-2}.*

According to Poiseuille's formula,

$$\left.\begin{array}{l}\text{Volume of liquid flowing}\\ \text{out of a tube per second}\end{array}\right\} = V = \frac{\pi\, pa^4}{8\,\eta l} \qquad ...(1)$$

Let A be the area of cross-section of the vessel and h the height of water above the capillary tube. Then, $p = h\rho g$. Suppose in a small time dt the level of water in the vessel falls by dh. Then, volume of water flowing out in time $dt = A\,dh$.

$\therefore$ Rate of flow of water $= A\,dh/dt$ ...(2)

From (1) and (2), $-A\dfrac{dh}{dt} = \dfrac{\pi\,(h\rho g)\,a^4}{8\,\eta l}$

or $$dt = -\frac{8\,A\eta l}{\pi\,\rho g a^4} \times \frac{dh}{h}$$

Integrating, $$t = -\frac{8\,A\eta l}{\pi\,\rho g a^4}\,[\log_e h]_{h_1}^{h_2}$$

$$= \frac{8A\eta l}{\pi \rho g a^4}\log_e\left(\frac{h_1}{h_2}\right)$$

Here, $A = 0.002\ \text{m}^2$; $\eta = 10^{-3}\ \text{Nsm}^{-2}$, $l = 0.1$ m; $\rho = 1000\ \text{kg m}^{-3}$
$g = 9.8\ \text{ms}^{-2}$, $a = 0.0005$m; $h_1 = 0.2$m and $h_2 = 0.1$m.

$$\therefore \qquad t = \frac{8 \times 0.002 \times 10^{-3} \times 0.1}{\pi \times 1000 \times 9.8\,(0.0005)^4}\log_e 2 = 576.2s.$$

Example 3 : *Calculate the mass of water flowing in 10 minutes through a tube 0.001m diameter and 0.4m long if there is a constant pressure head of 0.2m of water. The coefficient of viscosity of water is 0.00082 Nsm^{-2}.*

$$\left.\begin{array}{l}\text{Mass of water flowing}\\ \text{in 10 minutes}\end{array}\right\} = M = V \times t \times \rho = \frac{\pi\, pa^4 t}{8\,\eta\, l}\rho$$

Here, $p = h\,\rho g = 0.2 \times 1000 \times 9.8 = 1960\ \text{Nm}^{-2}$; $a = 0.0005$ m.
$t = 10 \times 60 = 600s$; $\eta = 0.00082\ \text{Nsm}^{-2}$ and $l = 0.4$ m.

$$\therefore \qquad M = \frac{\pi \times 1960 \times (0.0005)^4 \times 600}{8 \times 0.00082 \times 0.4} \times 1000 = 0.08795\ \text{kg}$$

Example 4 : *Calculate the maximum velocity with which water of coefficient of viscosity 0.001 Nsm^{-2} flows through a tube of radius 6 × 10^{-4} m without turbulence being produced. Reynolds number is 1000. Find the rate of flow of water through the tube at this velocity and the pressure head required to maintain it if the length of the tube is 0.2m.*

$$\text{Critical velocity} = v_c = \frac{k\eta}{\rho r} = \frac{1000 \times 0.001}{1000 \times 6 \times 10^{-4}} = 1.667\text{ms}^{-1}$$

Rate of flow of water through the tube

$$V = \text{Area} \times \text{Velocity} = \pi.\, r^2 \times v_c = \pi\,(6 \times 10^{-4})^2\ 1.667$$
$$= 1.884 \times 10^{-6}\ \text{m}^3/\text{sec.}$$

$$V = \frac{\pi p\,.\,a^4}{8\,\eta\,.\,l} = \frac{\pi\, h\,\rho g\,.\,a^4}{8\,\eta\,.\,l} \text{ or } h = \frac{8\,\eta l\,.\,V}{\pi\,\rho g\,.\,a^4}$$

$$\therefore \qquad h = \frac{8 \times 0.001 \times 0.2 \times (1.884 \times 10^{-6})}{\pi \times 1000 \times 9.8 \times (6 \times 10^{-4})^4} = 0.7559\text{m}.$$

Example 5 : *Assuming the Poiseuille's formula for the rate of flow of liquid through a capillary, show that if two capillaries of radii a_1 and a_2 having lengths l_1 and l_2 respectively are set in series, the rate of flow is given by* $V = \frac{\pi . p}{8\,\eta}\left[\frac{l_1}{a_1^4} + \frac{l_2}{a_2^4}\right]^{-1}$ *where p is the pressure difference across the arrangement and η is coefficient of viscosity. Also show that with the same two capillaries connected in parallel,* $V = \frac{\pi . p}{8\,\eta}\left(\frac{a_1^4}{l_1} + \frac{a_2^4}{l_2}\right)$.

Consider the arrangement consisting of two capillaries of radii a_1 and a_2 having lengths l_1 and l_2 respectively joined in series. Let p_1 and p_3 be the pressures at the extreme ends and let p_2 be the pressure at the junction. Then, pressure difference across the ends of the first tube = $p_1 - p_2$ and pressure difference across the ends of the second tube = $p_2 - p_3$.

Since there is no accumulation of the liquid at the junction of the two capillaries, the *rate of flow* (*V*) through the whole arrangement must be the same.

$$\therefore \qquad V = \frac{\pi (p_1 - p_2)\, a_1^4}{8\,\eta . l_1} = \frac{\pi (p_2 - p_3)\, a_2^4}{8\,\eta . l_2}$$

or
$$p_1 - p_2 = \frac{8\,\eta\, V . l_1}{\pi . a_1^4} \quad ...(1) \text{ and } p_2 - p_3 = \frac{8\,\eta\, V . l_2}{\pi\, a_2^4} \quad ...(2)$$

Adding (1) and (2), $p_1 - p_3 = \frac{8\,\eta\, V}{\pi}\left[\frac{l_1}{a_1^4} + \frac{l_2}{a_2^4}\right]$

Now, $p_1 - p_3$ = pressure difference across the composite tube = p

Hence
$$p = \frac{8\,\eta\, V}{\pi}\left[\frac{l_1}{a_1^4} + \frac{l_2}{a_2^4}\right]$$

or
$$V = \frac{\pi\, p}{8\,\eta}\left[\frac{l_1}{a_1^4} + \frac{l_2}{a_2^4}\right]^{-1}$$

When the two capillary tubes are connected in parallel, the volume of liquid flowing per second through them is given by

$$V = V_1 + V_2 = \frac{\pi\, p\, a_1^4}{8\,\eta\, l_1} + \frac{\pi\, p a_2^4}{8\,\eta l_2} = \frac{\pi\, p}{8\eta}\left[\frac{a_1^4}{l_1} + \frac{a_2^4}{l_2}\right]$$

Example 6 : *Two tubes A and B of lengths 1m and 0.5 m have radii 10^{-4}m and 2×10^{-4}m respectively. If a liquid is passing through the two tubes, entering A at a pressure of 0.8m of mercury and leaving B at a pressure of 0.76 m of mercury, find the pressure at the junction of A and B.*

For a liquid which is incompressible, the volume of the liquid that flows out per second (*V*) will be same in *A* and *B*.

$$\therefore \qquad V = \frac{\pi\, p_1 . a_1^4}{8\,\eta . l_1} = \frac{\pi\, p_2 . a_2^4}{8\,\eta . l_2}$$

where p_1 is the pressure difference between the two ends of the tube *A* of length l_1 and radius a_1 and p_2 is the pressure difference between the two ends of the tube *B* of length l_2 and radius a_2.

Hence, $$\frac{p_1}{p_2} = \left(\frac{a_2}{a_1}\right)^4 \frac{l_1}{l_2}$$

Let the pressure at the junction of A and B be x. Then, pressure difference between the two ends of tube $A = p_1 = (0.8 - x)$; pressure difference between the two ends of tube B.

$$= p_2 = x - 0.76$$

$$\frac{a_2}{a_1} = \frac{2 \times 10^{-4}}{10^{-4}} = 2 \text{ and } \frac{l_1}{l_2} = \frac{1}{0.5} = 2.$$

$$\therefore \quad \frac{0.8 - x}{x - 0.76} = 2^4 \times 2 = 32 \text{ or } 32x - 32 \times 0.76 = 0.8 - x$$

or $\quad 32x - 24.32 = 0.8 - x$ or $33x = 25.12$.

$\therefore \quad x = 0.7612\ m$ of mercury.

MOTION IN A VISCOUS MEDIUM

2.8. TERMINAL VELOCITY AND STOKES' FORMULA

Let us consider an infinite column of a highly viscous liquid like castor oil contained in a tall jar. If a steel ball is dropped into the liquid, it begins to move down with acceleration under gravitational pull. But its motion in the liquid is opposed by viscous forces in the liquid. These viscous forces increase as the velocity of the ball increases. Finally a velocity will be attained when the apparent weight of the ball becomes equal to the retarding viscous forces acting on it. At this stage, the resultant force on the ball is zero. Therefore the ball continues to move down with the same velocity thereafter. This uniform velocity is called the *terminal velocity.*

Stokes' Formula : The viscous force F experienced by a falling sphere must depend on (*i*) the terminal velocity v of the ball, (*ii*) the radius (r) of the ball and (*iii*) the coefficient of viscosity (η) of the liquid. We can write $F = k\, v^a\, r^b\, \eta^c$ where k is a dimensionless constant. The dimensions of these quantities are $F = MLT^{-2}$; $v = LT^{-1}$; $r = L$; $\eta = ML^{-1}\,T^{-1}$; (k is a number; it has no dimensions).

$$\therefore \quad MLT^{-2} = (LT^{-1})^a\, L^b\, (ML^{-1}\,T^{-1})^c$$

$$MLT^{-2} = M^c\, L^{a+b-c}\, T^{-a-c}$$

Equating the powers of M, L and T on either side,

$c = 1$; $a + b - c = 1$ and $-a - c = -2$.

Solving, $a = 1$; $b = 1$ and $c = 1$. $\therefore F = k\, v\, r\, \eta$.

Stokes experimentally found the value of k to be 6π.

$$\therefore \quad F = 6\,\pi\, v\, r\, \eta.$$

Expression for terminal velocity. Let ρ be the density of the ball and ρ' the density of the liquid. Then,

the weight of the ball $= \frac{4}{3}\pi r^3 \rho g$.

The weight of the displaced liquid or the upthrust on the ball $= \frac{4}{3}\pi r^3 \rho' g$

The apparent weight of the ball $= \frac{4}{3}\pi r^3 \rho g - \frac{4}{3}\pi r^3 \rho' g = \frac{4}{3}\pi r^3 (\rho - \rho') g$.

When the ball attains its terminal velocity v,

the apparent weight of the ball = viscous force F.

$$\therefore \quad 6\pi v r \eta = \frac{4}{3}\pi r^3 (\rho - \rho')g$$

or
$$v = \frac{2}{9}\frac{r^2}{\eta}(\rho - \rho')g$$

Assumptions made by Stokes while deriving the formula:

(1) The medium through which the body falls is *infinite in extent.*

(2) The moving body is *perfectly rigid* and *smooth*.

(3) There is *no slip between the moving body and the medium.*

(4) There are no eddy currents or waves set up in the medium due to the motion of the body through it. In other words, the body is moving very slowly through it.

2.9. STOKES' METHOD FOR THE COEFFICIENT OF VISCOSITY OF A VISCOUS LIQUID

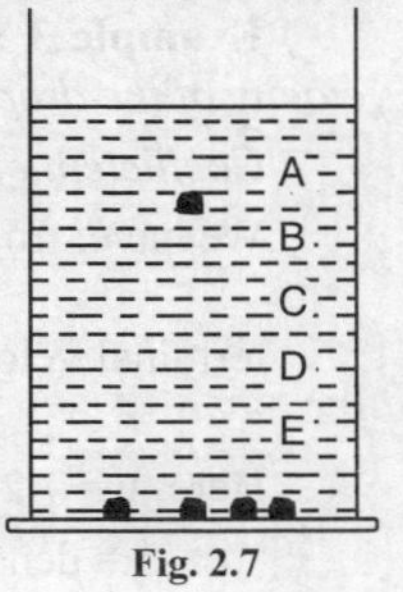

Fig. 2.7

Stokes' method is suitable for highly viscous liquids like castor oil and glycerine. The experimental liquid is taken in a tall and wide jar (Fig. 2.7). Four or five marks A, B, C, D... are drawn on the outside of the jar at intervals of 5 cm. A steel ball is gently dropped centrally into the jar. The time taken by the ball to move through the distances AB, BC, CD, ... are noted. When the times for two consecutive transits are equal, the ball has reached terminal velocity. Now another ball is gently dropped into the jar. When the ball just reaches a mark below the terminal stage, the time (t) taken by the ball to move through a definite distance (x) is noted.

$\therefore$ Terminal velocity $= v = x/t$.

The experiment is repeated for varying distances and the mean value of v is found.

The radius of the ball is measured accurately with a screw gauge. The density of the ball ρ and the density of the liquid ρ' are found by the principle of Archimedes. η is calculated using the formula

$$\eta = \frac{2}{9}\frac{r^2}{v}(\rho - \rho')\, g$$

Example 1 : *Assuming that when a spherical body moves in a viscous fluid under the action of a force F, the resultant force is given by* $ma = F - 6\,\pi r\,\eta\, v$*, calculate the terminal velocity of a rain drop of diameter* 10^{-3} *m. Density of air relative to water is* 1.3×10^{-3}; $\eta = 1.81 \times 10^{-5}$ $N\,sm^{-2}$.

When the body attains the terminal velocity, the acceleration of the body is zero. The body continues moving in the direction of the force with a constant velocity (v).

$$v = \frac{2}{9}\frac{r^2(\rho - \rho')g}{\eta};$$

Here
$$r = \frac{10^{-3}}{2} = 5 \times 10^{-4}\,\text{m}$$

$$\rho = \text{Density of water} = 1000 \text{ kg m}^{-3}.$$

ρ' = Density of air = Density of air relative to water × Density of water = $(1.3 \times 10^{-3}) \times 1000$ = 1.3 kg m^{-3}; $\eta = 1.81 \times 10^{-5}$ N sm^{-2} and $g = 9.8$ ms^{-2}.

$$\therefore \quad v = \frac{2}{9}\frac{(5\times10^{-4})^2\,(1000 - 1.3)\,9.8}{1.81\times10^{-5}} = 30 \text{ ms}^{-1}$$

Example 2 : *Two drops of water of the same size are falling through air with terminal velocities of 0.1 ms^{-1}. If the two drops combine to form a single drop, what will be the new terminal velocity?*

Let r be the radius of each of the original drops. Let R be the radius of the combined drop. Then

$$\frac{4}{3}\pi R^3 = 2 \times \frac{4}{3}\pi r^3 \text{ or } R^3 = 2r^3$$

Let v and v_1 be the velocities of old and new drops, respectively.

$$6\pi\eta r v = \frac{4}{3}\pi r^3 (\rho - \rho')g \text{ and } 6\pi\eta R v_1 = \frac{4}{3}\pi R^3 (\rho - \rho')g$$

Dividing, $rv/Rv_1 = r^3/R^3 = \frac{1}{2}$.

or $$\frac{v_1}{v} = \frac{2r}{R} = \frac{2}{2^{1/3}} = 2^{2/3}$$

$\therefore$ $$v_1 = 2^{2/3}\, v = 2^{2/3} \times 0.1 = 0.159 \text{ ms}^{-1}$$

Example 3 : *Determine the radius of the drop of water falling through air, if the terminal velocity of the drop is 1.2 × 10^{-2} ms^{-1}.*

Coefficient of viscosity for air = 18 × 10^{-6} Ns m^{-2} and density of air = 1.21 kg m^{-3}.

We have, from Stokes formula,

terminal velocity of the drop $= v = \frac{2}{9}\frac{r^2}{\eta}(\rho - \rho')g.$ ...(i)

Here, $v = 1.2 \times 10^{-2}$ ms^{-1}, $\eta = 18 \times 10^{-6}$ N sm^{-2}

ρ = density of water drop = 1000 kg m^{-3};

ρ' = density of air = 1.21 kg m^{-3}; $g = 9.8$ ms^{-2}, $r = ?$

From (*i*), $$r^2 = \frac{9\eta v}{2(\rho - \rho')g} \text{ or } r = \sqrt{\frac{9\eta v}{2(\rho - \rho')g}}$$

$$= \sqrt{\frac{9(18 \times 10^{-6})(1.2 \times 10^{-2})}{2(1000 - 1.21)\,9.8}} = 9.968 \times 10^{-6} \text{ m}$$

2.10. VARIATION OF VISCOSITY WITH TEMPERATURE AND PRESSURE

The viscosity of liquids decreases with temperature. For example, the viscosity of water decreases from 10^{-3} Nsm^{-2} at 20°C to 0.65 × 10^{-3} Nsm^{-2} at 40° C. However, there is no definite relation which expresses the variation of viscosity with temperature accurately. Slotte suggested the empirical formula $\eta_t = \frac{\eta_0}{1 + at + bt^2}$ where η_t and η_0 are the viscosities of the liquid at t°C and 0°C respectively and a and b are constants.

Andrade, on the basis of his theory of viscosity of liquids, derived the formula $\eta v^{\frac{1}{3}} = Ae^{c/vT}$ where A and c are constants, v is the *specific volume* of the liquid and T is the absolute temperature. This formula is in close agreement with the experimental results for all liquids so far examined except water and some tertiary alcohols.

Pressure produces a much smaller effect than temperature on the viscosity of liquids. Thus the viscosity of water decreases only marginally with increase of pressure up to a few hundred atmospheres.

The viscosity of ether at 20°C increases by only about 60% for an increase of 500 atmospheres of pressure. The change in viscosity with pressure is very much greater in the case of liquids with high viscosity. Barring the case of water whose viscosity decreases with pressure, the viscosity of all other liquids increases with pressure.

2.11. FRICTION AND LUBRICATION

When two solid surfaces in contact move relative to each other, friction opposes the relative motion. One of the most difficult problems of a machine is how to overcome friction. Friction reduces efficiency of a machine by converting mechanical energy into heat and causes much wear and tear of the moving parts. Friction is reduced by using lubricants.

A lubricant is a substance (a liquid or a solid) which is used to reduce friction. The lubricant forms a thin layer between the two surfaces in contact. It also fills the depressions present in the surfaces of contact and reduces friction considerably. In light machinery, thin oils (*e.g.*, clock oil) with low viscosity are used. In heavy and fast moving machinery, solids or thick highly viscous oils (*e.g.*, grease or heavy oils) are used.

A good lubricant should have the following properties :

(1) It should be able to spread and fill up the minute depressions in the surfaces.

(2) It should be chemically inert and should not undergo any decomposition at high temperatures.

(3) It should be capable of conducting away heat produced by friction.

2.12. SEARLE'S VISCOMETER:ROTATING CYLINDER METHOD OF FINDING η.

This method is suitable for finding the coefficient of viscosity of a very viscous liquid like castor oil, glycerine etc.

Theory : Consider two coaxial cylinders, one suspended inside the other. The space between the two cylinders is filled with the liquid whose coefficient of viscosity (η) is to be determined. The outer cylinder *B* (of radius *a*) is fixed and the inner cylinder *A* (of radius *b*) is rotated with a constant angular velocity ω_1 [Fig. 2.8]. The layer of liquid in contact with the inner cylinder will have an angular velocity ω_1 and the layer of liquid in contact with the outer cylinder will be at rest. The velocity of the liquid increases from zero at the outer cylinder to a maximum at the inner cylinder and thus a velocity gradient is set up. Let us consider a co-axial cylindrical layer of liquid of radius *r*, thickness *dr* and length *l*. If ω be its angular velocity, the linear velocity of the layer $v = r\omega$.

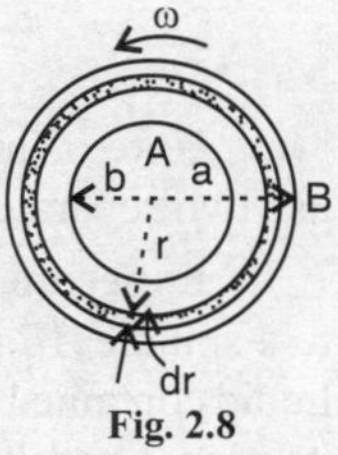

Fig. 2.8

Now, velocity gradient $= \frac{dv}{dr} = \frac{d}{dr}(r\,\omega) = \omega + r\frac{d\omega}{dr}$.

The angular velocity ω of the layer is necessary to prevent slipping of the layer and is not responsible for viscous drag.

Thus the effective velocity gradient is $r\,(d\,\omega/dr)$.

According to Newton's formula,

$$\left.\begin{array}{l}\text{the viscous force acting on the}\\ \text{cylinderical layer of radius } r\end{array}\right\} = F = \eta\, A\,\frac{dv}{dr}$$

$$= \eta\, 2\pi r l r\frac{d\omega}{dr} \qquad ...(1)$$

$$\left.\begin{array}{l}\text{The moment of this force (or torque)}\\ \text{about the axis of rotation}\end{array}\right\} = C = \eta\, 2\,\pi\, r l r\,\frac{d\omega}{dr} \times r$$

$$= 2\pi l \eta r^3 \frac{d\omega}{dr}$$

or $$C \frac{dr}{r^3} = 2\,\pi\, l\, \eta\, d\omega.$$

Integrating, $$C \int_b^a \frac{dr}{r^3} = 2\pi\, l\, \eta \int_0^{\omega_1} d\omega$$

$$C\left[-\frac{1}{2r^2}\right]_b^a = 2\,\pi\, l\,\eta\,\omega_1 \text{ i.e., } C\left[\frac{1}{b^2}-\frac{1}{a^2}\right] = 4\pi\, l\,\eta\,\omega_1$$

$$C = \frac{4\pi\,\eta\,\omega_1\, a^2 b^2}{(a^2-b^2)}\, l \quad ...(2)$$

Here we have omitted the torque due to viscosity between the bottoms of the two cylinders. Let it be c'. This effect can be eliminated by performing the experiment with two different heights of the liquid l_1 and l_2 without changing the distance between the bottoms of the cylinders.

Then, the total torque when the height of the liquid is l_1,

$$C_1 = \frac{4\,\pi\,\eta\,\omega_1\, a^2 b^2}{(a^2-b^2)}\, l_1 + c'$$

The total torque when the height of the liquid is l_2 $$= C_2 = \frac{4\,\pi\,\eta\,\omega_1 a^2 b^2}{(a^2-b^2)}\, l_2 + c'$$

$$\therefore \quad C_1 - C_2 = \frac{4\,\pi\,\eta\,\omega_1 a^2 b^2}{(a^2-b^2)}\,(l_1 - l_2) \quad ...(3)$$

Note : This expression is valid for *gases* also. Only change in such a case is that we have to use two separate inner cylinders of lengths l_1 and l_2.

Experiment : The apparatus consists of an inner cylinder *I* fixed to a spindle *AB* (Fig. 2.9). The inner cylinder *I* is surrounded by the liquid contained in the fixed outer cylinder *R*. The spindle is pivoted at its ends *A* and *B* so that it can be free to rotate about its axis *AB*. The disc *F* and the drum *D* are also attached to the same spindle. The disc *F*, in combination with the index *E*, measures the period of rotation of the cylinder. A string is wound over the drum *D*. The ends of the string are passed over pulleys P_1 and P_2. Two weights *mg*, *mg* are suspended from the ends of the string. As the weights fall, a steady condition will be reached when the couple due to the weights will be balanced by the couple due to the viscous forces and the inner cylinder rotates with uniform angular velocity. If the period of rotation of the inner cylinder is *T*, then

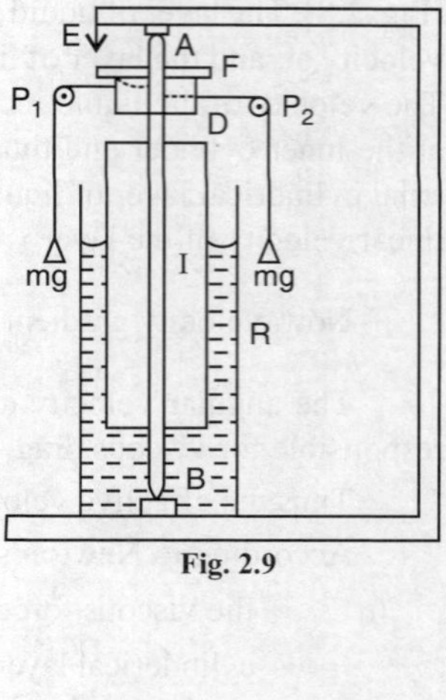

Fig. 2.9

$$\omega_1 = \frac{2\pi}{T}$$

From (2), $$\omega_1 = \frac{C\,(a^2-b^2)}{4\,\pi\,\eta\, a^2\, b^2 l} \quad ...(4)$$

Here $C = mgd$, where *d* is the diameter of the drum.

Substituting for ω_1 and *C* in Eq. (4), we get,

$$\frac{2\pi}{T} = \frac{mgd\,(a^2 - b^2)}{4\,\pi\,\eta\,a^2b^2 l} \text{ or } \eta = \frac{gd\,(a^2 - b^2)}{8\,\pi^2\,a^2b^2} \times \frac{mT}{l}$$

The experiment is repeated for different heights of the liquid and correspondingly adjusting m to get the same period of rotation T. A graph is plotted between mT and l. The graph will be a straight line. The intercept on the negative side of the length axis gives the bottom correction k. Thus the bottom correction can be accounted for by taking l as $l + k$.

Hence $$\eta = \frac{gd\,(a^2 - b^2)}{8\,\pi^2\,a^2b^2} \times \frac{mT}{(l + k)}$$

VISCOSITY OF GASES

2.13 MODIFICATION OF POISEUILLE'S FORMULA FOR GASES

Poiseuille's formula does not apply to gases since gases are, unlike liquids, highly compressible. The density of a gas varies directly with pressure ($\rho \propto P$). Under steady conditions, there is no accumulation of gas anywhere and so the mass (and not the volume) of the gas crossing any section of the tube per second must be constant. *i.e.*, PV = constant or ρV constant ($\because \rho \propto P$).

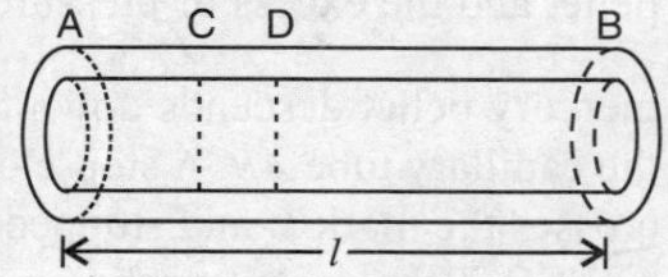

Fig. 2.10

Consider the flow of a gas through a capillary tube AB of length l and radius a [Fig. 2.10].

Consider an intermediate element CD of the tube of length dx. Let dp be the difference of pressure between C and D. Let p be the average pressure of the gas in CD.

The volume of gas flowing through CD per second $\Big\} = V = -\dfrac{\pi\,a^4}{8\,\eta}\dfrac{dp}{dx}$ ...(1)

The –ve sign indicates that the pressure decreases as x increases.

Let p_1 and p_2 be the pressures at the inlet and outlet of the tube. Let V_1 and V_2 be the volumes of the gas entering and leaving the tube respectively.

Now $$p_1V_1 = p_2V_2 = pV \qquad ...(2)$$

$$\therefore \quad p_1V_1 = pV = p\left[\frac{-\pi a^4}{8\eta}\frac{dp}{dx}\right]$$

or $$p_1V_1 dx = \frac{-\pi\,a^4}{8\,\eta}\,p\,dp$$

Integrating over the whole length of the tube,

$$p_1V_1 \int_0^l dx = -\frac{\pi\,a^4}{8\,\eta}\int_{p_1}^{p_2} p\,dp$$

$$p_1V_1\,(x)_0^l = -\frac{\pi\,a^4}{8\,\eta}\left(\frac{p^2}{2}\right)_{p_1}^{p_2}$$

$$p_1V_1\,l = \frac{-\pi\,a^4}{16\eta}\left[p_2^{\,2} - p_1^{\,2}\right]$$

$$p_1V_1 = \frac{\pi\,a^4}{16\,\eta l}\left(p_1^{\,2} - p_2^{\,2}\right)$$

Hence, $$p_1 V_1 = \frac{\pi a^4}{16\,\eta l}(p_1^2 - p_2^2) = p_2 V_2 \qquad \text{[From } Eq.\text{ (2)]}$$

This equation is called *Meyer's Formula* for the rate of flow of a gas through a capillary tube.

2.14. RANKINE'S METHOD FOR DETERMINATION OF η OF A GAS

His apparatus consists of a closed vessel *ABCD* [Fig. 2.11]. Between *A* and *B* there is a capillary tube of length l and radius a. In the opposite branch, there is a mercury pellet of mass m. There are two fixed marks *C* and *D* such that the volume v of the upper portion *BC* is equal to the volume of the lower portion *AD*. Let the volume of the gas in the whole tube *ABCD* be V. T_1 and T_2 are taps for filling the vessel with the experimental gas. Let α be the area of cross-section of the tube *CD*. Then the pressure difference between the ends of the capillary tube is caused by the mercury pellet and the excess of pressure $= \frac{mg}{\alpha}$. When the apparatus is held vertical, the mercury pellet descends down the tube, forcing a certain amount of gas through the capillary tube *AB*. A stop-clock is started just when the upper end of the pellet crosses the mark *C* and stopped just when the lower end of the pellet crosses the mark *D*. The time interval t is noted. Then η is calculated using the formula,

$$\eta = \frac{\pi a^4 \; mgt}{8\,\alpha l\,(V - 2v)}$$

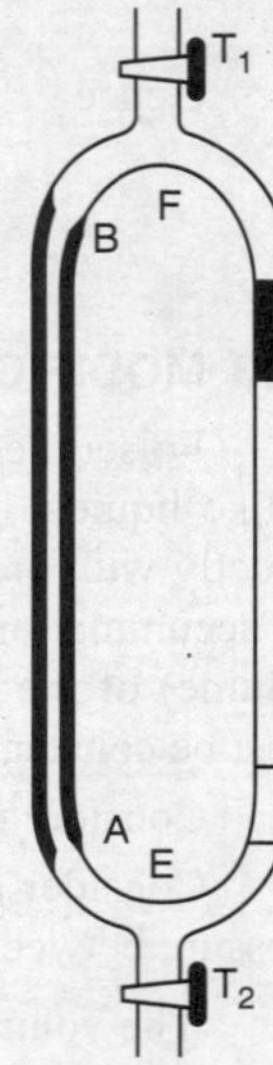

Fig. 2.11

Theory of the Experiment : When the apparatus is placed horizontally, the pressure p is the same throughout and the density of the gas is $p\,\rho$, where ρ is the density per unit pressure. The mass of the gas in the whole tube $= p\,\rho V$ and remains constant.

At the beginning of the experiment :

$$\left.\begin{matrix}\text{Pressure at}\\ \text{the outlet } B\end{matrix}\right\} = \left.\begin{matrix}\text{Pressure above}\\ \text{mercury pellet at } C\end{matrix}\right\} = p_1 \text{ (say)}$$

Pressure at the inlet A = Pressure below the mercury pellet at C

$$= p_1 + \frac{mg}{\alpha} = P_1$$

At the end of the experiment :

Pressure at the outlet B = Pressure above $D = p_2$ (say).

Pressure at the inlet A = Pressure below mercury pellet at D.

$$= p_2 + \frac{mg}{\alpha} = P_2.$$

Since the total mass of the gas remains constant, we can write two equations representing the beginning and the end of the experiment.

$$p\rho V = p_1\,\rho v + \left(p_1 + \frac{mg}{\alpha}\right)\rho\,(V - v) \qquad \text{...(1) at the beginning.}$$

$$p\rho V = p_2\rho\,(V - v) + \left(p_2 + \frac{mg}{\alpha}\right)\rho\,v \qquad \text{....(2) at the end}$$

From (1),

$$pV = p_1 v + p_1 V - p_1 v + \frac{mg}{\alpha}V - \frac{mg}{\alpha}v$$

or
$$p_1 V = pV - \frac{mg}{\alpha}V + \frac{mg}{\alpha}v.$$

or
$$p_1 = p - \frac{mg}{\alpha} + \frac{mg}{\alpha}\frac{v}{V}$$

$\therefore$
$$P_1 = p_1 + \frac{mg}{\alpha} = p + \frac{mg}{\alpha}\frac{v}{V}$$

From (2),
$$pV = p_2 V - p_2 v + p_2 v + \frac{mg}{\alpha}v.$$

or
$$p_2 V = pV - \frac{mg}{\alpha}v.$$

or
$$p_2 = p - \frac{mg}{\alpha}\frac{v}{V}.$$

$\therefore$
$$P_2 = p_2 + \frac{mg}{\alpha} = p - \frac{mg}{\alpha}\frac{v}{V} + \frac{mg}{\alpha}$$
$$= p + \frac{mg}{\alpha}\left(1 - \frac{v}{V}\right).$$

At the beginning of the experiment, the mass of the gas enclosed between the first position of the mercury pellet and the inlet of the capillary $= P_1\rho(V - v)$.

At the end of the experiment, the mass of the gas enclosed between the inlet and the second position of the mercury pellet $= P_2\rho v$.

Hence the mass of the gas that goes through the capillary tube in time $t = P_1\rho(V - v) - P_2\rho v$.
$$= \left(p + \frac{mg}{\alpha}\frac{v}{V}\right)\rho(V - v) - [p + \frac{mg}{\alpha}\left(1 - \frac{v}{V}\right)]\rho v$$
$$= p\rho(V - 2v).$$

$$\left.\begin{array}{l}\text{The average rate of flow}\\ \text{in kilogram per second}\end{array}\right\} = p\rho\frac{(V - 2v)}{t}$$

Let us find the average rate of flow from Meyer's formula.

At the beginning of the experiment, pressure at the inlet $A = P_1$. Pressure at the outlet $B = p_1$.

Let V_1 be the volume of the gas flowing through the capillary tube per second. From Meyer's formula,
$$P_1 V_1 = \frac{\pi a^4}{16\eta . l}(P_1^2 - p_1^2)$$

Similarly, at the end of the experiment, let V_2 be the volume of the gas flowing through the capillary tube per second. Then,
$$P_2 V_2 = \frac{\pi a^4}{16\eta l}(P_2^2 - p_2^2)$$

$$\left.\begin{array}{l}\text{The average rate of flow}\\ \text{in kilogram per second}\end{array}\right\} = \frac{\rho P_1 V_1 + \rho P_2 V_2}{2}$$
$$= \frac{\frac{\rho\pi a^4}{16\eta l}(P_1^2 - p_1^2) + \rho\frac{\pi a^4}{16\eta l}(P_2^2 - p_2^2)}{2}$$

$$= \frac{\rho \pi a^4}{32 \eta l}[(P_1^2 - p_1^2) + (P_2^2 - p_2^2)]$$

Now,

$$P_1^2 - p_1^2 = (P_1 + p_1)(P_1 - p_1) = \left(2p - \frac{mg}{\alpha} + \frac{2mg}{\alpha}\frac{v}{V}\right)\frac{mg}{\alpha}$$

$$P_2^2 - p_2^2 = (P_2 + p_2)(P_2 - p_2) = \left(2p + \frac{mg}{\alpha} - \frac{2mg}{\alpha}\frac{v}{V}\right)\frac{mg}{\alpha}$$

$\therefore$ Average rate of flow $= \dfrac{\rho \pi a^4}{32 \eta l} 4p \dfrac{mg}{\alpha}$

$$= p\rho \frac{\pi a^4}{8 \eta l}\frac{mg}{\alpha}$$

Equating the two expressions for the rate of flow, we have

$$\frac{p\rho (V - 2v)}{t} = p\rho \frac{\pi a^4}{8 \eta l}\frac{mg}{\alpha}$$

$$\therefore \quad \eta = \frac{\pi a^4 \, mgt}{8 \alpha l (V - 2v)}$$

Notes : (1) However pure the mercury of the pellet may be, some of it does stick to the wall of the tube and the effect of sticking is to reduce the effective weight of the pellet. In the actual experiment a correction for the sticking of the pellet is necessary.

(2) Rankine's method is used to find the variation of η with pressure and temperature. Rankine determined the coefficient of viscosity of gases under various pressures and found that coefficient of viscosity is quite independent of pressure. It was found that the coefficient of viscosity of a gas increases with temperature.

(3) According to the kinetic theory of gases, the coefficient of viscosity of a gas is given by

$$\eta = \frac{1}{3} m . n \bar{c} \lambda$$

where m is the mass, $\bar{c}$ the mean speed and λ the mean free path of the molecules of the gas. n is the number of molecules per unit volume. Determination of η enables us to estimate the mean free path. Further,

$$\lambda = \frac{1}{\sqrt{2} \pi \sigma^2 n}$$

where σ is the molecular diameter. By measuring η, the molecular diameter σ can be estimated.

EXERCISE II

1. Derive Poiseuille's formula for the rate of flow of a liquid through a capillary tube. Describe a laboratory method for determining the coefficient of viscosity of a liquid at room temperature.
2. Describe, with theory, Stokes method of determining the viscosity of a highly viscous fluid.
3. Describe the rotating cylinder method of determining the viscosity of a liquid. Deduce the necessary formula.
4. Describe Rankine's experiment for the measurement of the viscosity of a gas.
5. A cylindrical vessel of radius 0.05 m has at its bottom a horizontal capillary tube of length 0.2 m and internal radius 0.4mm. The vessel is completely filled with water and then the water is allowed to flow out of the capillary. How long will it take for the water level to fall to half its initial height in the vessel? The viscosity of water is 0.001 Nsm^{-2}. [**Ans.** 1.1×10^4 *s*]

6. A cylindrical glass vessel held vertically is fitted with a capillary glass tube of length 0.4 m and diameter 6×10^{-4} m at a depth of 0.5 m below its open end. If the vessel is completely filled with water, find the rate of flow of the liquid. [η for water = 1.257×10^{-3} Nsm^{-2}] **[Ans.** 3.06×10^{-8} m^3/s]

7. Water is conveyed through a horizontal tube 0.08 m in diameter and 4000 m in length at the rate of 20 litres per second. Assuming only viscous resistance, calculate the pressure required to maintain the flow. $\eta = 0.001$ *SI* units.

$$\textbf{Hint. } p = \frac{8\,V\,\eta l}{\pi\,a^4} = \frac{8 \times 0.02 \times 0.001 \times 4000}{\pi\,(0.04)^4} = 7.962 \times 10^4 \text{ Nm}^{-2}$$

8. A horizontal capillary of 1mm diameter and 0.1m length is connected to a tank of water and 1 cc of water flows out per minute. How much water will flow out per minute if another capillary of 0.5 mm diameter and 0.01 m length is connected in series with the first capillary ? **[Ans.** 3.8×10^{-6} m^3]

9. Water flows through a horizontal tube of length 0.2 m and internal radius 8.1×10^{-4} m under a constant head of the liquid 0.2 m high. In 12 minutes, 8.64×10^{-4} m^3 of liquid issues from the tube. Calculate the coefficient of viscosity of water and verify that the conditions for streamline flow exist. The density of water = 1000 kgm^{-3} and $g = 9.8$ ms^{-2}

[**Hint.** $\eta = 1.2 \times 10^{-3}$ Nsm^{-2}. The maximum velocity of flow which is along the axis of the tube = v_{maxm}.

$$= \frac{pa^2}{4\,\eta l} = 1.165 \text{ ms}^{-1};\ v_c = \frac{k\eta}{\rho a} = 1.705 \text{ ms}^{-1}.$$

$v_{maxm} < v_c$ and hence streamline flow exists]

10. Two equal drops of water are falling through air with a steady terminal velocity of 5×10^{-2} ms^{-1}. If the drops coalesce, what will be the new terminal velocity ? **[Ans.** 7.98×10^{-2} ms^{-1}]

11. A water drop of radius 10^{-5} m is falling through air; Find its terminal velocity. Neglect the density of air. (η for air = 18×10^{-6} Nsm^{-2}). **[Ans.** 1.2×10^{-2} ms^{-1}]

12. Castor oil at 20° C has a coefficient of viscosity 2.42 Nsm^{-2} and a density 940 kgm^{-3}. Calculate the terminal velocity of a steel ball of radius 2×10^{-2} m falling under gravity in the oil, taking the density of steel as 7800 kg m^{-3}. **[Ans.** 0.025 ms^{-1}]

13. A small steel ball-bearing falling through glycerine has a terminal velocity of 2×10^{-2} ms^{-1}. Find the terminal velocity of an air bubble of the same size rising through glycerine at the same temperature. Neglect the weight of air in the bubble and assume that the viscous force on each sphere is proportional to its terminal velocity. (Density of steel = 7700 kg m^{-3}. Density of glycerine = 1260 kg m^{-3}]. **[Ans.** 39×10^{-3} ms^{-1}].

SHORT QUESTIONS AND ANSWERS

1. *Bring out the analogy between liquid flow and current flow.*

(*i*) As a liquid flows through a tube, mass of the liquid gets displaced.

$$\left.\begin{array}{l}\text{Volume of liquid}\\ \text{flowing per second}\end{array}\right\} = V = \frac{\pi\,pa^4}{8\,\eta\,l}$$

In a current carrying conductor, charges get displaced.

Electric current is $I = E/R$.

$\therefore$ Volume of liquid flow/sec is analogous to charge flow/sec.

(*ii*) $V = \dfrac{p}{8\,\eta\,l\,/\,(\pi\,a^4)}$

Pressure head is analogous to *EMF* or potenmtial difference. $8\,\eta\,l/(\pi a^4)$ is called the *viscous resisitance.* Thus viscous resistance is analogous to electrical resistance.

2. *On what factors does the terminal velocity of a small body depend ? What will happen if the density of the medium is greater than that of the body ?*

Consider a small sphere of radius r and density ρ falling freely from rest under gravity through a fluid of density ρ' and coefficient of viscosity η. Then

$$\left.\begin{array}{l}\text{the resultant downward driving}\\ \text{force on the sphere}\end{array}\right\} = \frac{4}{3}\pi r^3 (\rho - \rho')\, g$$

When the sphere attains a terminal velocity v, the retarding force $6\,\pi\,\eta\,r\,v$ becomes equal to the resultant driving force.

$$\therefore \qquad 6\,\pi\,\eta\,r\,v = \frac{4}{3}\pi\,r^3\,(\rho - \rho')g$$

or
$$v = \frac{2}{9}\frac{r^2(\rho - \rho')\,g}{\eta}$$

Thus the terminal velocity depends on r, ρ, ρ' and η. The radius of the sphere should be quite small. Otherwise, terminal velocity will be quite high as $v \propto r^2$.

If the density of the medium is greater than the density of the body (*i.e.*, $\rho' > \rho$) the terminal velocity becomes negative and the sphere moves upward with a constant velocity. Slow bubbling out of CO_2 from a soda water bottle is an example of this phenomenon.

3. *Is it possible to determine the viscosity of water by Stokes' method ?*

Stokes' method is applicable for determining the coefficient of viscosity of highly viscous liquids like castor oil. In the case of water (medium), there is slip between the body and medium. Hence this method is not suitable for liquids like water.

4. *Mention some practical applications of the knowledge of viscosity.*

(*i*) **Lubrication :** The knowledge of viscosity has wide applications in the field of lubrication. The knowledge of viscosity and its variation with temperature helps to use a suitable lubricant for a certain machine. Liquids with moderate viscosities are good lubricants for light machinery. Highly viscous oils are used as lubricants for heavy machinery.

(*ii*) **Damping of instruments :** The viscosity of air and some liquids is used for damping the motion of some instruments. Paddles or vanes of such instruments move in the liquid or air. The viscous force opposes the motion. Thus the moving parts of the instruments attain the final position quickly.

5. *Which rain drops fall faster, big ones or small ones ? Why ?*

$$v = \frac{2}{9}\frac{r^2}{\eta}(\rho - \rho')\,g\;.\; i.e.,\; v \propto r^2$$

Hence, big rain drops fall faster.

3 Surface Tension

CHAPTER

3.1 INTRODUCTION

Any liquid in small quantity, so that gravity influence is negligibly small, will always assume the form of a spherical drop *e.g.*, rain drops, small quantities of mercury placed on a clean glass plate etc. So a liquid must experience some kind of force, so as to occupy a minimum surface area. This contracting tendency of a liquid surface is known as surface tension of liquid. This is a fundamental property of every liquid.

The following experiment illustrates the tendency of a liquid to decrease its surface area.

When a camel hair brush is dipped into water, the bristles spread out [Fig. 3.1 (a)]. When the brush is taken out, the bristles cling together on account of the films of water between them contracting [Fig. 3.1. (b)]. This experiment clearly shows that the surface of a liquid behaves like an elastic membrane under tension with a tendency to contract. This tension or pull in the surface of a liquid is called its *surface tension*.

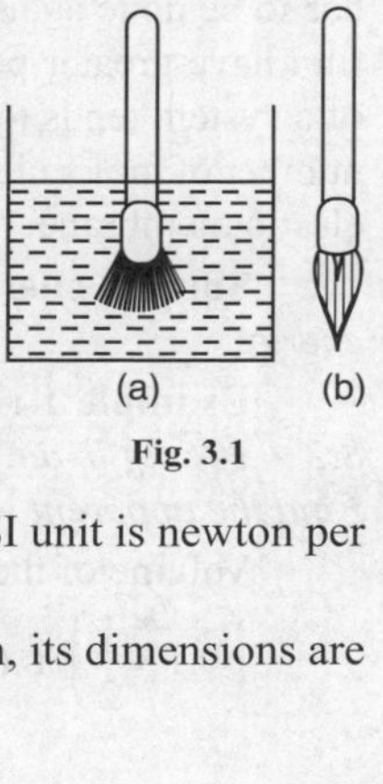

(a) (b)

Fig. 3.1

Definition : *It may be defined as the force per unit length of a line drawn in the liquid surface, acting perpendicular to it at every point and tending to pull the surface apart along the line.*

Unit of surface tension. Surface tension being force per unit length, its SI unit is newton per metre (Nm^{-1}).

Dimensions of surface tension : Since it is the ratio of a force to a length, its dimensions are $MLT^{-2}/L = MT^{-2}$.

Molecular forces : There are two kinds of molecular forces :

(*i*) adhesive forces and (*ii*) cohesive forces.

(*i*) Forces of attraction between molecules of different substances are known as adhesive forces. For example, the force of attraction between the glass molecules of a beaker and molecules of water contained in it is an adhesive force. Adhesive force is different for different pairs of substances.

(*ii*) Force of attraction between molecules of the same substance is called cohesive force. This force varies inversely probably as the eighth power of the distance between two molecules. Hence, it is very appreciable when the distance between two molecules is small. It is the greatest in solids, less in liquids and the least in gases. Therefore, a solid has a definite shape, a liquid has a definite free surface and a gas has neither.

The maximum distance up to which a molecule exerts a force of attraction on another is called the range of *molecular attraction* and is generally of the order of 10^{-9} m. A sphere with the molecule as centre and the range of molecular attraction as radius is called the *sphere of influence* of the molecule. The molecule attracts and is, in turn, attracted by the molecules present inside this sphere.

3.2 EXPLANATION OF SURFACE TENSION ON KINETIC THEORY

Consider three molecules A, B and C of a liquid [Fig. 3.2]. The circles around them indicate their respective spheres of influence.

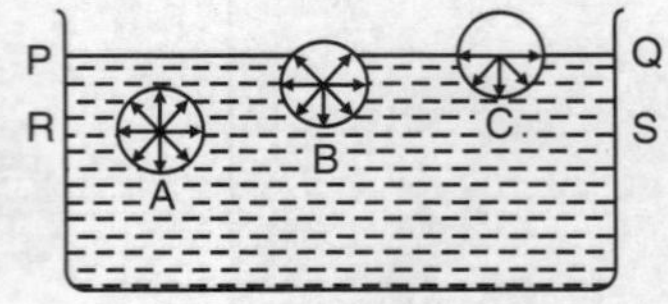

Fig. 3.2

(*i*) The molecule A is well within the liquid and it is attracted equally in all directions by the other molecules lying within its sphere of influence. Therefore, it does not experience any resultant force in any direction. This happens only as long as the sphere of influence is well within the liquid.

(*ii*) The sphere of influence of molecule B lies partly outside the liquid. The upper half of the sphere contains fewer molecules attracting the molecule B upwards, than the lower half attracting it downwards. Hence, there is a resultant downward force acting on B.

(*iii*) The molecule C lies on the surface of the liquid. Half of its sphere of influence lies above the surface of the liquid and contains only a few vapour molecules, whereas there are many liquid molecules in its entire lower half. Thus the resultant downward force in this case is the maximum. If a plane RS is drawn parallel to the free surface PQ of the liquid at a distance equal to the molecular range, then the layer of the liquid between the planes PQ and RS is called the *surface film*. Hence all the molecules in the surface film are pulled downward due to the cohesive force between molecules.

If a molecule is to be brought from the interior of the liquid to the surface of the liquid, work has to be done against the downward cohesive force acting upon it. Hence molecules in the surface film have greater potential energy than the molecules inside the liquid. Since the potential energy of a system tends towards a minimum, the surface film tends to contract, so as to contain minimum number of molecules in it. Thus the surface of the liquid is under tension and behaves like a stretched elastic membrane.

Surface energy : *The potential energy per unit area of the surface film is called its surface energy.*

Example 1 : *A glass plate of length 0.1 m, breadth 0.0154 m and thickness 0.002 m weighs 8.2×10^{-3} kg in air. It is held vertically with the long side horizontal and the lower half under water. Find the apparent weight of the plate. Surface tension of water $=73 \times 10^{-3}$ N/m, $g = 9.8$ m/s^2.*

Volume of the portion of the plate immersed in water is

$$\left(0.1 \times \frac{1}{2}\right)(0.0154)\,(0.002) = 1.54 \times 10^{-6} \text{ m}^3$$

$$\begin{aligned} \text{Upthrust} &= \text{Weight of water displaced} \\ &= (1.54 \times 10^{-6}) \times 1000 \times 9.8 \\ &= 1.509 \times 10^{-2} \text{ N} = 0.01509 \text{ N} \end{aligned}$$

Now, total length of the plate in contact with the water surface is $2(0.1 + 0.002) = 0.204$ m.

$\therefore$ Downward pull upon the plate due to surface tension is

$0.204 \times (73 \times 10^{-3}) = 0.0149$ N

$$\begin{aligned} \therefore \quad \text{Resultant upthrust} &= 0.01509 - 0.0149 = 0.00019 \text{ N} \\ &= 0.00019/9.8 \text{ kg wt} = 1.939 \times 10^{-5} \text{ kg wt} \end{aligned}$$

$\therefore$ Apparent weight of the plate in water = Weight of the plate in air – resultant upthrust

$$= 8.2 \times 10^{-3} - 1.939 \times 10^{-5} = 8.18 \times 10^{-3} \text{ kg}$$

3.3. WORK DONE IN INCREASING THE AREA OF A SURFACE

Take a rectangular framework of wire $ABCD$, with a horizontal wire EF placed across it, free to move up and down [Fig. 3.3]. Form a soap film across $BCFE$ by dipping it in a soap solution. The wire EF will now be pulled upwards by the surface tension of the film acting in the plane of the film and perpendicular to EF. Therefore, to keep the wire EF in equilibrium, a force F (including the weight of the wire EF) has to be applied downwards. If σ = the force per unit length of the film (*i.e.*, S.T.) and l = the length of the wire EF, then the upward force due to S.T. = $2l\sigma$ [∵ The film has two surfaces and each has a S.T. σ]

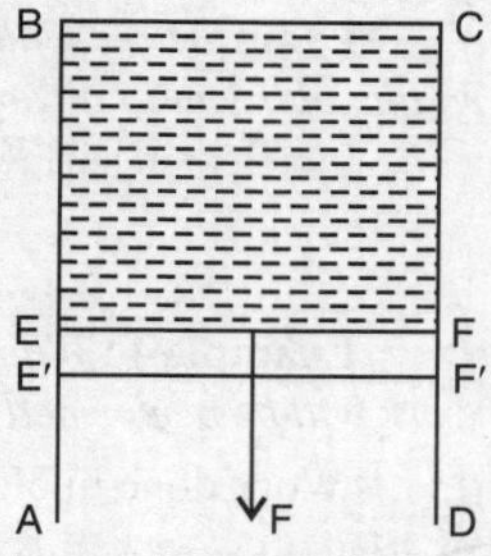

Fig. 3.3

Hence $$F = 2\,l\,.\,\sigma$$

If the wire is pulled downwards through a small distance x to the position $E'\,F'$,

$$\text{the work done} = Fx = 2\,l\,\sigma\,.\,x = \sigma\,2\,l\,.\,x$$
$$= \text{S.T.} \times \text{Increase in surface area}$$

$$\left.\begin{array}{l}\because \text{ the work done in increasing the} \\ \text{surface area of the liquid film by unity}\end{array}\right\} = \frac{\sigma\,2\,l\,.\,x}{2\,l\,.\,x} = \sigma$$

∴ The surface tension of a liquid may be defined *as the amount of work done in increasing the surface area of the liquid film by unity.*

The work required to enlarge the surface of separation between two immiscible or partially miscible liquids is called the interfacial surface energy, and is often expressed as the *interfacial tension* in newton per metre.

Example 2 : *Calculate the work done in spraying a spherical drop of water of 10^{-3} m radius into million droplets, all of the same size, the surface tension of water being 72×10^{-3} Nm^{-1}.*

Breaking the liquid drop means an increase in surface area. Therefore work has to be done for this purpose.

Work done = Surface tension × Increase in surface area.

Let us calculate the increase in surface area.

Let R be the radius of the larger drop and r the radius of the smaller droplets.

Volume of the original drop = Volume of 10^6 droplets.

$$\frac{4}{3}\pi R^3 = 10^6 \times \frac{4}{3}\pi r^3$$

$$\frac{4}{3}\pi (10^{-3})^3 = 10^6 \times \frac{4}{3}\pi r^3 \qquad [\because R = 10^{-3}\text{ m}]$$

or $$r^3 = 10^{-15}\text{ m} \quad \text{or} \quad r = 10^{-5}\text{ m}$$

Area of original large drop = $4\,\pi\,R^2 = 4\,\pi\,(10^{-3})^2 = 4\,\pi \times 10^{-6}\text{ m}^2$

Area of 10^6 droplets $= 10^6 \times 4\,\pi\,r^2 = 10^6 \times 4\,\pi\,(10^{-5})^2$
$= 4\,\pi \times 10^{-4}\text{ m}^2$

Hence increase in surface area $= 4\,\pi \times 10^{-4} - 4\,\pi \times 10^{-6} = 1.244 \times 10^{-3}\text{ m}^2$

∴ Work done = Surface tension × Increase in surface area.
$= (72 \times 10^{-3}) \times (1.244 \times 10^{-3})$
$= 8.956 \times 10^{-5}$ joules

3.4. WORK DONE IN BLOWING A BUBBLE

Let the radius of the bubble blown be r. A bubble has two surfaces, an inner and an outer one, each of surface area $4\,\pi\,r^2$.

∴ The surface area of the film forming the bubble $= 2 \times 4\,\pi\,r^2 = 8\,\pi\,r^2$

Work done in blowing the bubble = Surface tension × surface area of the film formed
$= \sigma \times 8\,\pi\,r^2 = 8\,\pi\,r^2\,\sigma$

Example 3 : *Calculate the amount of work done if a soap bubble is slowly enlarged from a radius of 0.1 m to a radius of 0.2 m.* $\sigma = 30 \times 10^{-3}\ Nm^{-1}$.

Increase in surface area $= 2[4\pi\,(r_2^2 - r_1^2)] = 8\pi\,[(0.2)^2 - (0.1)^2] = 0.7536\ m^2$

$$\therefore \quad \text{Work done} = \text{Increase in surface area} \times \sigma = 0.7536 \times (30 \times 10^{-3}) = 0.02261\ J$$

Example 4 : *What is the work done in blowing a soap bubble of radius 0.1 m? What additional work will be performed in further blowing it, so that its radius becomes 0.15 m?* $\sigma = 30 \times 10^{-3}\ Nm^{-1}$.

$$\left.\begin{array}{l}\text{Work done in blowing a soap}\\ \text{bubble of radius 0.1 m}\end{array}\right\} = \text{Surface tension} \times \text{Increase in surface area}$$

$$W = \sigma \times 8\pi r_1^2 = (30 \times 10^{-3})\,8\pi\,(0.1)^2 = 7.536 \times 10^{-3}\ J$$

$$\left.\begin{array}{l}\text{Work done in increasing the radius of}\\ \text{the soap bubble from 0.1 m to 0.15 m}\end{array}\right\} = \sigma \times 8\pi\,(r_2^2 - r_1^2)$$

i.e.,

$$W = (30 \times 10^{-3})\,8\pi\,[(0.15)^2 - (0.1)^2] = 9.42 \times 10^{-3}\ J$$

3.5. FORMS OF LIQUID DROPS

When a quantity of liquid rests upon a horizontal solid plate, which it does not wet, the shape of the drop is determined by surface tension and gravity. For extremely small drops, the S.T. effects are great and the gravitational effects small. So S.T. determines the shape of the drop. It is therefore spherical. Rain drops, a small quantity of mercury on a glass plate, water drops on leaves, all assume spherical shapes on account of this.

On increasing the size of the drop, the effect of gravitation becomes greater and that of S.T. less. Now, the effect of gravitation alone would be to make the drop spread out, so that its centre of gravity may be the lowest. Hence, a large drop of a heavy liquid spreads out when placed on a glass plate. Therefore, a large drop of mercury is always flat.

3.6. ANGLE OF CONTACT

When a glass plate is dipped in water, the water molecules cling to the surface of glass and the water molecules rise along the plate. The shape of water is as shown in Fig. 3.4 (*i*). When the glass plate is dipped in mercury, the mercury molecules cling to the surface and the liquid is depressed along the plate as shown in Fig. 3.4 (*ii*).

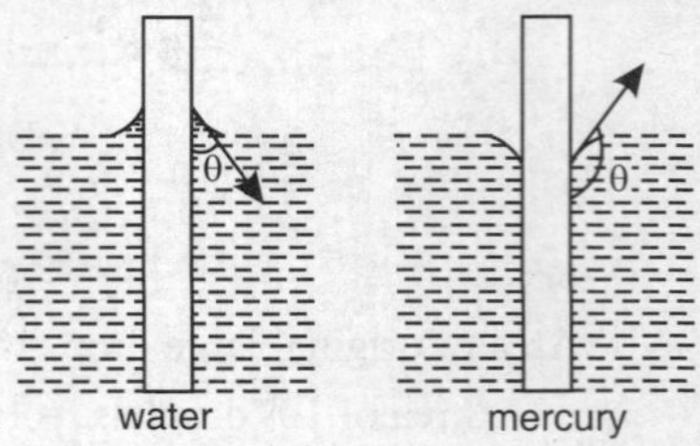

Fig. 3.4

The angle of contact θ ***is defined as the anlge made by the tangent*** *at the point of contact of the liquid surface with the glass surface inside the liquid.*

This angle may have any value between 0° and 180°. For most liquids and glass, it is less than 90°; for mercury and glass, it is about 140°. It really depends upon the nature of the liquid and the solid. It is quite independent of the angle of inclination of the solid to the liquid surface.

3.7. SPREADING OF ONE LIQUID OVER ANOTHER

Let a liquid (liquid *A*) be in contact with another liquid (liquid *B*) as shown in Fig. 3.5. The free surfaces of both liquids are in contact with air. Three forces are acting at the point of contact *P* of the three substances. (1) S.T. σ_1 between liquid *A* and air, (2) S.T. σ_2 between liquid *B* and air and (3) S.T. σ_3 between liquids *A* and *B*. The directions of these tensions are along the tangents at the lines of common contact. If equilibrium is possible, we should be able to represent the three forces acting at the

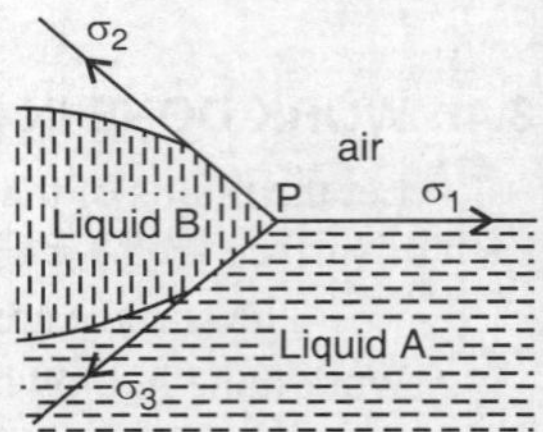

Fig. 3.5

point P by the three sides of a triangle, taken in order. Such a triangle is known as the *Neumann's triangle*. The sum of any two sides of a triangle is always greater than the third side. Therefore, the sum of any two of the three surface tensions σ_1, σ_2 and σ_3 should be greater than the third. No two pure liquids satisfy this condition. Thus it is not possible to have a drop of one liquid remaining in equilibrium over another liquid. The liquid B, if it is lighter than the liquid A, will spread over the surface of A. Quincke showed that pure water spreads over pure mercury. But, if the mercury surface is contaminated with grease, the water will form a drop on the mercury surface. For contaminated surfaces of liquids, the construction of *Neumann's triangle* can be possible.

3.8. PRESSURE DIFFERENCE ACROSS A LIQUID SURFACE

(*a*) If the free surface of the liquid is plane, as in Fig. 3.6 (*a*), the resultant force due to S.T. on a molecule on the surface is zero.

(*b*) If the free surface of the liquid is concave, as in Fig. 3.6 (*b*), the resultant force due to S.T. on a molecule on the surface acts vertically upwards.

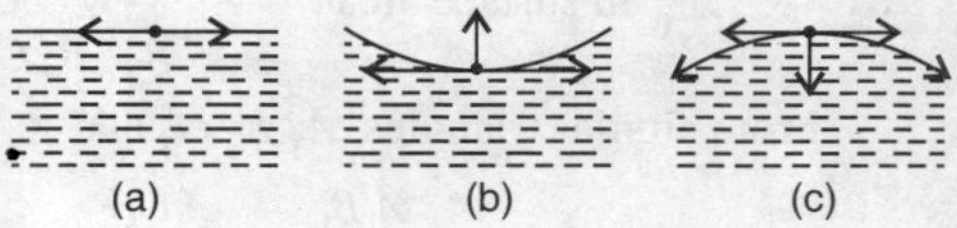

Fig. 3.6

(*c*) If the free surface of the liquid is convex, as in Fig. 3.6 (*c*), the resultant force due to S.T. on a molecule on the surface acts vertically downwards (into the liquid).

Excess pressure inside a liquid drop : A spherical liquid drop has a convex surface, as in Fig. 3.7 (*i*). The molecules near the surface of the drop experience a resultant force, acting inwards due to surface tension. Therefore, the pressure inside the drop must be greater than the pressure outside it. Let this excess pressure inside the liquid drop over the pressure outside it be p.

Imagine the drop to be divided into two exactly equal halves. Consider the equilibrium of the upper-half (or the upper hemisphere) of the drop as shown in Fig. 3.7 (*ii*). If r is the radius of the drop, and σ its S.T.

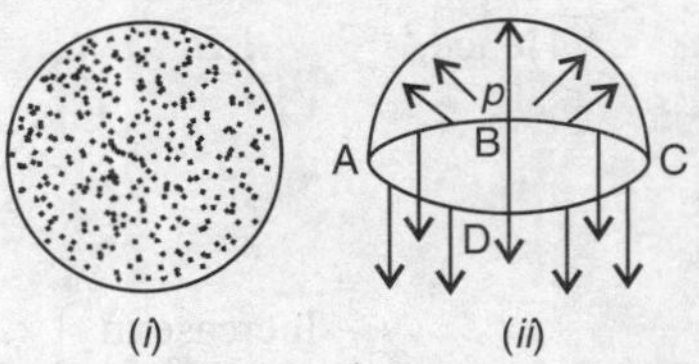

Fig. 3.7

$$\left.\begin{array}{l}\text{the upward force on the plane face}\\ ABCD \text{ due to the excesss pressure } p\end{array}\right\} = p\,\pi\, r^2$$

$$\left.\begin{array}{l}\text{the downward force due to surface tension acting}\\ \text{along the circumference of the circle } ABCD\end{array}\right\} = \sigma\, 2\pi\, r$$

Since the hemisphere is in equilibrium, the two forces are equal.

$$\therefore \qquad p\,\pi\, r^2 = \sigma\, 2\pi\, r \quad \text{or} \quad p = 2\sigma/r$$

Excess pressure inside a soap bubble : A soap bubble has two liquid surfaces in contact with air, one inside the bubble and the other outside the bubble.

The force due to S.T. in this case $= 2 \times \sigma\, 2\pi\, r = 4\pi\, r\, \sigma$

Therefore, for equilibrium of the hemisphere,

$$p\pi\, r^2 = 4\pi\, r\, \sigma \quad \text{or} \quad p = 4\sigma/r$$

Thus the excess pressure inside a drop or a bubble is inversely proportional to its radius (*i.e.*, $p \propto 1/r$). Since $p \propto 1/r$, the pressure needed to form a very small bubble is high. This explains why one needs to blow hard to start a balloon growing. Once the balloon has grown, less air pressure is needed to make it expand more.

3.9. EXCESS PRESSURE INSIDE A CURVED LIQUID SURFACE

When the pressure on both sides of a liquid surface is same, then the surface is flat, without any curvature. But when it is curved convex upwards, then the pressure inside must be greater than the pressure outside and the excess pressure inside is balanced by the force of S.T.

To find the excess pressure, consider a small curvilinear rectangular element $A_1 B_1 C_1 D_1$ of a liquid surface [Fig. 3.8]. $A_1 B_1$ has a radius of curvature R_1 with centre at O_1. $B_1 C_1$ has a radius of curvature R_2 with centre at O_2. Let p be the excess of pressure inside the surface over that outside. Then the outward thrust on the surface $A_1 B_1 C_1 D_1 = p \times$ Area of the element $A_1 B_1 C_1 D_1 = p \times A_1 B_1 \times B_1 C_1$.

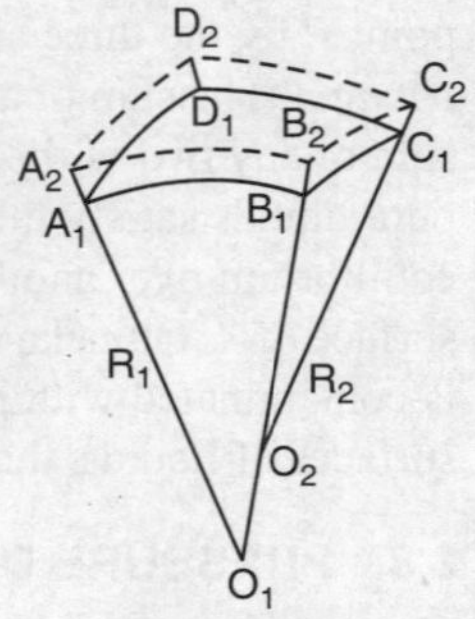

Fig. 3.8

Now, let the surface be moved outward through a very small distance δx.

Let the new position of the surface be $A_2 B_2 C_2 D_2$.

Work done in the displacement $= p \cdot A_1 B_1 \cdot B_1 C_1 \cdot \delta x$ ...(1)

$$\left.\begin{array}{l}\text{Now, increase}\\ \text{in surface area}\end{array}\right\} = \text{Area of } A_2 B_2 C_2 D_2 - \text{Area of } A_1 B_1 C_1 D_1$$

$$= A_2 B_2 \cdot B_2 C_2 - A_1 B_1 \cdot B_1 C_1$$

From similar triangles, $A_1 B_1 O_1$ and $A_2 B_2 O_1$,

$$\frac{A_1B_1}{A_2B_2} = \frac{A_1O_1}{A_2O_1} \quad \text{or} \quad \frac{A_1B_1}{A_2B_2} = \frac{R_1}{(R_1 + \delta x)}$$

$$\therefore \quad A_2B_2 = \frac{A_1B_1\,(R_1 + \delta x)}{R_1} = A_1B_1\left[1 + \frac{\delta x}{R_1}\right]$$

Similarly, $$B_2C_2 = B_1C_1\left(1 + \frac{\delta x}{R_2}\right)$$

Hence $$A_2B_2 \cdot B_2C_2 = A_1B_1\left(1 + \frac{\delta x}{R_1}\right)B_1C_1\left(1 + \frac{\delta x}{R_2}\right)$$

$$= A_1B_1 \cdot B_1C_1\left(1 + \frac{\delta x}{R_1} + \frac{\delta x}{R_2}\right)$$

$$\therefore \quad \left.\begin{array}{l}\text{Increase in}\\ \text{surface area}\end{array}\right\} = A_1B_1 \cdot B_1C_1\left(1 + \frac{\delta x}{R_1} + \frac{\delta x}{R_2}\right) - A_1B_1 \cdot B_1C_1$$

$$= A_1B_1 \cdot B_1C_1 \cdot \delta x\left(\frac{1}{R_1} + \frac{1}{R_2}\right)$$

$$\left.\begin{array}{l}\text{Work done in increasing}\\ \text{the area of a surface}\end{array}\right\} = \{\text{Surace tension} \times \text{increase in surface area}\}$$

$$= \sigma \cdot A_1B_1 \cdot B_1C_1 \cdot \delta x\left(\frac{1}{R_1} + \frac{1}{R_2}\right) \quad ...(2)$$

Equating (1) and (2) we have,

$$p \cdot A_1B_1 \cdot B_1C_1 \cdot \delta x = \sigma \cdot A_1B_1 \cdot B_1C_1 \cdot \delta x\left(\frac{1}{R_1} + \frac{1}{R_2}\right)$$

or $$p = \sigma\left(\frac{1}{R_1} + \frac{1}{R_2}\right)$$

The excess of pressure inside any particular surface can be deduced from the above expression.

1. **Spherical liquid drop** (an air bubble in a liquid).

It has only one surface and the radius of curvature is the same everywhere, *i.e.*, $R_1 = R_2 = r$.

$$\therefore \quad p = 2\sigma/r$$

2. **Spherical soap bubble.** Here there are two surfaces having the same radius of curvature.

$\therefore \qquad p = 4\sigma/r$

3. **Cylindrical drop.** $R_1 = r$ = Radius of the cylinder and $R_2 = \infty$.

$\therefore \qquad p = \sigma/r$

4. **Cylindrical bubble.** $p = 2\sigma/r$ since it has two surfaces.

Note: We have, $p = \sigma\left(\frac{1}{R_1} + \frac{1}{R_2}\right)$. This expression holds good for surfaces such as spherical or ellipsoidal, for which the principal radii of curvature are on the same side. Such surfaces are called synclastic surfaces [Fig. 3.9 (a)]. But in cases where the two radii are in opposite directions, the surface is called anticlastic [Fig. 3.9 (b)]. The expression for such surfaces is

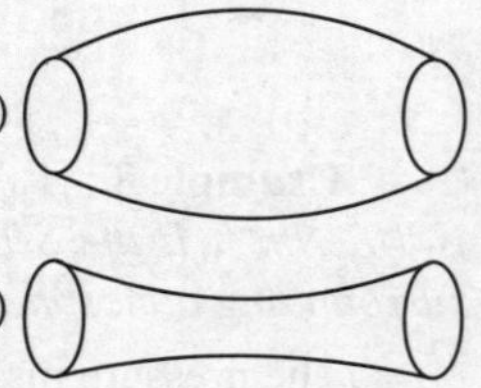

Fig. 3.9

$$p = \sigma\left(\frac{1}{R_1} - \frac{1}{R_2}\right)$$

Combining the two cases, the general relation is

$$p = \sigma\left(\frac{1}{R_1} \pm \frac{1}{R_2}\right)$$

Example 5 : *There is a minute circular hole at the bottom of a small hollow vessel. The vessel has to be immersed in water to a depth of 0.4 m, before any water penetrates inside. Find the radius of the hole, if the surface tension and density of water be 73×10^{-3} Nm^{-1} and 1000 kg m^{-3} respectively.*

Water cannot penetrate till the hydrostatic pressure is greater than the excess of pressure that comes into play due to surface tension. At the limiting condition, when water just penetrates, the two pressures should be equal; *i.e.,* $h\rho g = 2\sigma/r$ where r is the radius of the hole.

$\therefore \qquad r = 2.\sigma/(h\rho g)$

Here, $\qquad \sigma = 73 \times 10^{-3}$ Nm^{-1}; $h = 0.4$ m; $\rho = 1000$ kgm^{-3}; $r = ?$

$$r = \frac{2 \times (73 \times 10^{-3})}{0.4 \times 1000 \ 9.8} = 3.724 \times 10^{-5} \text{ m}$$

Example 6 : *What would be the pressure inside a small air bubble of 10^{-4} m radius, situated just below the surface of water? S.T. of water may be taken to be 70×10^{-3} Nm^{-1} and the atmospheric pressure to be 1.012×10^5 Nm^{-2}.*

$\left.\begin{array}{l}\text{Excess of pressure inside the spherical} \\ \text{air bubble over that of the atmosphere}\end{array}\right\} = p = \frac{2\sigma}{r}$

Here, $\sigma = 70 \times 10^{-3}$ Nm^{-1}; $r = 10^{-4}$ m

$$\left.\begin{array}{r}\text{Excess} \\ \text{pressure}\end{array}\right\} = p = \frac{2\sigma}{r} = \frac{2 \times (70 \times 10^{-3})}{10^{-4}} = 1400 \text{ Nm}^{-2}$$

$$\therefore \left.\begin{array}{l}\text{Total pressure inside} \\ \text{the air bubble}\end{array}\right\} = \left\{\begin{array}{l}\text{Atmospheric pressure} \\ + \text{Excess pressure}\end{array}\right\}$$

$$= 1.012 \times 10^5 + 1400 = 1.026 \times 10^5 \text{ Nm}^{-2}$$

Example 7 : *The pressure of air in a soap bubble of 7×10^{-3} m diameter is 8×10^{-3} m of water above the atmospheric pressure. Calculate the S.T. of the soap solution.*

$\left.\begin{array}{l}\text{Excess of pressure inside a soap} \\ \text{bubble over that outside it}\end{array}\right\} = p = \frac{4\sigma}{r}$

Here, $p = 8 \times 10^{-3}$ m of water

$= (8 \times 10^{-3}) \times 1000 \times 9.81$ Nm^{-2}

$= 78.48$ Nm^{-2}

$r = (7 \times 10^{-3})/2 = 3.5 \times 10^{-3}$ m

$$\therefore \quad \sigma = \frac{pr}{4} = \frac{78.48 \times (3.5 \times 10^{-3})}{4} = 68.67 \times 10^{-3} \text{ Nm}^{-1}$$

Example 8 : *A spherical bubble of radius 0.001 m is blown in an atmosphere whose pressure is 10^5 Nm^{-2}. If the S.T. of the liquid comprising the film is 0.05 Nm^{-1}, to what pressure must the surrounding atmosphere be brought in order that the radius of the bubble may be doubled ?*

The pressure inside the bubble initially is

$$p_i = \frac{4\sigma}{r} + P = \frac{4 \times 0.05}{0.001} + 10^5 = 1.002 \times 10^5 \text{ Nm}^{-2}$$

The volume of the bubble $= V_i = \frac{4}{3}\pi r^3 = \frac{4}{3}\pi \times 10^{-9}$ m^3

In the second case, the radius of the bubble is 0.002 m. Let the required pressure be P_f. Then the total pressure inside the bubble,

$$p_f = \frac{4\sigma}{0.002} + P_f = \frac{4 \times 0.05}{0.002} + P_f = (100 + P_f) \text{ Nm}^{-2}$$

In the second case, volume of the bubble $= V_f = \frac{4}{3}\pi (0.002)^3$

Applying Boyle's law, $p_i V_i = p_f V_f$,

$$(1.002 \times 10^5)\left(\frac{4}{3}\pi \times 10^{-9}\right) = (100 + P_f)\frac{4}{3}\pi (0.002)^3$$

Simplifying, $P_f = 1.252 \times 10^4$ Nm^{-2}

Example 9 : *Two soap bubbles of radii r_1 and r_2 coalesce to form a single bubble of radius r. If the external pressure is P, prove that the S.T. of the solution from which the soap bubble is formed is given by $\sigma = \frac{1}{4}P(r^3 - r_1^3 - r_2^3)/(r_1^2 + r_2^2 - r^2)$.*

Pressure inside the first bubble $= p_1 = (4\sigma/r_1) + P$

Pressure inside the second bubble $= p_2 = (4\sigma/r_2) + P$

Let the bubbles coalesce into a larger bubble of radius r.

Then, pressure inside this larger bubble $= p_3 = (4\sigma/r) + P$

Let V_1, V_2 be the volumes of the two bubbles before they coalesce and V_3 the volume of the large bubble formed. Then by Boyle's law,

$$p_1V_1 + p_2V_2 = p_3V_3$$

$$\left(\frac{4\sigma}{r_1} + P\right)\frac{4}{3}\pi r_1^3 + \left(\frac{4\sigma}{r_2} + P\right)\frac{4}{3}\pi r_2^3 = \left(\frac{4\sigma}{r} + P\right)\frac{4}{3}\pi r^3$$

or $4\sigma(r_1^2 + r_2^2 - r^2) = P(r^3 - r_1^3 - r_2^3)$

or $$\sigma = \frac{1}{4}P(r^3 - r_1^3 - r_2^3)/(r_1^2 + r_2^2 - r^2)$$

Example 10 : *A minute spherical air bubble is rising slowly through a column of mercury contained in a deep jar. If the radius of the bubble at a depth of 1 m is 0.0001 m, calculate its depth when its radius is 0.000126 m. Given : S.T. of mercury = 567 × 10^{-3} Nm^{-1}. Atmospheric pressure = 0.76 m of Hg.*

Total pressure inside the bubble at a depth of 1m = p_i = Atmospheric pressure + pressure due to 1 m of mercury column + excess pressure inside the bubble (2 σ/r)] *i.e.*,

$$p_i = 0.76 \times 13600 \times 9.81 + 1 \times 13600 \times 9.81 + (2 \times 567 \times 10^{-3}/0.0001)$$
$$= 2.461 \times 10^5 \text{ Nm}^{-2}$$

$$\text{Volume of the bubble} = V_i = \frac{4}{3}\pi r_2^3 = \frac{4}{3}\pi (0.0001)^3$$
$$= 4.187 \times 10^{-12} \text{ m}^3$$

Let h be the depth at which radius of the bubble is 0.000126 m.

Total pressure inside the bubble at a depth of h metre is

$$p_f = 0.76 \times 13600 \times 9.81 + h \times 13600 \times 9.81 + (2 \times 567 \times 10^{-3}/0.000126)$$
$$= 1.014 \times 10^5 + 1.334 \times 10^5 h + 9000 = 133400h + 110400$$

$$\text{Volume of the bubble} = V_f = \frac{4}{3}\pi r_2^3 = \frac{4}{3}\pi (0.000126)^3$$
$$= 8.373 \times 10^{-12} \text{ m}^3$$

Applying Boyle's law, $p_i V_i = p_f V_f$

$(2.461 \times 10^5)(4.187 \times 10^{-12}) = (133400h + 110400)(8.373 \times 10^{-12})$

Simplifying, $h = 0.0945$m

3.10. FORCE BETWEEN TWO PLATES SEPARATED BY A THIN LAYER OF A LIQUID

Let PQ and RS be two equal horizontal plates separated by a very thin layer of liquid of thickness d. Let the thin layer of liquid wet the plates over an almost circular area A which is concave outwards, as shown in Fig. 3.10. The radius of curvature of the two concave edges of the liquid layer is nearly $d/2$.

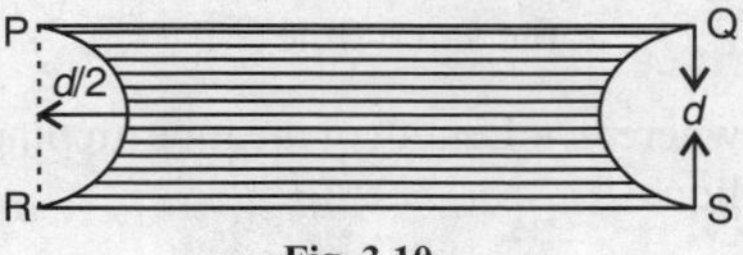

Fig. 3.10

$$\left.\begin{array}{l}\text{Excess pressure inside the liquid-film}\\ \text{over the outside atmospheric pressure}\end{array}\right\} = p = \sigma\left(\frac{1}{r} - \frac{2}{d}\right)$$

where σ = S.T. of the liquid-film and r = radius of the circular area of the liquid film.

or $\quad p = -2\sigma/d \quad$ since $r >> d$.

The negative sign indicates that the pressure inside the film is less than the atmospheric pressure by 2 σ/d. Therefore the excess pressure of the atmosphere on the two plates pushes them closer together. If the plates are to be separated by pulling them normally apart, a force 2 A σ/d should be applied.

Note : This explains why two plates with moistened surfaces stick to each other strongly.

Example 11 : *Calculate the force required to separate two plates of glass of area 10^{-3} m^2 each with a layer of water 10^{-6} m thick. S.T. of water = 75 × 10^{-3} Nm^{-1}.*

Here, $\quad A = 10^{-3}$ m², σ = 75 × 10^{-3} Nm⁻¹, $d = 10^{-6}$ m, F = ?

$$F = \frac{2A\sigma}{d} = \frac{2 \times 10^{-3} \times (75 \times 10^{-3})}{10^{-6}} = 150 \text{ N}$$

EXPERIMENTAL DETERMINATION OF SURFACE TENSION

3.11. JAEGAR'S METHOD

Principle : The experiment is based on the principle that the pressure inside an air bubble in a liquid is greater than the pressure outside it by 2 σ/r. Here σ is the S.T. of the liquid and r the radius of the air bubble. This excess pressure can be directly found and hence σ can be calculated.

Apparatus : An aspirator A is closed with a two-holed rubber stopper through which pass two glass tubes [Fig.3.11]. One of these is connected to a water reservoir through a stopcock B and the other is joined through a tap C to a manometer M and a vertical tube DE. The tube DE ends in a narrow orifice at E and dips into the experimental liquid contained in a beaker.

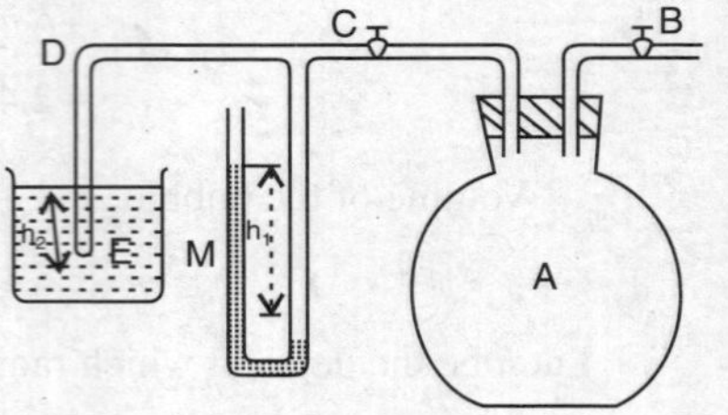

Fig. 3.11

Experimental details : If the stopcock B is opened, water flows into the aspirator and the air in the aspirator is displaced. The displaced air forces its way through the tube DE and forms air bubbles at E. The size of each air bubble gradually grows. When its radius becomes equal to the radius of the tube at E, it becomes unstable and breaks away. During the growth of the bubble, the pressure inside increases and reaches a maximum value at the instant of detachment. The difference in manometer levels h_1 is noted just when the bubble detaches itself. At the moment of detachment,

the pressure inside the bubble $= p_1 = H + h_1 \rho_1 g$, where

H = atmospheric pressure,

h_1 = the difference in manometer levels

and ρ_1 = density of the manometric liquid

$$\left.\begin{array}{l}\text{The pressure outside the bubble}\\ \text{at the same time}\end{array}\right\} = p_2 = H + h_2 \rho_2 g$$

where h_2 = Length of the tube dipping in the experimental liquid and ρ_2 = Density of the experimental liquid.

$$\therefore \quad \left.\begin{array}{l}\text{Excess pressure}\\ \text{inside the bubble}\end{array}\right\} = p = (H + h_1 \rho_1 g) - (H + h_2 \rho_2 g)$$

$$= (h_1 \rho_1 - h_2 \rho_2)\, g$$

But the excess pressure inside the bubble $= 2\sigma/r$

Hence $\quad 2\sigma/r = (h_1\rho_1 - h_2\rho_2)g \quad$ or $\quad \sigma = \dfrac{1}{2} rg\,(h_1\rho_1 - h_2\rho_2)$

Advantages of the method : (1) The angle of contact need not be known.

(2) The continual renewal of the liquid air interface helps in avoiding contamination.

(3) The experiment does not require a large quantity of liquid.

(4) The liquid in the beaker may be heated to various temperatures. Hence the S.T. of a liquid can be determined at various temperatures.

Drawbacks: (1) The exact value of the radius of the bubble when it breaks away cannot be ascertained.

(2) The drop may not be hemispherical and of quite the same radius as the aperture at E.

(3) The calculations are based on the assumption of static conditions but the phenomenon is not entirely statical.

For these reasons, this method does not give very accurate results for the surface tension.

Example 12. *In a Jaegar type of experiment to measure the S.T. of a liquid, the vertical capillary tube of radius 0.0005 m was dipped inside a liquid of density 1100kg/m³ to a depth 0.04 m below its surface. When an air bubble was formed at the end of the capillary tube dipping inside the liquid, it was observed from manometric reading that the pressure inside the bubble exceeded the atmospheric pressure by 0.00547m of mercury. Calculate the S.T. of the liquid.*

$$\sigma = \frac{rg}{2}(h_1\rho_1 - h_2\rho_2)$$

Here, $r = 0.0005$ m, $g = 9.8$ m/s², $h_1 = 0.00547$ m, $\rho_1 = 13600$ kg/m³, $h_2 = 0.04$ m and $\rho_2 = 1100$ kg/m³, $\sigma = ?$

$$\sigma = \frac{0.0005 \times 9.8}{2}(0.00547 \times 13600 - 0.04 \times 1100)$$

$$= 0.07446 \text{ Nm}^{-1}$$

3.12. VARIATION OF SURFACE TENSION WITH TEMPERATURE

Liquids are of two types, *viz.*, (*i*) unassociated liquid and (*ii*) associated liquid. An unassociated liquid contains the individual molecules of that liquid. **Example:** Benzene and carbon tetrachloride. An associated liquid contains groups of molecules of quite another type. These groups, however, tend to break up into single molecules with a rise in temperature. At the ordinary temperatures, water is known to consist of groups, consisting of two H_2O molecules, in addition to ordinary single H_2O molecules. Thus water is an associated liquid at these temperatures.

The S.T. of an unassociated liquid is found to decrease with rise of temperature, according to the simple formula $\sigma_t = \sigma_0(1 - \alpha t)$ where σ_t is the S.T. at t°C, σ_0 at 0°C and α, the temperature coefficient of S.T. for the liquid. Van der Waals and Ferguson suggested other relations from which it could be easily deduced that the S.T. is zero at the critical temperature. The best relation connecting S.T. and temperature, for both associated and unassociated liquids, is due to Eotvos. This formula was later modified by Ramsay and Shields.

This is represented by $\sigma(Mvx)^{3/2} = k(\theta_c - \theta - d)$ where σ = Surface tension at θ K, θ_c = Critical temperature, d = a constant, varying from 6 to 8 for most of the liquids, k = another constant having the value 2.12 for associated liquids and 2.22 for unassociated liquids.

x = Coefficient of association

$$= \frac{\text{effective molecular weight of associated liquid}}{\text{mol wt of the unassociated liquid with the same molecules}}$$

M = molecular weight of the unassociated liquid and v its specific volume.

This shows that the S.T. is zero, when $\theta = (\theta_c - d)$ *i.e.*, at a temperature a little below the critical temperature.

Note 1 : At higher temperature, the molecules themselves possess greater average kinetic energy and are moving about more rapidly. This reduces the effect of molecular attractions. So less work is needed to bring a molecule from the interior of the liquid to the surface. Hence the S.T. of all liquids decreases as the temperature rises.

Note 2 : Impurities, contaminations and dissolved substances all lower the S.T. of a liquid.

3.13. QUINCKE'S METHOD

The shape of a drop depends on the combined action of S.T. and gravity. When a large drop of mercury is placed on a clean horizontal glass plate, the drop flattens out until its top becomes perfectly horizontal as shown in Fig. 3.12 (a).

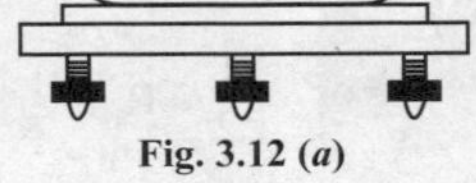

Fig. 3.12 (*a*)

Imagine that the drop is cut into two halves by a vertical plane *ABCD* [Fig. 3.12 (*b*)]. Let the drop be further cut by two vertical planes *BCG* and *ADH* at a distance *l* from each other and perpendicular to *ABCD*. *GH* is the most protruding portion of the drop. Let h_1 be the height of the flat top *AB* above the horizontal plane *EFGH* and h_2 the total height *BC* of the drop.

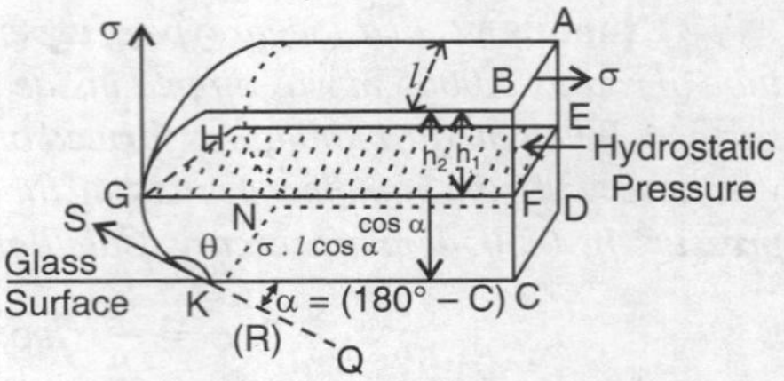

Fig. 3.12 (*b*)

The portion of the drop lying above the horizontal plane *EFGH* is in equilibrium under the action of the following forces:

1. Force due to S.T. acting at right angles to *AB* from left to right horizontally = σ. *l*.
2. Hydrostatic thrust acting horizontally from right to left on the plane *ABEF* of the drop due to part of the liquid on the right. The hydrostatic pressure is zero at *AB* and increases to $h_1 \rho g$ at *EF*.

$$\therefore \quad \text{Average pressure} = (0 + h_1 \rho g)/2 = \frac{1}{2} h_1 \rho g$$

$$\therefore \quad \text{Total hydrostatic thrust on the area } ABEF = \frac{1}{2} h_1 \rho g \times h_1 l$$

$$= \frac{1}{2} h_1^2 \rho g l$$

3. The S.T. at *G*, which acts vertically upwards, has no component along the horizontal.

Since the drop is in equilibrium, these horizontal forces must balance. Hence $\sigma . l = \frac{1}{2} h_1^2 \rho g l$,

or $\sigma = \frac{1}{2} h_1^2 \rho g$

Determination of the angle of contact : Consider the equilibrium of the whole drop. The forces acting on the face *ABCD* are :

Pull due to S.T. acting horizontally from left to right and perpendicular to $AB = \sigma . l$.

Hydrostatic thrust acting horizontally from right to left on the face *ABCD* $\Big\} = \frac{1}{2} h_2 \rho g \times h_2 l = \frac{1}{2} h_2^2 \rho g l$

The S.T. pull due to mercury on glass at *K* acting tangentially to the slice in the direction *KS* $\Big\} = \sigma . l$

The glass plate exerts an equal and opposite reaction along *KQ*.

The horizontal component of this reactional force due to glass along *KC*, acting from left to right $\Big\} = \sigma . l \cos \alpha$

where $\alpha = (180 - \theta)$

In equilibrium, $\sigma . l + \sigma . l \cos \alpha = \frac{1}{2} h_2^2 \rho g . l$

$$\text{or} \quad \sigma (1 + \cos \alpha) = \frac{1}{2} h_2^2 \rho . g$$

$$\text{or} \quad \frac{1}{2} h_1^2 \rho . g (1 + \cos \alpha) = \frac{1}{2} h_2^2 \rho . g \qquad \left[\because \sigma = \frac{1}{2} h_1^2 \rho g \right]$$

or $$(1 + \cos\alpha) = \frac{h_2^2}{h_1^2} \text{ or } \cos\alpha = \frac{h_2^2}{h_1^2} - 1$$

or $$\cos\alpha = (h_2^2 - h_1^2)/h_1^2$$

The value of α can be determined from this relation.

The angle of contact $\theta = 180 - \alpha$.

Experiment : Place a clean glass plate on a small table provided with levelling screws. Arrange it horizontally with the help of a spirit level. Form a large drop of mercury on the glass plate so that its upper surface is perfectly plane.

Sprinkle a fine layer of lycopodium powder at the top of the drop. Adjust the microscope so that the horizontal cross-wire is in line with the lycopodium powder. Lower the microscope and focus it on the line of separation of mercury and glass. The difference between the two readings gives h_2.

Move the microscope to one edge of the drop. Focus the microscope so that the horizontal cross-wire is in line with the most protruding part of the drop. To find this position very accurately, light from an incandescent lamp S is focussed by a lens L and the plane glass plate P, acting as a mirror, on to the edge of the drop (Fig.3.13). A bright, thin, horizontal line is seen at G, where the drop protrudes out most. The horizontal cross-wire is placed coinciding with this bright horizontal line and the reading is noted. The difference between the top surface reading and this reading gives h_1. Using the relations given above, S.T. of mercury and angle of contact are determined.

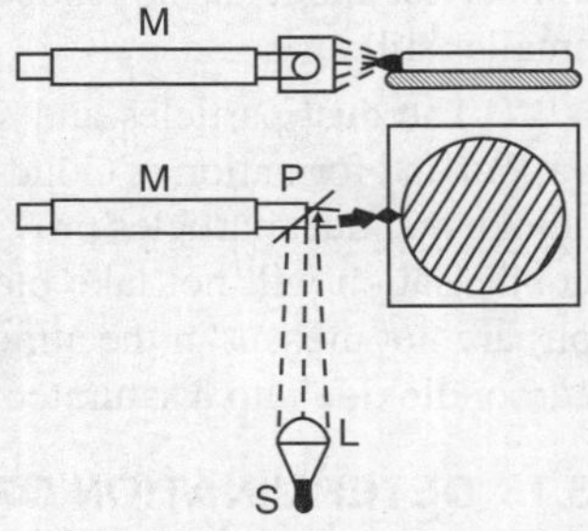

Fig. 3.13

3.14 VAPOUR PRESSURE OVER FLAT AND CURVED SURFACES

When a liquid is contained in a closed vessel, the space above the liquid is filled with vapour and a stage is reached when evaporation and condensation go on at equal rates. At this stage, the vapour is in dynamic equilibrium with the liquid and the pressure exerted by it is called the saturation vapour pressure of the liquid at that temperature.

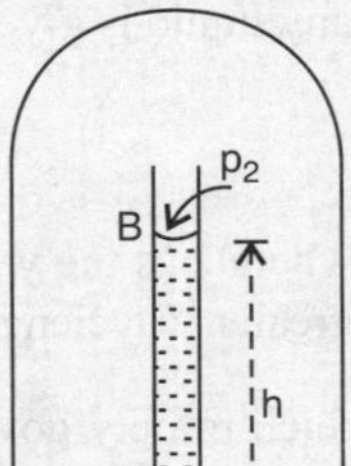

Fig. 3.14

Let a capillary tube of radius r be dipped vertically in a liquid (which wets the tube) of S.T. σ and density ρ (Fig. 3.14). Let the whole arrangement be enclosed in an exhausted bell jar, so that the effect of atmosphere may be neglected. The liquid evaporates till a steady state is reached and the space within the bell-jar is saturated with the vapour. The liquid rises to a height h given by

$$\sigma = rh\rho g/2 \qquad ...(i)$$

If p_1 is the vapour pressure at the horizontal surface A, p_2 is the vapour pressure above the concave surface B (Fig. 3.14) of the liquid in the tube and ρ' is the density of the vapour supposed to be uniform between A and B of height h, then

$$p_1 - p_2 = h\rho' g \qquad ...(ii)$$

Dividing (*ii*) by (*i*), $$\frac{p_1 - p_2}{\sigma} = \frac{2\rho'}{r\rho} \text{ or } p_1 - p_2 = \frac{2\sigma}{r} \times \frac{\rho'}{\rho}$$

Therefore, the saturation vapour pressure over the concave surface is less than that over a flat surface by $\frac{2\sigma}{r} \times \frac{\rho'}{\rho}$.

In the case of a liquid like mercury, there is a capillary depression and the meniscus is convex. Then the saturation vapour pressure over the convex surface is more than that over the flat surface by, $\frac{2\sigma}{r} \times \frac{\rho'}{\rho}$.

Formation of drops. The saturated vapour pressure over a convex surface is greater than that over a flat surface by $\frac{2\sigma}{r} \times \frac{\rho'}{\rho}$. Let us place a drop of water in a space in which the vapour pressure is at the saturation value for a plane surface. The drop is spherical and has a convex surface. The maximum vapour pressure over a convex surface must be more than that on a plane surface. Hence the drop evaporates in order to increase the vapour pressure to its own saturation value.

This will result in a further decrease in the radius of the drop and the value of saturation vapour pressure over it must rise further. Therefore, the drop evaporates more and more rapidly to increase the vapour pressure. The space then contains more vapour than that required for saturation at that temperature, without condensation taking place and the vapour becomes super-saturated.

If dust particles or charged ions are present in the atmosphere, vapour condenses on them and the size of the drop formed, even in the beginning, is sufficiently large. The surface of a large drop is almost flat and it has no tendency to evaporate. As its size grows, its tendency to evaporate becomes smaller still.

The dust particles and smoke particles play an important role in the condensation of water vapour and formation of clouds. That is why the weather looks foggy in large industrial towns, where smoke and dust particles provide nuclei for condensation to set in. Even in a super-saturated cloud, condensation will not take place if dust particles (which act as nuclei for the vapour to condense on) are not present in the atmosphere. Artificial rain is sometimes brought about by injecting solid carbondioxide into a saturated cloud to act as nuclei for the water vapour to condense on.

3.15 DETERMINATION OF S.T. BY RIPPLES METHOD (RAYLEIGH'S METHOD)

Waves and Ripples : The velocity of waves travelling over the surface of a liquid is governed by the force of gravity (g) and S.T. (σ). The velocity, v, of a harmonic disturbance on the surface of any liquid is given by

$$v = \sqrt{\frac{\lambda g}{2\pi} + \frac{2\pi\sigma}{\lambda\rho}} \qquad ...(1)$$

where λ is the wavelength of the waves, σ, the S.T. and ρ density of the liquid. In the case of the greater wavelengths, the velocity of the wave is found to be predominantly dependent on gravity. Such gravity governed waves are called *waves*. Therefore, for long waves the term $\frac{2\pi\sigma}{\lambda\rho}$ can be neglected, and the velocity of propagation of the wave due to gravity only is $\sqrt{\frac{\lambda g}{2\pi}}$.

Equation (1) can be written as

$$v = \sqrt{\frac{\lambda}{2\pi}\left(g + \frac{4\pi^2}{\lambda^2\rho}\sigma\right)} \qquad ... (2)$$

From equation (2) it appears that the effect of S.T. is to increase the effective value of g by $\frac{4\pi^2}{\lambda^2\rho}\sigma$.

If the wavelength is smaller than a particular wavelength λ_c, the critical wavelength, it is found that the velocity of wave is chiefly governed by S.T. Such waves are called *ripples*. The first term in Eq.(1) becomes negligible in comparison with the second when $\lambda < \lambda_c$. Therefore, the velocity of ripples is given by

$$v = \sqrt{\frac{2\pi\sigma}{\lambda\rho}}$$

Eq. (1) shows that $v = \infty$ when $\lambda = 0$ and when $\lambda = \infty$. Therefore λ has a critical value λ_c for which v is minimum. For v_{minm} the two terms should be equal.

Note : The product of the two terms $\frac{\lambda g}{2\pi}$ and $\frac{2\pi\,\sigma}{\lambda\,.\,\rho}$ is $\frac{g\sigma}{\rho}$ and is a constant. Therefore, their sum will be minimum when they are equal.)

i.e., $$\frac{\lambda_c g}{2\pi} = \frac{2\pi\,\sigma}{\lambda_c\rho} \text{ or } \lambda_c = 2\pi\sqrt{\frac{\sigma}{\rho\,.\,g}}$$

For water, $\sigma = 75 \times 10^{-3}$ Nm^{-1}; $\rho = 1000$ kg m^{-3}, $g = 9.8$ ms^{-2}.

$$\therefore \qquad \lambda_c = 2\pi\sqrt{\frac{75 \times 10^{-3}}{1000 \times 9.8}} = 0.017 \text{ m}$$

Waves having a smaller wavelength than this critical value are called ripples.

Principle of the experiment. This method depends on a measurement of the wavelength of ripples formed on the surface of the liquid whose S.T. is to be determined.

EXPERIMENTAL ARRANGEMENT

The perfectly clean liquid is placed in a large flat dish (*A*), such as a photographic developing dish (Fig. 3.15). An electrically maintained tuning fork (*F*) of large frequency (100) is held in position above the liquid. The tuning fork is held horizontally with the prongs vibrating in the vertical plane. A dipper (*D*) of polished silver or aluminium is attached to the lower prong and its lower end touches the liquid surface in the dish.

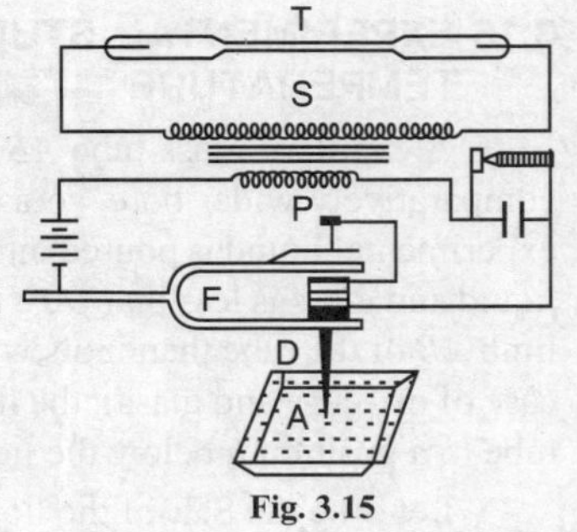

Fig. 3.15

When the prong is set into vibration, the vertical to and fro motion of the dipper produces ripples in the liquid. The ripples formed are reflected from the walls of the trough and give rise to stationary waves.

To observe the ripples and to measure their wavelength, the liquid surface must be properly illuminated. This is done by completing the tuning fork circuit through the primary *P* of a small induction coil. The neon discharge tube (*T*) is connected to the secondary *S* of the induction coil. Every time the tuning fork circuit is made, an electric discharge passes through the neon discharge tube. Then the light from the neon tube brilliantly illuminates the liquid surface in the trough.

During one vibration of the fork, there is one alternate make and break of the tuning fork circuit, giving us one view of the ripples on the liquid surface. The same occurs again when the fork has made one vibration during which the ripples have moved through one full wavelength. So when the liquid surface is lit up again as before, we get a second view of the ripples which is identical with the first. Hence the ripples appear to be stationary in their earlier position due to persistence of vision.

The frequency with which the liquid surface is lit up is the same as that of the tuning fork and, therefore, of the ripples.

A graph paper is placed at the bottom of the trough and the distance between a large number of ripples is measured. From this the mean wavelength λ is determined.

Calculation: Let n be the frequency of the fork and λ the wavelength of ripples. Then, we have $v = n\lambda$ where v is the velocity of the ripples. Substituting this value of v in Eq. (1),

$$v = n\lambda = \sqrt{\frac{\lambda g}{2\pi} + \frac{2\pi\,\sigma}{\lambda\,.\,\rho}}$$

$$\text{or} \qquad n^2\,.\,\lambda^2 = \frac{\lambda\,g}{2\pi} + \frac{2\pi\,\sigma}{\lambda\,.\,\rho} \text{ or } \frac{2\pi\sigma}{\lambda\,.\,\rho} = n^2\,.\,\lambda^2 - \frac{\lambda\,.\,g}{2\pi}$$

or $$\sigma = \frac{\lambda . \rho}{2\pi} n^2 . \lambda^2 - \frac{\lambda \rho}{2\pi} \times \frac{\lambda . g}{2\pi}$$

or $$\sigma = \frac{n^2 . \lambda^3 \rho}{2\pi} - \frac{\lambda^2 \rho . g}{4\pi^2}$$

Since λ, n and ρ are known, the S.T. of the given liquid can be calculated.

Example 13 : *A light stylus attached to the prong of a tuning fork vibrating 100 times per second dips slightly in a trough full of mercury and produces ripples of wavelength 0.00278m. Calculate the S.T. of mercury.*

We have, $$\sigma = \frac{n^2 . \lambda^3 \rho}{2\pi} - \frac{\lambda^2 \rho . g}{4\pi^2}$$

Here, $n = 100$ Hz. $\lambda = 0.00278$ m. $\rho = 13600$ kg m^{-3} and $g = 9.8$ ms^{-2}.

$$\sigma = \frac{(100)^2 \ (0.00278)^3 \ 13600}{2\pi} - \frac{(0.00278)^2 \ (13600) \ 9.8}{4\pi^2}$$

$$= 0.4649 - 0.02602 = 0.4389 \text{ Nm}^{-1}.$$

3.16 EXPERIMENTAL STUDY OF THE VARIATION OF SURFACE TENSION WITH TEMPERATURE

A capillary glass tube *AB* is connected to another glass tube *CD*, of a comparatively wider bore, so as to form a complete *U* tube (Fig. 3.16). The experimental liquid is poured into this *U* tube. If the angle of contact (θ) for the liquid and glass is less than 90°, the liquid stands much higher in the narrower limb *AB* of the tube than in its wider limb *CD*. If θ > 90° (for example, in the case of mercury and glass), the liquid column in *AB* will be depressed into the tube to a position *h* below the liquid column in *CD*.

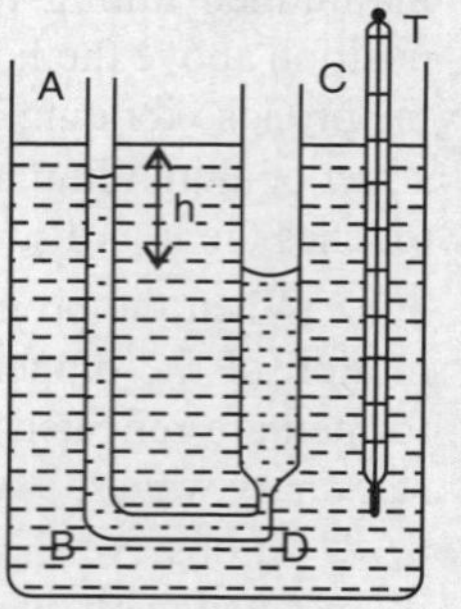

Fig. 3.16

Let σ be the S.T. of the liquid and θ, its angle of contact with glass. Let r_1 and r_2 be the radii of the bores of the two limbs respectively. Let *P* be the atmospheric pressure. Then,

$$\left.\begin{array}{l}\text{pressure just below the} \\ \text{liquid meniscus in } AB\end{array}\right\} = P - \frac{2\sigma \cos\theta}{r_1}$$

$$\left.\begin{array}{l}\text{pressure just below the} \\ \text{liquid meniscus in } CD\end{array}\right\} = P - \frac{2\sigma \cos\theta}{r_2}$$

$$\therefore \left.\begin{array}{l}\text{Difference of pressure} \\ \text{in the two limbs}\end{array}\right\} = 2\sigma \cos\theta \left(\frac{1}{r_1} - \frac{1}{r_2}\right)$$

This is balanced by the hydrostatic pressure $h \rho g$ due to the liquid column in the *U*-tube. Here, *h* is the difference of liquid levels in the two limbs. ρ is the density of the liquid.

$$\therefore \quad 2\sigma \cos\theta \left(\frac{1}{r_1} - \frac{1}{r_2}\right) = h \rho g$$

or $$\sigma = \frac{h \rho g}{2\left(\frac{1}{r_1} - \frac{1}{r_2}\right) \cos\theta}$$

The experiment is repeated with the liquid maintained at different temperatures by means of suitable temperature baths.

3.17. DROP-WEIGHT METHOD OF DETERMINING THE SURFACE TENSION OF A LIQUID

Experiment : A short glass tube is connected to the lower end of a burette (or funnel) clamped vertically, by means of a rubber tube [Fig. 3.17]. The funnel is filled with the liquid whose S.T. is to be determined. A beaker is arranged under the glass tube to collect the liquid dropping from the funnel. The stopcock is adjusted so that the liquid drops are formed slowly. In a previously weighed beaker a known number of drops (say, 50) are collected.

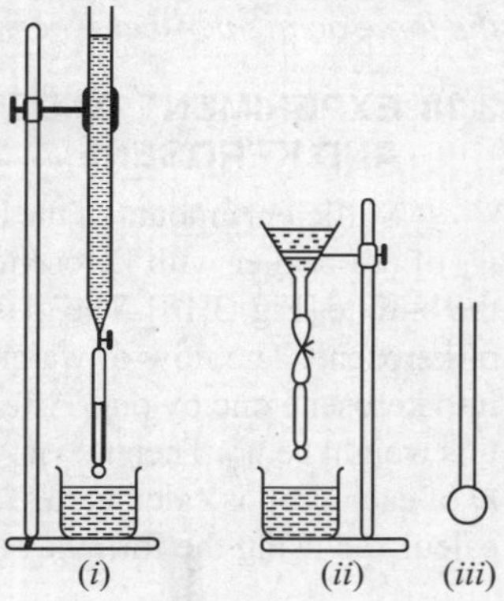

Fig. 3.17

The beaker is again weighed. The difference between this weight and the weight of the empty beaker gives the weight of 50 drops of the liquid. From this the mass m of each drop is calculated. The inner radius r of the tube is determined using a vernier calipers. The S.T. of the liquid at the room temperature is calculated using the formula,

$$\sigma = \frac{m \cdot g}{3.8\, r}$$

Theory : Here, we consider the vertical forces that keep a small drop of liquid in equilibrium, just before it gets detached from the end of a vertical glass tube of circular aperture. At the instant the drop gets detached, it assumes a cylindrical shape at the orifice of the tube [Fig. 3.17(*iii*)]. Let σ = S.T. of the liquid and r = radius of the orifice.

$$\left.\begin{array}{l}\text{Excess pressure } (p) \text{ inside the drop over} \\ \text{the outside atmospheric pressure}\end{array}\right\} = \frac{\sigma}{r}$$

The area of the section is πr^2. Therefore,

$$\left.\begin{array}{l}\text{downward force on the drop} \\ \text{due to this excess of pressure}\end{array}\right\} = \pi r^2 \frac{\sigma}{r}$$

The weight mg of the drop also acts vertically downwards.

∴ Total downward force on the drop = $(\pi r^2 \sigma/r) + mg$

This downward force is balanced by the upward pull due to surface tension $2\pi r \sigma$ acting along a circle of radius r. Therefore

$$2\pi r\,\sigma = \frac{\pi r^2 \sigma}{r} + mg \quad \text{or} \quad 2\pi r\,\sigma = \pi r \sigma + mg$$

$$\therefore \qquad \sigma = \frac{m \cdot g}{\pi \cdot r}$$

But the equilibrium of the drop at the instant of its detachment is dynamic and not static. Lord Rayleigh, taking dynamical aspect into account, showed that

$$\sigma = \frac{m \cdot g}{3.8\, r}$$

Example 14 : *In a drop weight method for the determination of S.T. between water and air, a glass tube of external diameter 2 mm is used, and 100 drops of water are collected. The mass of these drops is 2.8 gms. Find the S.T. of water in air.*

Here, $r = 10^{-3}$ m. $m = 2.8 \times 10^{-3}/100 = 2.8 \times 10^{-5}$ kg

$$\therefore \qquad \sigma = \frac{m \cdot g}{3.8\, r} = \frac{(2.8 \times 10^{-5}) \times 9.8}{3.8 \times 10^{-3}} = 0.07221 \text{ Nm}^{-1}$$

Interfacial Tension: At the surface of separation between two immiscible liquids there is a tension similar to surface tension. It is called the interfacial tension.

Definition: *When one liquid rests on another without mixing with it, the interface between the two liquids possesses energy just like the surface of a liquid. The interfacial tension is the value of the force acting per metre normal to a line drawn on the interface.*

3.18. EXPERIMENT TO DETERMINE THE INTERFACIAL TENSION BETWEEN WATER AND KEROSENE

Sufficient amount of the lighter liquid (kerosene) is taken in a beaker. The weight w_1 of the beaker with kerosene is determined. The heavier liquid (water) is taken in the burette [Fig. 3.18]. The glass tube is fixed vertically with its end under the surface of kerosene. The flow of water is regulated so that drops of water detach themselves into kerosene one by one. After collecting 50 drops, the beaker is again weighed. Let this weight be w_2. Then $w_2 - w_1$ gives the mass of 50 drops. From this the average mass m of each drop is calculated. The interfacial tension σ between water and kerosene is calculated using the formula

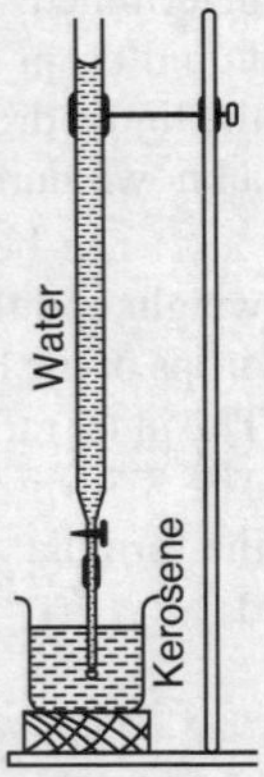

Fig. 3.18

$$\sigma = \frac{m \cdot g}{3.8\,r}\left(1 - \frac{\rho_2}{\rho_1}\right)$$

No.	Mass of beaker + lighter liquid		Mean w_1 kg	Mass of beaker + lighter liquid + 50 drops		Mean w_2 kg	Mass of 50 drops = $(w_2 - w_1)$ kg
	Trial I	Trial II		Trial I	Trial II		

Theory : Let ρ_1 and ρ_2 be the densities of water and kerosene respectively. Let m be the mass of water drop in air.

Volume of water drop $= m/\rho_1$

$\therefore$ volume of kerosene displaced by the water drop $= \dfrac{m}{\rho_1}$

mass of kerosene displaced by the water drop $= \dfrac{m\,\rho_2}{\rho_1}$

$\therefore$ Apparent weight of the water drop in kerosene $= mg - \dfrac{m\,\rho_2\,g}{\rho_1}$

Let σ be the S.T. at the interface between the two liquids.

Then,
$$2\pi\, r\, \sigma = \frac{\pi r^2 \sigma}{r} + m \cdot g - \frac{m\,\rho_2 \cdot g}{\rho_1}$$

$$\therefore \quad \sigma = \frac{m \cdot g}{\pi r}\left(1 - \frac{\rho_2}{\rho_1}\right)$$

Again the more accurate equation will be

$$\sigma = \frac{m \cdot g}{3.8\,r}\left(1 - \frac{\rho_2}{\rho_1}\right)$$

Example 15 : *100 drops of water falling down a tube of external diameter 3.5 mm are collected under coconut oil of specific gravity 0.8. Calculate the interfacial tension between water and oil if the water collected weighs 12.35 gm.*

Here, $m = (12.35 \times 10^{-3})/100 = 1.235 \times 10^{-4}$ kg

$r = 1.75 \times 10^{-3}$ m; $\rho_1 = 1000$ kg/m³ and $\rho_2 = 800$ kg/m³ . σ = ?

$$\sigma = \frac{m \cdot g}{3.8\, r}\left(1 - \frac{\rho_2}{\rho_1}\right)$$

$$= \frac{(1.235 \times 10^{-4})\, 9.8}{3.8 \times (1.75 \times 10^{-3})} \times \left(1 - \frac{800}{1000}\right) = 0.0364 \text{ Nm}^{-1}$$

EXERCISE III

SECTION – A

1. Define S.T. and angle of contact.
2. What are the units and dimensions of surface tension ?
3. Explain S.T. from the point of view of molecular theory of matter.
4. What is surface energy ? How is it related to S.T.?
5. It is easier to spray water when soap is added to it than when it is pure. Why ?

   [**Hint.** *When soap is added to water, the surface tension decreases. Therefore less work is required to spray water.*]
6. Several fog particles merge to form a single particle. Has the potential energy decreased or increased ? Explain your answer. [**Hint.** *Area decreases → P.E. decreases.*]
7. In the case of mercury, there is capillary depression. Why ?

   [**Ans.** *This is because the cohesive forces in the liquid molecules are greater than the forces of attraction between the glass and liquid molecules*]
8. Water wets the glass surface while mercury does not. Why ?

   [**Hint.** *Because the forces of adhesion between water and glass are greater than the forces of cohesion between water molecules.*]
9. If two bubbles of different size are blown at opposite ends of a pipe, what will happen ?

   [**Ans.** *The smaller bubble will contract and the larger bubble will expand, showing that the air flows from smaller bubble which has more pressure to larger bubble which has less pressure.*]
10. Obtain the relation between the vapour pressure over a curved surface and that over a flat surface.
11. Obtain an expression for the excess of pressure inside (*i*) a spherical soap bubble and (*ii*) a spherical liquid drop.

SECTION – B

12. Prove that the excess of pressure on one side of a soap film of surface tension σ over that on the other side is given by $p = 2\sigma\left(\frac{1}{R_1} + \frac{1}{R_2}\right)$.
13. Describe Jaegar's method of studying the variation of S.T. of water with temperature.
14. Describe Quincke's method of finding S.T. Derive the formula employed.
15. Deduce the relation between S.T. and vapour pressure at a curved surface and discuss its effects on evaporation and condensation.
16. How do you distinguish between ripples and waves ? Describe Rayleigh's method for measuring the surface tension of a liquid. Derive the necessary formula used.
17. Explain the drop-weight method experiment to determine the surface tension of a liquid.

SECTION – C

18. A drop of water of radius 10^{-4} m is split into 1000 equal tiny droplets. Find the mechanical work done. Calculate the excess of pressure inside each droplet, if the S.T. of water is 75×10^{-3} Nm^{-1}.

    [**Ans.** $W = 8.482 \times 10^{-8}$ J; $P = 1.5 \times 10^4$ Nm^{-2}]

19. Calculate the excess pressure inside a soap bubble of radius 3×10^{-3} m. Surface tension of soap solution is 20×10^{-3} N/m. Calculate also the surface energy of the soap bubble.

[**Ans.** 26.66 N/m^2. 4.525×10^{-6} *J*]

20. What amount of energy will be liberated, if a thousand droplets of water each 10^{-8} m in diameter coalesce to form one large spherical drop? S.T. of water $=75 \times 10^{-3}$ Nm^{-1}. [**Ans.** 2.12×10^{-14} *J*]

21. A liquid drop of radius R breaks into 64 tiny drops. Find the resulting change in energy. Surface tension of the liquid is σ. [**Ans.** $12\pi\, \sigma\, R^2$]

22. Calculate the amount of work done in blowing a soap bubble from a radius of 0.1 m to 0.15 m. S.T. of soap solution $= 35 \times 10^{-3}$ Nm^{-1}. [**Ans.** 1.1×10^{-2} *J*]

23. The difference of pressure between the inside and outside of a soap bubble of diameter 3×10^{-3} m is 10^{-2} m of liquid of density 800 kg m^{-3}. Calculate the S.T. of the soap solution. ($g = 9.8$ ms^{-2}).

[**Ans.** 29.4×10^{-3} Nm^{-1}]

24. If a number of little droplets of water, all of the same radius r, coalesce to form a single drop of radius R, show that the rise in temperature of water will be given by $\theta = 3\sigma\left(\frac{1}{r} - \frac{1}{R}\right)$ where σ is the S.T. of water.

25. Prove that the air inside a soap bubble of sufficiently large radius R would at atmospheric pressure P fill a sphere of radius $R + \frac{4\sigma}{3P}$, where σ is the S.T.

[**Hint.** Pressure inside the soap bubble $= P + \frac{4\sigma}{R}$.

Suppose x is the radius of the bubble when the pressure inside is atmospheric. Then by Boyle's law,

$$P \times \frac{4}{3}\pi x^3 = \left(P + \frac{4\sigma}{R}\right)\frac{4}{3}\pi R^3, \quad \text{or} \quad x = R\left(1 + \frac{4\sigma}{RP}\right)^{\frac{1}{3}}$$

$$= R\left(1 + \frac{4\sigma}{3RP}\right) \text{ approximately, } \therefore\ x = R + \left(\frac{4\sigma}{3P}\right)$$

26. Calculate the amount of energy evolved when 1000 droplets of water of radius 0.0002 m combine to form a single drop. S.T. of water $= 72 \times 10^{-3}$ Nm^{-1} [**Ans.** 3.257×10^{-5} *J*]

27. Two spherical soap bubbles of diameter 0.1 m and 0.06 m respectively are formed at the ends of a narrow horizontal tube. What is the difference of pressure between the two ends of the tube ? (S.T. of soap solution $= 3 \times 10^{-2}$ Nm^{-1}) [**Ans.** 1.6 Nm^{-2}]

28. A spherical soap bubble of radius 0.01 m is formed inside another soap bubble of radius 0.02 m. Calculate the radius of a single soap bubble which will have an excess of pressure equal to the difference in pressure between the inside of the inner bubble and the outside of the larger bubble. [**Ans.** 6.67×10^{-3} m]

29. A minute spherical air bubble is formed in water at a depth of 1 m from its surface. The diameter of the bubble is 1.4×10^{-4} m and the S.T. of water is 70×10^{-3} Nm^{-1}. Calculate the pressure inside the bubble. (Pressure of air over the surface of water = 0.75 m of Hg]. [**Ans.** 0.8385 m of Hg]

30. Calculate the difference in vapour pressure of water for a plane surface and that for a drop of radius 2×10^{-4} m. Density of water vapour = 0.6 kg/m^3 and surface tension of water $= 70 \times 10^{-3}$ N/m.

[**Ans.** $p_1 - p_2 = \frac{2\sigma}{r} \times \frac{\rho'}{\rho} = \frac{2 \times (70 \times 10^{-3})}{2 \times 10^{-4}} \times \frac{0.6}{1000} = 0.42 \text{ N/m}^2$]

CHAPTER 4

Hydrodynamics

4.1 EQUATION OF CONTINUITY

The equation of continuity is an expression of the law of conservation of mass in fluid mechanics. Fig. 4.1 represents a tube of varying cross-section through which a non-viscous incompressible fluid of density ρ flows. Let a_1 and a_2 be the cross-sectional areas of the tube at the points A and B. Let the velocity of the fluid at A and B be v_1 and v_2 respectively. Since the fluid is incompressible, in the steady state, mass of fluid entering the tube per second through the section A = mass of fluid leaving the tube per second through the section B.

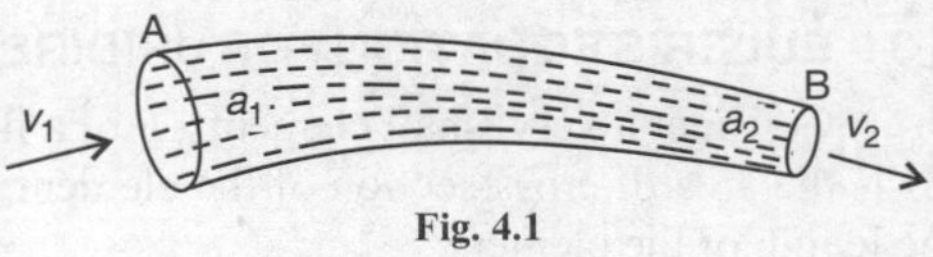

Fig. 4.1

Mass of fluid entering the tube per second across the section $A = a_1 v_1 \rho$

Mass of fluid leaving the tube per second across the section $B = a_2 v_2 \rho$

$$\therefore \qquad a_1 v_1 \rho = a_2 v_2 \rho \text{ or } a_1 v_1 = a_2 v_2.$$

Thus the product av is constant along any given flow tube. It follows that the speed of flow through a tube is inversely proportional to the cross-sectional area of the tube. It means that where the area of cross-section of the tube is large, the velocity is small and *vice versa.*

Example. *Water flowing with a velocity of 3 m/s in a 4 cm diameter pipe enters a narrow pipe having a diameter of only 2 cm. Calculate the velocity in the narrow pipe.*

Here $a_1 = \pi\,(0.02)^2;\ v_1 = 3\text{ m/s};\ a_2 = \pi\,(0.01)^2\ ;\ v_2 = ?$

$$v_2 = a_1 v_1 / a_2 = \pi\,(0.02)^2 \times 3 / \{\pi\,(0.01)^2\} = 12 \text{ m/s}.$$

4.2 ENERGY OF THE LIQUID

A liquid in motion possesses three types of energy, viz., (*i*) potential energy, (*ii*) kinetic energy and (*iii*) pressure energy.

(*i*) Potential energy. If we have a mass m of the liquid at a height h above the earth's surface, its P.E. = mgh.

P. E. per unit mass of the liquid = gh.

P. E. per unit volume of the liquid = ρgh

(*ii*) Kinetic energy. The K. E. is the energy possessed by the liquid by virtue of its motion. The K. E. of a mass m of a liquid flowing with a velocity v is $\frac{1}{2}mv^2$.

K. E. per unit mass of the liquid = $\frac{1}{2}v^2$

K. E. per unit volume of the liquid = $\frac{1}{2}\rho v^2$

***(iii)* Pressure energy.** Consider an incompressible non-viscous liquid contained in a tank (Fig. 4.2). The tank has a side tube at an axial depth h below the free surface of the liquid in the tank. The side tube is fitted with a smooth piston. Let a be the area of cross-section of the side tube. Let ρ be the density of the liquid. Hydrostatic pressure of the liquid on the piston $= p = h\rho g$. Force on the piston $= pa$.

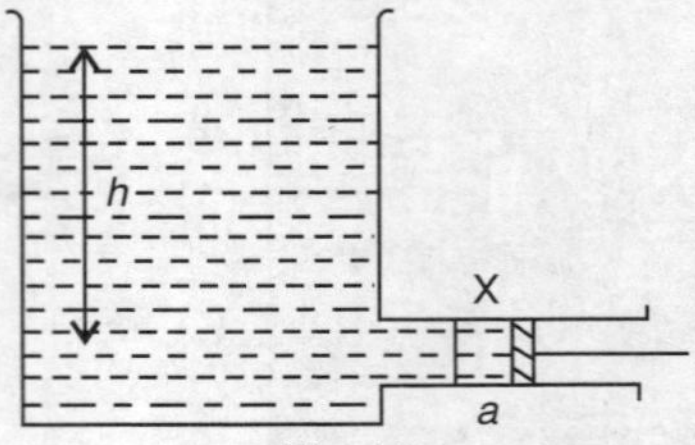

Fig. 4.2

If the piston is moved inwards through a distance x, the work done $= pax$. This will force a mass ρax of the liquid into the tank. This work done on the mass ρax accounts for an expenditure of energy pax. This is stored up as pressure energy of the same mass without imparting any velocity to it.

$$\text{Pressure energy per unit mass} = \frac{pax}{\rho\, ax} = \frac{p}{\rho}$$

Pressure energy per unit volume of the liquid $= p$

The three forms of energy possessed by a liquid under flow are mutually convertible.

4.3 EULER'S EQUATION FOR UNIDIRECTIONAL FLOW

Consider a very small element AB of a flowing liquid (Fig. 4.3). ρ is the density of the liquid. dA is the area of cross-section of the element, ds is the length of the element.

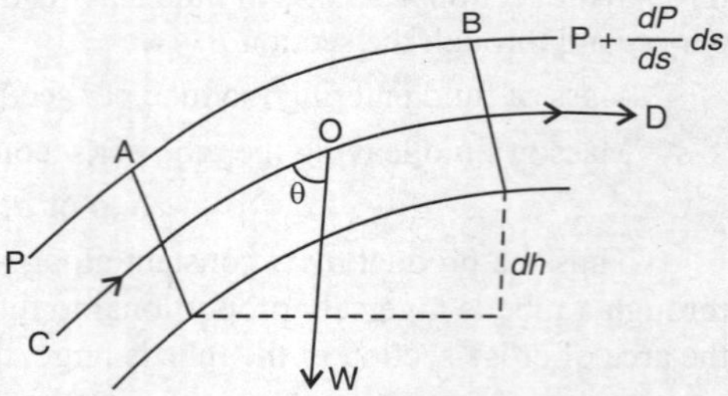

Fig. 4.3

Volume of the element $= dA.ds$

Mass of the element $= \rho\, dA.ds$.

Weight of the element $= \rho dA.ds.g$.

Let the streamline CD make an angle θ with the vertical at O.

Component of the weight of the element along the streamline

$$= \rho dA.ds.g.\cos\theta.$$

Let the pressure of the liquid at A be P.

$$\text{Then, the pressure at } B = \left(P + \frac{dP}{ds}ds\right)$$

$$\text{Resultant force on the element} = PdA - \left(P + \frac{dP}{ds}ds\right)dA = -\frac{dP}{ds}ds\,dA.$$

This force is directed from B to A. The component of the weight of the element is also directed from B to A.

$$\text{The net force acting on the element} = -\frac{dP}{ds}ds\,dA - \rho\, ds\, dA\, g\cos\theta$$

Let dh be the difference in heights between A and B. Then $\cos\theta = dh/ds$.

$\therefore$ the resultant force on the element is

$$F = -\frac{dP}{ds}ds\,dA - \rho\, ds\, dA\, g\,\frac{dh}{ds} = -\frac{dP}{ds}ds\,dA - \rho\, dA\, dh\, g$$

Let the velocity of the element of liquid be v. Then its acceleration is

$$a = \frac{dv}{dt} = \frac{ds}{dt}\frac{dv}{ds} = v\frac{dv}{ds}$$

According to Newton's II law of motion, $F = ma$.

$$-\frac{dP}{ds}\,ds\,dA - \rho\,dA\,dh\,g = \rho\,dA\,ds\,v\frac{dv}{ds}$$

or $$dP + \rho g\,dh = -\rho\,v\,dv$$

or $$\frac{dP}{\rho g} + dh + \frac{v\,dv}{g} = 0$$

This equation is called Euler's equation of motion.

Integrating, $$\frac{P}{\rho g} + \frac{v^2}{2g} + h = constant.$$

or $$\frac{P}{\rho} + \frac{1}{2}v^2 + gh = constant.$$

This is Bernoulli's equation.

4.4 BERNOULLI'S THEOREM

Statement. *The total energy of an incompressible non-viscous fluid flowing from one point to another, without any friction, remains constant throughout the motion.*

Explanation. According to this theorem, the sum of kinetic, potential and pressure energies of any element of an incompressible fluid in streamline flow remains constant. Suppose the height of an element of fluid of density ρ above ground level is h. Let it be moving with a velocity v. Let it have a pressure p. Then, its total energy per unit volume is

$$E = \rho v^2/2 + \rho\,gh + p.$$

Bernoulli's theorem states that E is a constant.

If at two points in the fluid the velocities are v_1, v_2, the heights are h_1, h_2 and the pressures are p_1, p_2, then,

$$\rho v_1^2/2 + \rho\,gh_1, + p_1 = \rho\,v_2^2/2 + \rho\,gh_2 + p_2$$

The K. E. per unit weight is called *velocity head* and is equal to $v^2/2g$. The P. E. per unit weight is called the *gravitational head* and is equal to h. The pressure energy per unit weight is called the *pressure head* and is equal to $p/\rho g$. Bernoulli's equation can be written as

$$v^2/(2g) + h + p/(\rho g) = constant$$

i.e., velocity head + gravitational head + pressure head = constant.

In the case of liquid flowing *along a horizontal pipe,* the gravitational head h is a constant.

$$\therefore \quad \frac{v^2}{2g} + \frac{p}{\rho g} = \text{constant or } \frac{v^2}{2} + \frac{p}{\rho} = \text{constant}$$

or $$p + \rho v^2/2 = \text{constant}$$

or static pressure + dynamic pressure = constant.

This expression shows that greater velocity corresponds to a decrease in pressure and vice versa, *i.e.,* points of maximum pressure correspond to those of minimum velocity and vice versa. This principle may be used to determine fluid speeds by means of pressure measurements.

Example. Venturimeter, Pitot tube, etc.

Proof. Consider a fluid in streamline motion along a non-uniform tube (Fig. 4.4). Let A and B be two transverse sections of the tube at heights h_1 and h_2 from a reference plane (the surface of the earth). Let a_1 and a_2 be the areas of cross-section at A and B. Let v_1 and v_2 be the velocities of the fluid at A and B. Let p_1 be the pressure at A due to the driving pressure head. Let p_2 be the pressure at B. Since $a_1 > a_2$, $v_2 > v_1$. Hence the fluid is accelerated as it flows from A to B.

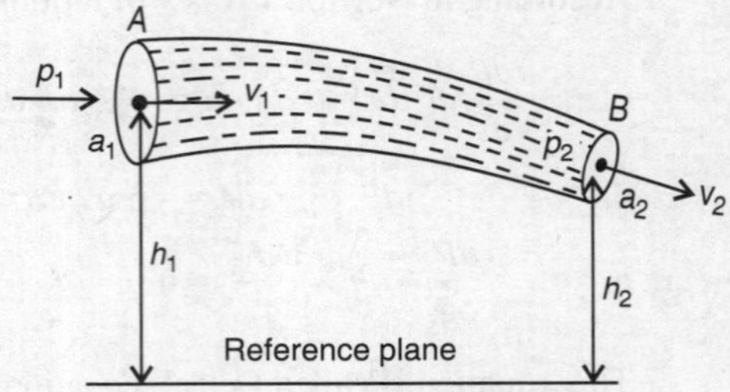

Fig. 4.4

Work done per second on the liquid entering at A is

$$W_1 = \text{Force at } A \times \text{Distance moved by the liquid in 1 second}$$

$$= p_1 a_1 \times v_1 = p_1 a_1 v_1$$

Work done per second by the liquid leaving the tube at B is

$$W_2 = p_2 a_2 v_2$$

$\therefore$ Net work done by the fluid in passing from A to B

$$W = W_1 - W_2 = p_1 a_1 v_1 - p_2 a_2 v_2$$

But $\quad a_2 v_2 = a_1 v_1$

$\therefore \quad W = (p_1 - p_2)\, a_1 v_1$

The work done on the liquid is used in changing its potential and kinetic energies.

$$\text{Decrease in P.E.} = (a_1\, v_1\, \rho) g\, (h_1 - h_2)$$

$$\text{Increase in K. E.} = \frac{1}{2}(a_1 v_1 \rho)\,(v_2^2 - v_1^2)$$

Hence, the total gain in the energy of the system when the liquid flows from A to B

$$= \frac{1}{2}(a_1 v_1 \rho)\,(v_2^2 - v_1^2) - (a_1 v_1 \rho g)\,(h_1 - h_2)$$

$$\therefore \quad (p_1 - p_2)\, a_1\, v_1 = \frac{1}{2}(a_1 v_1 \rho)(v_2^2 - v_1^2) - (a_1 v_1 \rho g)\,(h_1 - h_2)$$

or $\quad p_1 + \frac{1}{2}\rho v_1^2 + \rho g h_1 = p_2 + \frac{1}{2}\rho v_2^2 + \rho g h_2$

or $\quad p + \frac{1}{2}\rho v^2 + h\rho g = \text{constant}$

or $\quad \frac{p}{\rho} + \frac{1}{2}v^2 + hg = \text{constant}$

or $\quad \frac{p}{\rho g} + \frac{v^2}{2g} + h = \text{constant}$

$p/(\rho g)$ is called the pressure head, $v^2/(2g)$ is called the velocity head and h is called the gravitational head. Hence, Bernoulli's theorem may be stated as follows.

When an incompressible non-viscous fluid flows in streamline motion, the sum of the pressure head, velocity head and gravitational head remains constant throughout its motion.

Example 1. *Water flows through a horizontal tube of varying section. At a place the velocity is 0.3 m/s when the pressure is 0.016 m of mercury. Find the pressure when the velocity is 0.6 m/s.*

Sol. Let A and B be two points in the tube at the same horizontal level. The potential energies at these points are the same. Let p_1 and p_2 be the pressures at A and B. Let v_1 and v_2 be the velocities of the liquid at A and B. Then, according to Bernoulli's theorem,

$$p_1 + \frac{1}{2}\rho v_1^2 = p_2 + \frac{1}{2}\rho v_2^2.$$

Here, $p_1 = 0.016$ m of Hg $= 0.016 \times 13600 \times 9.8$ N/m²

$$v_1 = 0.3 \text{ m/s}; \ v_2 = 0.6 \text{ m/s}; \ p_2 = ?$$

$$p_2 = p_1 + \frac{1}{2}\rho(v_1^2 - v_2^2)$$

$$= (0.016 \times 13600 \times 9.8) + \frac{1}{2} \times 1000 \times (0.3^2 - 0.6^2)$$

$$= 1997 \ N/m^2$$

$$= \frac{1997}{13600 \times 9.8} \text{ m of Hg} = 0.015 \text{ m of Hg.}$$

Example 2. *Water flows along a horizontal tube of which the cross-section is not constant. Calculate the change in pressure when the velocity of flow changes from* 0.1 *m/s to* 0.2 *m/s*.

Sol. For horizontal flow Bernoulli's equation takes the form

$$p_1 + \rho\frac{v_1^2}{2} = p_2 + \rho\frac{v_2^2}{2}$$

$$\therefore \quad p_2 - p_1 = \frac{\rho}{2}(v_1^2 - v_2^2) = \frac{1000}{2}[(0.1)^2 - (0.2)^2].$$

$$= -15 \text{ N m}^{-2}.$$

The negative sign indicates that due to increase of velocity the pressure decreases.

Example 3. *A railway locomotive lowers a tube in a reservoir of water in between the rails. The immersed tube is horizontal and the upper end is 3 m above the surface of water in the reservoir. If the locomotive is running at 48 kmph, calculate the speed with which the water rushes out of the upper end.*

Sol. Let the lower end be h metre below the water surface.

The total energy at the upper end of the tube is equal to the total energy at the lower end.

$$\frac{1}{2}v_2^2 + \frac{p_2}{\rho} + h_2 g = \frac{1}{2}v_1^2 + \frac{p_1}{\rho} + h_1 g$$

Now, $v_1 = 48$ kmph $= 13.33$ m/s; $p_1 = p_0 + h\rho g$ where p_0 is the atmospheric pressure; $p_2 = p_0$; $h_1 = 0$ and $h_2 = (h + 3)$. $v_2 = ?$

$$\therefore \quad \frac{1}{2}v_2^2 + \frac{p_0}{\rho} + (h+3)g = \frac{1}{2}(13.33)^2 + \frac{p_0 + h\rho g}{\rho} + 0.$$

$$\therefore \quad v_2^2 = (13.33)^2 - 2 \times 3 \times 9.8 = 118.9 \text{ or } v_2 = 10.90 \text{ m/s.}$$

Example 4. *A fire engine pumps water from a hydrant at the rate of 1000 litres/s. It ejects it from a nozzle 5m above the surface of water in the hydrant with a velocity of* $10 \ ms^{-1}$. *Calculate (i) the pressure difference between the water at the pump and the nozzle and (ii) power of the engine.*

Sol. (*i*) $p_1 + \frac{1}{2}\rho v_1^2 + \rho g h_1 = p_2 + \frac{1}{2}\rho v_2^2 + \rho g h_2$

$$\therefore \quad p_1 - p_2 = \frac{1}{2}\rho\,(v_2^2 - v_1^2) + \rho\,g\,(h_2 - h_1)$$

$$= \frac{1}{2} \times 1000 \times [10^2 - 0^2] + 1000 \times 9.8 \times 5$$

$$= 9.9 \times 10^4 \text{ Nm}^{-2}.$$

(*ii*) Power of the engine = $(p_1 - p_2)$ (rate of flow)

$= (9.9 \times 10^4) \times (1)$ ($\because$ 1000 litres/s = 1m^3/s)

$= 9.9 \times 10^4$ W.

Applications of Bernoulli's theorem

(*i*) Torricelli's theorem

Consider a vessel *T* having a liquid of density ρ. Let the surface of the liquid be at a height *h* above the level of the orifice *O* [Fig. 4.5]. If the vessel is sufficiently wide, the velocity of the liquid on the free surface *A* may be taken as zero. The pressure on the free surface is atmospheric pressure. Since the pressure is also atmospheric at the orifice, where the liquid emerges, it plays no part in the flow of the liquid. Let *v* be the velocity of the liquid at the orifice. Consider a tube of flow beginning at *A* and ending at *O*.

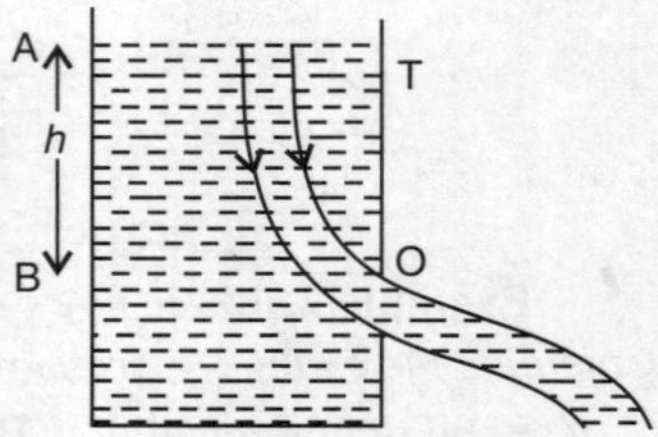

Fig. 4.5

Total energy at *A*= pressure energy + potential energy + kinetic energy

$= 0 + hg + 0 = hg.$

$$\text{Total energy at } O = 0 + 0 + \frac{1}{2}v^2 = \frac{1}{2}v^2.$$

Since total energy remains the same.

$$\frac{1}{2}v^2 = hg \text{ or } v = \sqrt{2gh}$$

This gives the velocity of efflux at the orifice. This result is known as Torricelli's theorem. It states that the *velocity of efflux of a liquid through an orifice is equal to that which a body attains in falling freely from the surface to the orifice*.The liquid that escapes through the orifice with the velocity $v = \sqrt{(2gh)}$ flows out along a parabolic path. Let h_1 be the height of the orifice above the bottom of the vessel. The liquid reaches the horizontal plane through the bottom of the vessel in time

$$t = \sqrt{2h_1 / g}$$

The escaping liquid will strike the horizontal plane through the bottom of the vessel at a distance *H,* where $H = v \times t$

$$H = \sqrt{(2gh)} \times \sqrt{2h_1 / g} = 2\sqrt{hh_1}$$

The range *H* is maximum for a given height $(h + h_1)$ if $h = h_1$.

Vena contracta. It is to be noted that the volume of the liquid that moves across the orifice in one second cannot be computed by multiplying the area of the orifice with the velocity of efflux. The streamlines converge as they approach the orifice. Therefore, the fluid velocities, as the jet leaves the hole, are not parallel to one another but have components inwards towards the centre of the stream. This inward momentum of the emergent fluid causes a contraction of the jet. After the jet has gone a little away, the contraction stops and the velocities become parallel. This point where the contraction of the jet stops and the fluid velocities become parallel is known as *vena contracta.* It is at this point that the velocity multiplied by area gives the rate of flow of the liquid. For an orifice of circular shape, the area of the *vena contracta* is about 65% of the area of the orifice.

Example 1. *A tank containing water has an orifice on one vertical side. If the centre of the orifice is 4.9 m below the surface level in the tank, find the velocity of discharge, assuming that there is no wastage of energy.*

$$v = \sqrt{2gh} = \sqrt{2 \times 9.8 \times 4.9} = 9.8 \text{ ms}^{-1}.$$

Example 2. *Calculate the velocity of efflux of kerosene oil from a tank in which the pressure is 35150 kg wt per m² above the atmospheric pressure. The density of kerosene is 800 kg m⁻³.*

Sol. Now,
$$h = \frac{p}{\rho} = \frac{35150}{800} = 43.94m.$$

$$\therefore \quad v = \sqrt{2gh} = \sqrt{2 \times 9.8 \times 43.94} = 29.35 \text{ms}^{-1}.$$

***(ii)* Venturimeter.** It is a device based on Bernoulli's principle. It is used for measuring the rate of flow of liquids in pipes. It consists of two wide conical tubes C_1 and C_2 with a constriction T between them. T is called the *throat.* Let the area of cross-section of C_1 and C_2 be A. Let a be the area of cross-section of the throat [Fig. 4.6].

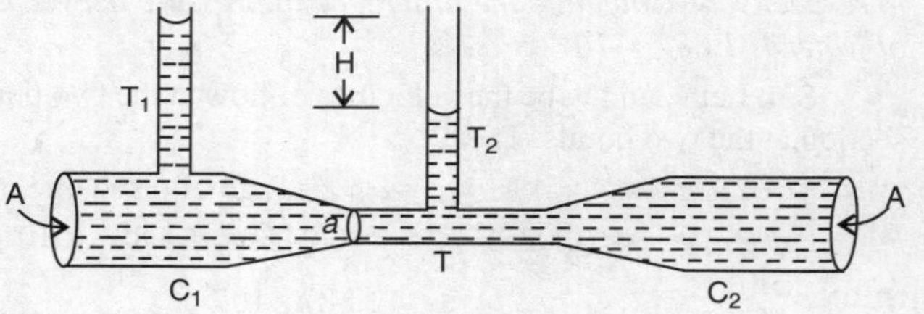

Fig. 4.6

When the flow is steady, let V be the volume of water flowing per second through the venturimeter.

Then, $V = Av_1 = av_2$

Here v_1 = velocity in C_1 or C_2 and v_2 = velocity in T.

$\therefore$ $v_1 = V/A$ and $v_2 = V/a$.

Hence velocity of water in T is greater than the velocity in C_1 and C_2. Consequently, the pressure in T is smaller than the pressure in C_1 and C_2. This difference in pressure H is measured by the difference of the water levels in the vertical glass tubes T_1 and T_2 connected to C_1 and T respectively. Let p_1 and p_2 be the pressures in the wider limb and throat respectively.

According to Bernoulli's equation for a horizontal flow,

$$\frac{p_1}{\rho g} + \frac{v_1^2}{2g} = \frac{p_2}{\rho g} + \frac{v_2^2}{2g}$$

or
$$\frac{p_1 - p_2}{\rho g} = \frac{v_2^2 - v_1^2}{2g}$$

The difference in pressure in C_1 and $T = p_1 - p_2 = H\rho g$.

Hence
$$\frac{H\rho g}{\rho g} = \frac{1}{2g}\left[\frac{V^2}{a^2} - \frac{V^2}{A^2}\right] = \frac{V^2}{2g}\left[\frac{A^2 - a^2}{A^2 a^2}\right]$$

$$\therefore \quad V = Aa\sqrt{\frac{2gH}{(A^2 - a^2)}}$$

The rate of flow of water through the pipeline can be determined by measuring H, and knowing the constants A, a, and g.

Example 1. *The diameter of the throat of a venturimeter is 0.06 m. When it is inserted in a horizontal pipe line of diameter 0.1 m, the pressure difference between the pipe and the throat equals 0.08 m of water. Calculate the rate of flow.*

Sol. Here $a = \pi\,(0.03)^2 = 2.826 \times 10^{-3}\ m^2\ ;\ A = \pi\,(0.05)^2$

$= 7.85 \times 10^{-3}\ m^2.\ H = 0.08\ m.\ V = ?$

$$V = Aa\sqrt{\frac{2gH}{A^2 - a^2}}$$

$$= (7.85 \times 10^{-3})\,(2.826 \times 10^{-3})\sqrt{\frac{2 \times 9.8 \times 0.08}{(7.85 \times 10^{-3})^2 - (2.826 \times 10^{-3})^2}}$$

$$= 3.793 \times 10^{-3}\ \text{m}^3/\text{s}.$$

Example 2. *The diameters of a horizontal water pipe line at two points are 0.05 m and 0.08 m respectively. Calculate the difference in pressure between the two points if the rate of discharge of water is 1.887×10^{-3} m³/s.*

Sol. Let v_1 and v_2 be the velocities of flow at the two points. Let a_1 and a_2 be the areas of cross-section at the two points. Then,

$$a_1 v_1 = a_2 v_2 = \text{Rate of discharge of water}$$

$$\therefore \quad \pi\,(0.025)^2\, v_1 = \pi\,(0.04)^2\, v_2 = 1.887 \times 10^{-3}.$$

Hence,
$$v_1 = \frac{1.887 \times 10^{-3}}{\pi\,(0.025)^2} = 0.9615\ \text{m}/s$$

$$v_2 = \frac{1.887 \times 10^{-3}}{\pi\,(0.04)^2} = 0.3756\ \text{m/s}$$

Applying Bernoulli's theorem

$$\frac{p_1}{\rho} + \frac{1}{2}v_1^{\,2} = \frac{p_2}{\rho} + \frac{1}{2}v_2^{\,2}$$

or
$$p_2 - p_1 = \frac{\rho}{2}(v_1^2 - v_2^2)$$

$$p_2 - p_1 = \frac{1000}{2}[(0.9615)^2 - (0.3756)^2] = 391.75\ \text{N/m}^2$$

Let H be the pressure difference in terms of metres of water.

$$H = \frac{p_2 - p_1}{\rho g} = \frac{391.75}{1000 \times 9.8} = 0.04\ \text{m}.$$

Therefore, the pressure difference between the two points is 0.04 metres of water.

Example 3. *A venturimeter has a pipe diameter of* 0.2 *m and a throat diameter* 0.15 *m. The levels of water column in the two limbs differ by* 0.1 *m. Calculate the amount of water discharged through the pipe in one hour. Density of water* = 1000 *kg m*$^{-3}$.

Sol.
$$A = \frac{\pi d_1^2}{4} = \frac{\pi\,(0.2)^2}{4} = 0.0314 m^2$$

$$a = \frac{\pi d_2^2}{4} = \frac{\pi\,(0.15)^2}{4} = 0.0177\text{m}^2$$

$$H = 0.1\ \text{m}.$$

$$\therefore \qquad V = Aa\sqrt{\frac{2gH}{A^2 - a^2}} = 0.0314 \times 0.0177\sqrt{\frac{2 \times 9.8 \times 0.1}{(0.0314)^2 - (0.0177)^2}}$$

$$= 0.03 \text{ m}^3 \text{ s}^{-1}.$$

Therefore, Volume of water flowing per hour = 0.03 × 3600 $m^3 hr^{-1}$ = 108 m^3/hr.

(*iii*) **Pitot tube.** It is an instrument used to measure the rate of flow of water through a pipeline. It is based on Bernoulli's principle. It consists of two vertical tubes *PQ* and *RS* with small apertures at their lower ends (Fig. 4.7). The plane of aperture of the tube *PQ* is parallel to the direction of flow of water. The plane of aperture of the tube *RS* faces the flow of water perpendicularly. The rise of the water column in the tube *RS*, therefore, measures the pressure at *S*.

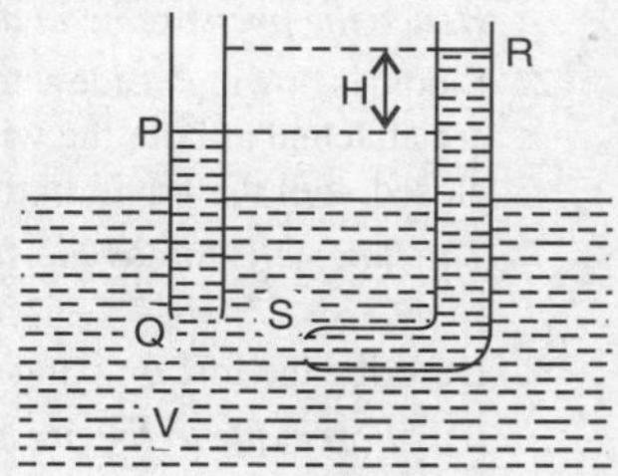

Fig. 4.7

Let p_1 and p_2 be the pressures of water at *Q* and *S* respectively. Let *v* be the velocity of water at *Q*. Since the water is stopped in the plane of the aperture *S* of the tube *RS*, its velocity at *S* becomes zero. Hence the pressure increases to p_2 at *S*. Let *H* be difference of level in the two tubes. Applying Bernoulli's theorem to the ends *Q* and *S*,

$$\therefore \qquad \frac{1}{2}v^2 + \frac{p_1}{\rho} = \frac{p_2}{\rho} \text{ or } v^2 = \frac{2}{\rho}(p_2 - p_1) = \frac{2}{\rho}H\rho g.$$

$$\therefore \qquad v = \sqrt{2gH}$$

Rate of flow of water = *av*

where *a* = area of cross-section of pipe.

Example 4. *A Pitot tube is fixed in a main of diameter* 0.15 *m and the difference of pressure indicated by the gauge is 0.04 m of water column. Find the volume of water passing through the main in a minute.*

Sol. Here, $\qquad v = \sqrt{2gH} = \sqrt{2 \times 9.8 \times 0.04} = 0.8854 \text{ m/s}$

∴ volume of water flowing per second across the section

= area of section × velocity = π (0.075)^2 × 0.8854.

Volume of water passing through the main in a minute

= [π (0.075)^2 × 0.8854] × 60 = 0.9383 m^3.

Wings of an aeroplane: Wings of an aeroplane are shaped as shown in Fig. 4.8, with the lower surface being flat and the upper surface being curved. The figure shows a cross-section of the wing perpendicular to its length. A stream of air flowing past the wing divides into two branches, one above and one below the wing. The streamlines far away from the wing are straight. The streamlines just above the wing get distorted due to the curvature of the upper surface of the wing. A tube of flow just above the wing will therefore get constricted.

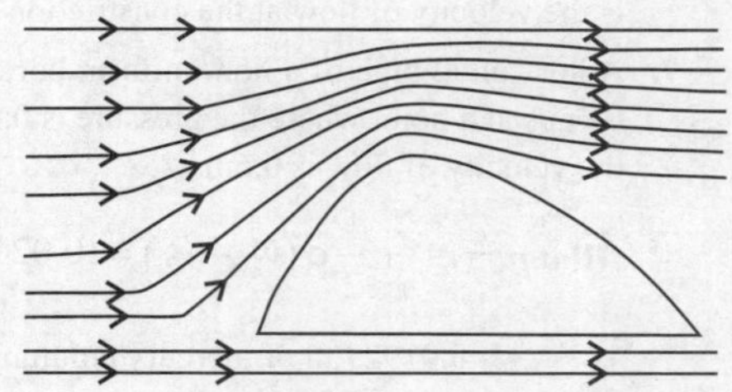
Fig. 4.8

Its area of cross-section decreases. According to the equation of continuity ($a_1v_1 = a_2v_2$), the velocity of air increases. According to Bernoulli's theorem, this causes a decrease in pressure above the wing. The tubes of flow below the wing are not affected much. The pressure below the wing continues to be atmospheric. Therefore, there is a lesser pressure above the wing than below it. This difference of pressure is helpful in giving a vertical lift to the plane.

SHORT ANSWER QUESTIONS

1. Two rowboats moving parallel to one another in the same direction are pulled toward one another. Two automobiles moving parallel are also pulled together. Explain such phenomena on the basis of Bernoulli's equation.

   In both the cases the velocity head of the air in between increases due to streamlines becoming denser. Hence the pressure head decreases causing the pull towards one another.

2. Liquid is flowing inside a horizontal pipe which has a constriction along its length. Vertical tube manometers are attached at both the wide portion and the narrow portion of the pipe. If a stopcock at the exit end is closed, will the liquid in the manometer tubes rise or fall ? Explain.

   The flow will stop as the stopcock at the exit end is closed. So levels in all the manometer tubes will be same.

3. Describe the forces acting on an element of fluid as it flows through a pipe of non-uniform cross-section.

   (*i*) *Force due to the pressure difference at the two sides of section of interest*

   (*ii*) *Viscous force*

   (*iii*) *Gravitational force if the pipe is not horizontal.*

4. The destructive effect of a tornado (twister) is greater near the centre of the disturbance than near the edge. Explain.

   The velocity in the tornado is the greatest at the centre and pressure minimum. So the destruction near the centre is maximum.

5. Does the difference in pressure between the lower and upper surfaces of an airplane wing depend on the altitude of the moving plane ? Explain.

   Yes. Density of air ρ *changes with altitude.*

$$p_1 - p_2 = \frac{1}{2}\rho\,(v_2^2 - v_1^2)$$

EXERCISE – IV

1. Obtain the equation of continuity for an incompressible fluid in streamline flow.
2. Derive Euler's equation of flow for liquids. Hence derive Bernoulli's equation.
3. State and explain Bernoulli's theorem. Obtain Torricelli's theorem.
4. State and prove Bernoulli's theorem. Discuss in detail two applications of this theorem.
5. Describe and explain the working of (*i*) venturimeter and (*ii*) Pitot tube.
6. The diameter of a main pipe line having constriction is 10cm and the velocity of flow is 0.5 ms^{-1}. What is the velocity of flow at the constriction if the diameter of the constriction is 2 cm.? [**Ans.** 12.5 ms^{-1}.]
7. A horizontal pipe of a non-uniform bore has water flowing through it such that the velocity of flow is 0.4m/s at a point where the pressure is 0.02 *m* of mercury column. What is the pressure at a point where the velocity of flow is 0.6m/s?

   [**Hint** $p_2 = p_1 + \frac{1}{2}\rho\,(v_1^2 - v_2^2) = (0.02 \times 13600 \times 9.8) + \frac{1}{2} \times 1000 \times (0.4^2 - 0.6^2) = 2565.6\ N/m^2$

   $= 0.01923$ m of mercury column]

8. The pressure of water in a horizontal pipe line of varying cross-section where the velocity is 0.24 ms^{-1} is 0.015 m of Hg. What is the pressure at a point where the velocity is 0.48 ms^{-1}?

   [**Ans.** 0.01435 m of Hg]

9. A railway locomotive lowers a tube which projects horizontally in a reservoir of water in between the rails. If the upper end of the tube is 4m above the surface of water in the reservoir, find the speed with which the water rushes out of the upper end if the locomotive is running at 72 km/hr. [**Ans.** 17.9 ms^{-1}]

10. A vertical tube of 8mm diameter at the bottom has water passing through it. If the pressure be atmospheric at the bottom, where the rate of flow is 50×10^{-6} m^3/s, what is the pressure at a point in the tube 0.5 m above the bottom, where the diameter is 4 mm ? [**Ans.** 0.737 m of mercury.]

11. An engine pumps water from a tank at the rate of 5 kg s^{-1} and ejects from a nozzle 6m above the surface of the tank with a velocity of 10ms^{-1}. Calculate the output power of the engine. [**Ans.** 544 W]

12. A horizontal tube has different areas of cross-section at two points P_1 and P_2. The diameter at P_1 is 0.04 m and the diameter at P_2 is 0.02 m . Two manometer limbs are fixed at P_1 and P_2. When a liquid of density 800 kg m^{-3} flows through, the difference in pressure between the manometer limbs is 0.08 m. Calculate the rate of flow of the liquid through the tube. [**Ans.** 4.062×10^{-4} m^3/s]

13. Water flows through a pipeline of varying cross-section. If the pressure of water equals 0.05 m of Hg at a point, where the velocity of flow is 0.25 m/s, what is the pressure at another point, where the velocity of flow is 0.4 m/s ? [**Ans.** 0.04963 m of Hg.]

14. If the diameter of the pipe and the throat of a venturimeter are 0.08 and 0.05 m respectively and the pressure difference equals 0.04 m of water column, find the volume of water flowing through the pipe per second. [**Ans.** 4.368×10^{-4} m^3/s]

15. A pipe is running full of water. At a certain point A it tapers from 0.6 m diameter to 0.2 m diameter at B. The pressure difference between A and B is 1 m of water column. Find the rate of flow of water through the pipe. [**Ans.** 0.1403m^3/s]

CHAPTER 5 Osmosis

5.1. INTRODUCTION

When a semi-permeable membrane separates a solution from its pure solvent, both being at the same hydrostatic pressure, a diffusion of the solvent takes place into the solution, causing a dilution of the solution. This is known as osmosis.

Osmosis may be demonstrated using the following simple experimental arrangement. A semi-permeable membrane is stretched across the mouth of a thistle funnel and tied firmly. A concentrated solution of sugar in water is poured into the funnel through the stem, keeping the funnel inverted. The funnel is then supported vertically in water kept in a beaker (Fig. 5.1).

The funnel is raised or lowered until the level of the solution in the funnel is the same as the level of water in the beaker. Water is observed to pass from the beaker into the funnel through the semi-permeable membrane. The level of the solution in the tube gradually rises. Therefore, the pressure on the solution side of the membrane gradually increases. This process continues until the excess of pressure on the solution side over the solvent side of the membrane is just sufficient to prevent further diffusion of the solvent into the solution.

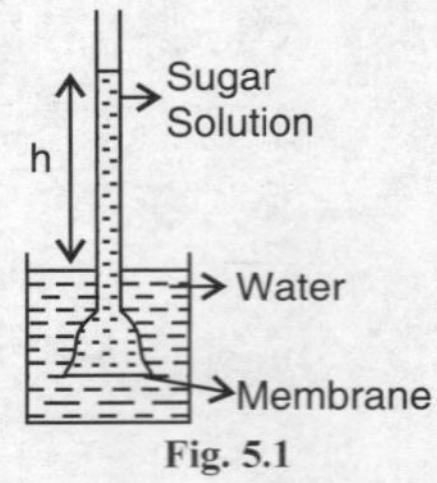

Fig. 5.1

The excess of pressure on the solution side over the solvent side of the membrane when there is no further diffusion of the solvent into the solution is called the osmotic pressure of the sugar solution.

5.2. EXPERIMENTAL DETERMINATION OF OSMOTIC PRESSURE [BERKELEY AND HARTLEY METHOD]

The semi-permeable membrane is in the form of a cylindrical porous pot *AB* with a deposit of copper ferro-cyanide in its pores (Fig. 5.2). The ends of *AB* are closed by air-tight one-holed rubber bungs. Through one of them a capillary tube *C*, closed at one end, is passed. Through the other end, a tube connected to a funnel *F* is passed. The pot is surrounded by a steel cylinder *D* containing the given solution. A side-tube in the cylinder is connected to a pump and a nitrogen manometer *M*.

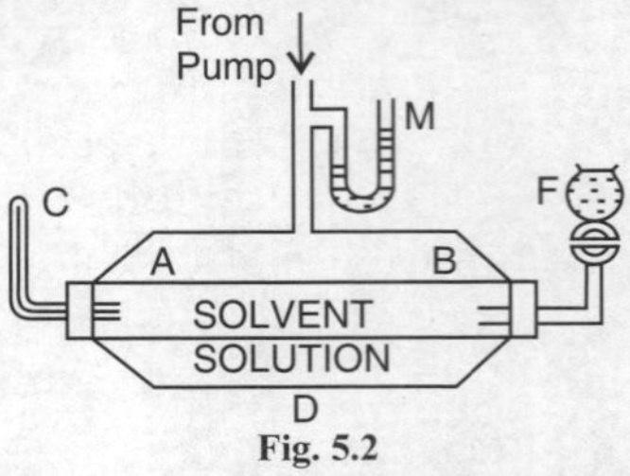

Fig. 5.2

The pure solvent is poured into the pot through the funnel *F*. Pressure is applied by means of the pump on the solution until the level of solvent in the capillary tube *C* remains steady. The solvent in the pot is then just prevented from entering the solution due to the pressure applied on the solution. This pressure is noted from the manometer *M*. It gives the osmotic pressure of the solution of the given concentration.

5.3. LAWS OF OSMOTIC PRESSURE

It is found that a dilute solution of a non-electrolyte behaves in the same way as a perfect gas. The following laws hold good for non-electrolytes.

1. *At constant temperature, the osmotic pressure of a dilute solution is directly proportional to the concentration of the solution.*

   If m kg of a solute are dissolved in V m^3 of a solvent, the concentration $C = m/V$.

   $\therefore \quad P \propto m/V$ or PV is constant. This is similar to Boyle's law.

2. *When the concentration of the solution is constant, its osmotic pressure is directly proportional to the absolute temperature,* ($P \propto T$). This is similar to Charles' law for gases.

3. *Solutions of non-electrolytes dissolved in the same solvent and having the same osmotic pressure at the same temperature, contain the same number of gram molecules of the solute in the same volume.* This is similar to Avogadro's law.

The first two laws can be combined and expressed as $PV = KT$ where K is a constant and V is the volume containing 1 gram molecule of the solute. $K = R$ for non-electrolytes.

The osmotic pressure of electrolytes is much higher than that of non-electrolytes, under the same temperature and concentration. This abnormal behaviour of electrolytes is attributed to the *dissociation* of the electrolyte in solution. In the case of electrolytes, the relation between P, V and T is $PV = i\,RT$ where i is a factor which depends on the degree of dissociation.

5.4. OSMOSIS AND VAPOUR PRESSURE OF A SOLUTION

The vapour pressure of a solution is always less than that of the pure solvent. *A* tube *AB* containing a solution is closed at *B* by a semi-permeable membrane. The tube *AB* dips vertically into the solvent contained in a beaker. The tube and the beaker are enclosed in an evacuated chamber (Fig. 5.3). The solution rises in the tube due to osmosis to the point *M*. Let the height of the solution in the tube above the level of the solvent be h. The vapour at *M* is in contact with the surface of the solution. The vapour at *N* is in contact with the solvent. Let p_1 and p_2 be the pressures of the vapour at *M* and *N* respectively. Let σ be the *uniform density* of the vapour in the enclosure. Then

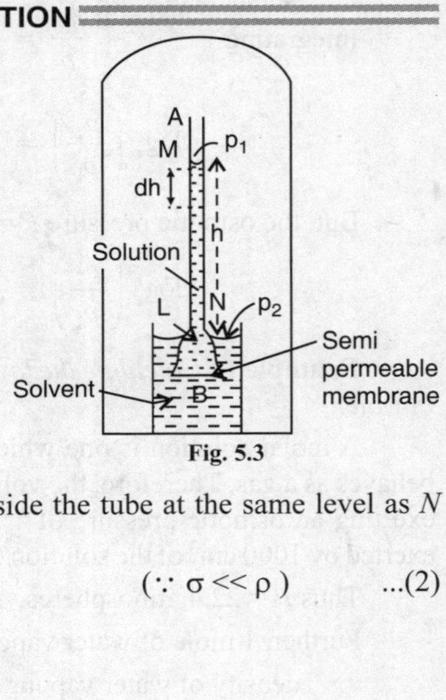

Fig. 5.3

$$p_2 - p_1 = h\sigma g \qquad ...(1)$$

Let ρ be the density of the solution. Let L be a point inside the tube at the same level as N outside. If P is the osmotic pressure, then

$$P = h\,g\,(\rho - \sigma) = hg\rho \qquad (\because \sigma << \rho) \qquad ...(2)$$

Dividing (1) by (2),

$$\frac{p_2 - p_1}{P} = \frac{\sigma}{\rho}$$

$$\therefore \qquad p_2 - p_1 = P\,\sigma/\rho \qquad ...(3)$$

or

$$p_2 - p_1 \propto P$$

$p_2 - p_1$ represents the lowering of vapour pressure. *The lowering of vapour pressure of the solvent is directly proportional to the osmotic pressure P.*

Now, σ is the density of the vapour at the pressure p_2. Let σ_0 be the density of the vapour at normal atmospheric pressure P_0. Then,

$$\frac{p_2}{\sigma} = \frac{P_0}{\sigma_0} \quad \text{or } \sigma = \frac{p_2}{P_0}\sigma_0$$

Substituting this value of σ in Eq. (3),

$$\frac{p_2 - p_1}{p_2} = \frac{P\sigma_0}{P_0\rho}$$

It is not correct to assume the density of the vapour to be uniform, if the osmotic pressure is high.

Let p be the pressure of the vapour at a height h above the surface and σ the density of the vapour at that height.

For a small height dh, let the pressure difference be dp.

$$dp = -g\sigma\, dh$$

Let σ_0 be the density of the solvent vapour at the normal atmospheric pressure P_0. Then, at constant temperature T,

$$\frac{p}{\sigma} = \frac{P_0}{\sigma_0} \text{ or } \sigma = \frac{p\sigma_0}{P_0}$$

$$\therefore \quad dp = -g\frac{p\sigma_0}{P_0}\, dh$$

$$\text{or} \quad \frac{dp}{p} = \frac{g\sigma_0}{P_0}\, dh$$

$$\text{Integrating,} \quad \int_{p_2}^{p_1} \frac{dp}{p} = -\frac{g\sigma_0}{P_0}\int_0^h dh$$

$$\log_e\left(\frac{p_2}{p_1}\right) = \frac{\sigma_0}{P_0}(hg)$$

But, the osmotic pressure $P = h\,g\,\rho$ or $h\,g = P/\rho$.

$$\therefore \quad \log_e\left(\frac{p_2}{p_1}\right) = \frac{\sigma_0 P}{P_0 \rho}$$

Example 1 : *Calculate the lowering of vapour pressure of a molar solution of a non-electrolyte in water.*

A molar solution is one which contains 1 mole of the solute in 1000 cm^3. A dilute solution behaves as a gas. Therefore, the volume of the solution containing 1 mole of the dissolved substance exerting an osmotic pressure of 1 atmosphere at 0°C is 22400 cm^3. Hence, the osmotic pressure exerted by 1000 cm^3 of the solution containing 1 mole of the dissolved substance is 22.4 atmospheres.

Thus $P = 22.4$ atmospheres. $P_0 = 1$ atmosphere.

Further, 1 mole of water vapour at N.T.P. occupies 22400 cm^3.

$\therefore$ density of water vapour at N.T.P.

$$= \sigma_0 = (18 \times 10^{-3})/(22400 \times 10^{-6})$$

$$= 0.8036 \text{ kg m}^{-3}.$$

Density of water $= \rho = 1000 \text{ kg m}^{-3}$.

We have the relation, $\dfrac{p_2 - p_1}{p_2} = \dfrac{P\sigma_0}{P_0\rho}$

$$\therefore \quad \frac{p_2 - p_1}{p_2} = \frac{22.4 \times 0.8036}{1 \times 1000} = 0.018 = 1.8\%$$

The percentage decrease of 1 mole dissolved in 100 c.c. of the solution would be about 18%.

6 Gravitation

CHAPTER

6.1. NEWTON'S LAW OF GRAVITATION

Statement : *Every particle of matter in the universe attracts every other particle with a force which is directly proportional to the product of their masses and inversely proportional to the square of the distance between them.*

Explanation : If m_1 and m_2 are the masses of two particles situated at a distance r apart, the force of attraction between them is given by

$$F \propto \frac{m_1 m_2}{r^2} \text{ or } F = \frac{G m_1 m_2}{r^2}$$

where G is a universal constant, called the *Universal gravitational constant*. The law of gravitation is universal. It holds from huge interplanetary distances to extremely small distances. The law does not hold good for interatomic distances, which are as small as 10^{-9} m. The force of attraction between any two bodies is not affected by the intervening medium. This force is also not affected by the nature, state or chemical structure of the bodies involved but depends only on their masses. Even temperature has no appreciable effect on gravitation.

Definition of G. If $m_1 = m_2 = 1$ kg and $r = 1$ m, then $F = G$. Thus, *the Gravitational constant is equal to the force of attraction between two unit masses of matter unit distance apart.*

Dimensions of G. $G = \dfrac{Fr^2}{m_1 m_2}$.

Dimensions of G are given by $[G] = \dfrac{MLT^{-2}\, L^2}{M^2} = M^{-1} L^3 T^{-2}$.

Mass and Density of earth : If m is the mass of a body and g the acceleration due to gravity, the force of attraction of the earth on the mass $m = mg$.

Let M = mass of the earth ; R = radius of the earth.

$$\left.\begin{array}{l}\text{Gravitational force of attraction between}\\ \text{a body of mass } m \text{ and earth}\end{array}\right\} = \frac{GMm}{R^2}$$

$$\therefore \quad \frac{GMm}{R^2} = m \,.\, g \text{ or } M = \frac{R^2 . g}{G}$$

Volume of the earth $= V = \dfrac{4}{3}\pi R^3$.

$$\therefore \quad \text{Density of the earth} = \rho = \frac{M}{V} = \frac{(R^2 g/G)}{\frac{4}{3}\pi R^3} = \frac{3g}{4\pi RG}$$

Inertial mass : The mass of a body may be determined by measuring the acceleration a produced on it by a known force F.

Thus, $m = F/a$. The mass m thus determined is called inertial mass.

Gravitational mass : The mass of a body may also be determined by measuring the gravitational force exerted on it by earth.

$$F = \frac{GMm}{R^2} \text{ or } m = \frac{FR^2}{GM}.$$

The mass m thus determined is called gravitational mass.

Example 1 : *Estimate the mass of the Sun, assuming the orbit of the earth round the Sun to be a circle. The distance between the Sun and the earth is 1.49 × 10^{11} m, and G = 6.66 × 10^{-11} Nm^2 kg^{-2}.*

$$\left.\begin{array}{l}\text{Force of attraction between}\\ \text{the sun and the earth}\end{array}\right\} = \frac{GMm}{R^2}$$

Here, M = mass of the Sun ; m = mass of the earth, and R = Radius of the earth's orbit round the sun.

This clearly supplies the centripetal force to the earth as it goes round the Sun in its orbit.

The centripetal force = mv^2/R.

Here, v = velocity of the earth in its circular orbit.

Now $v = 2\pi R/T$

where T = Time period of the earth's motion round the Sun.

$$\therefore \quad \text{Centripetal force} = \frac{m\,(2\pi R/T)^2}{R} = \frac{m 4\pi^2 R}{T^2}$$

$$\text{For equilibrium,} \quad \frac{GMm}{R^2} = \frac{m\,4\,\pi^2 R}{T^2} \text{ or } M = \frac{4\,\pi^2 R^3}{T^2 . G}$$

Here, $R = 1.49 \times 10^{11}$ m, T = 365 days = 365 × 24 × 60 × 60 s, $G = 6.66 \times 10^{-11}$ Nm^2 kg^{-2}

$$\therefore \quad M = \frac{4\,\pi^2\,(1.49 \times 10^{11})^3}{(365 \times 24 \times 60 \times 60)^2\,(6.66 \times 10^{-11})} = 1.971 \times 10^{30} \text{ kg}$$

Example 2 : *Calculate the mass of the earth and the mean density of the earth from the following data :*

Radius of the earth = 6.4 × 10^6 m ; g = 9.8 ms^{-2} ; G = 6.67 × 10^{-11} SI units.

We have, mass of the earth = $M = R^2 . g/G$.

Here, $R = 6.4 \times 10^6$ m ; $g = 9.8$ ms^{-2} ; $G = 6.67 \times 10^{-11}$ SI units

$$\therefore \quad M = \frac{(6.4 \times 10^6)^2 (9.8)}{(6.67 \times 10^{-11})} = 6.017 \times 10^{24} \text{ kg}$$

Density of the earth = $\rho = 3g/(4\pi\,RG)$

$$= \frac{3 \times 9.8}{4\pi\,(6.4 \times 10^6)\,(6.67 \times 10^{-11})} = 5480 \text{ kg m}^{-3}$$

Example 3 : *A body weighs 900 kg on the surface of the earth. How much will it weigh on the surface of Mars whose mass is one-ninth and radius one-half that of the earth ?*

Let M and R be the mass and radius of the earth. Let m be the mass of the body. Then, the force of attraction which the earth exerts on this body

$$F = \frac{GMm}{R^2} \text{ or } 900 = \frac{GMm}{R^2} \quad ...(1)$$

Let W be the weight of the body on the surface of Mars.

Then, $$W = \frac{G.(M/9)m}{(R/2)^2} \text{ or } W = \frac{GMm}{R^2} \times \frac{4}{9} \quad ...(2)$$

Dividing (2) by (1), $\frac{W}{900} = \frac{4}{9}$ or $W = 400$ kgf.

Example 4 : *Show how the mass of the earth may be compared with that of the sun from a knowledge of the time-periods of the earth round the sun and of the moon round the earth, together with the radii of their respective orbits, (taken to be circular).*

Let M_s and M_e be the masses of the sun and the earth respectively. Let R_e be the radius of the earth's orbit round the sun. Let ω_e be the angular velocity of the earth.

Then, $$\frac{G.M_s M_e}{R_e^2} = M_e R_e \omega_e^2$$

or $$GM_s = R_e^3 \omega_e^2 \quad ...(i)$$

Similarly, $$\frac{G M_e M_m}{R_m^2} = M_m R_m \omega_m^2$$

where M_m = mass of the moon, R_m = radius of moon's orbit round the earth and ω_m = angular velocity of the moon.

$$\therefore \quad G.M_e = R_m^3 . \omega_m^2 \quad ...(ii)$$

From (*i*) and (*ii*), $$\frac{M_e}{M_s} = \left(\frac{R_m}{R_e}\right)^3 \left(\frac{\omega_m}{\omega_e}\right)^2 \quad ...(iii)$$

Let T_e and T_m be the time periods of the earth and the moon respectively. Then

$$\left(\frac{\omega_m}{\omega_e}\right)^2 = \left(\frac{T_e}{T_m}\right)^2 \qquad \left(\because \ \omega = \frac{2\pi}{T}\right)$$

$\therefore$ From (*iii*), $$\frac{M_e}{M_s} = \left(\frac{R_m}{R_e}\right)^3 \left(\frac{T_e}{T_m}\right)^2$$

Thus knowing R_m, R_e, T_e and T_m, the mass of the earth can be compared with that of the sun.

6.2. KEPLER'S LAWS OF PLANETARY MOTION

(1) *Every planet moves in an elliptical orbit around the sun, the sun being at one of the foci.*

(2) *The radius vector, drawn from the sun to a planet sweeps out equal areas in equal times i.e., the areal velocity of the radius vector is constant (dA/dt = constant).*

(3) *The square of the period of revolution of the planet around the sun is proportional to the cube of the semi-major axis of the ellipse ($T^2 \propto a^3$).*

Deduction of Newton's Law of Gravitation from Kepler's Laws

Consider two planets of masses m_1 and m_2. Let r_1 and r_2 be the radii of their circular orbits. Let T_1 and T_2 be their periods of revolution round the sun.

The centrifugal force acting on the first planet,

$$F_1 = m_1 r_1 \cdot \omega^2 = m_1 r_1 \left(\frac{2\pi}{T_1}\right)^2$$

Similarly, the centrifugal force acting on the second planet

$$F_2 = m_2 r_2 \left(\frac{2\pi}{T_2}\right)^2$$

Now, $$\frac{F_1}{F_2} = \frac{m_1 r_1}{m_2 r_2}\left(\frac{T_2}{T_1}\right)^2$$

But according to Kepler's third law, $$\left(\frac{T_2}{T_1}\right)^2 = \left(\frac{r_2}{r_1}\right)^3$$

or $$\frac{F_1}{F_2} = \frac{m_1 \cdot r_1}{m_2 \cdot r_2}\left(\frac{r_2}{r_1}\right)^3 = \frac{m_1 \cdot r_2^2}{m_2 \cdot r_1^2}$$

i.e., the force on the planet is directly proportional to $\frac{m}{r^2}$ or $F \propto \frac{m}{r^2}$. Therefore, the force is proportional to the mass of the planet. Since the attraction is mutual, the force is also proportional to the mass of the sun M. Hence $F \propto \frac{Mm}{r^2}$ or $F = \frac{GMm}{r^2}$ which is Newton's Law of Gravitation.

6.3. DETERMINATION OF G–BOYS' EXPERIMENT

The apparatus consists of two co-axial glass tubes T_1 and T_2 mounted on a platform provided with levelling screws (Fig. 6.1). The inner tube T_1 is fixed, while the outer tube T_2 can be rotated about the common axis. A small mirror, RS, is suspended in the inner tube by a fine quartz fibre f from a torsion head H. From the two ends of the mirror, two gold spheres A and B are suspended, such that the spheres are at different depths below the mirror. In the outer co-axial tube T_2, two large lead balls C and D are suspended from its revolving lid such that the centre of C is in level with that of A, the centre of D is in level with that of B and the distance $AC = BD$. Two rubber pads P_1 and P_2 are placed below the two lead spheres, as a safeguard against damage, in case they should fall accidentally.

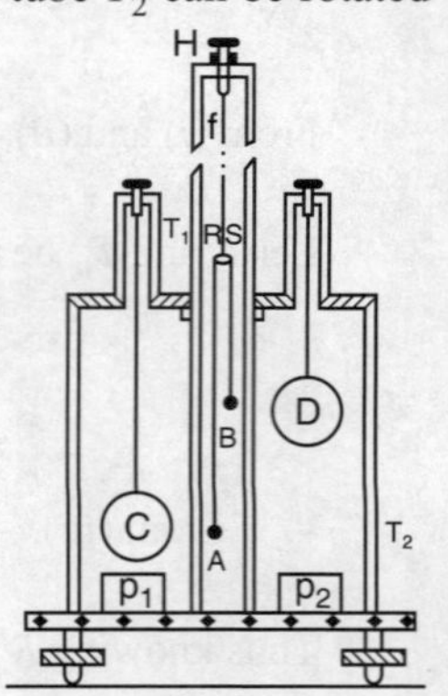

Fig. 6.1

The experiment is performed by rotating the outer glass tube until the large lead spheres lie on the opposite sides of the two gold balls, so as to exert the maximum moment on the suspended system. In this position, the angle through which the mirror (RS) turns is maximum. The outer glass tube is then rotated so that the lead spheres now lie on the other sides of the gold balls, in an exactly similar position, producing the greatest deflection. The mean of these two observations gives the deflection of the mirror θ.

A lamp and scale arrangement is used to measure θ.

Force of attraction between spheres A and $C = GMm/(AC)^2$

Force of attraction between spheres B and $D = GMm/(BD)^2$

Since $AC = BD$, the two forces are equal, parallel and act in opposite directions, thus constituting a couple.

$$\therefore \quad \left.\begin{array}{l}\text{The moment of the}\\ \text{deflecting couple}\end{array}\right\} = \frac{GMm}{(AC)^2} \times 2l = \frac{GMm}{d^2} \times 2l$$

(where $2\,l$ = the length of the mirror strip RS and $AC = d$).

The deflection of the mirror strip under this couple is resisted by the torsion or twist set up in the suspension fibre. The mirror strip comes to rest when the deflecting couple due to gravitational pull is balanced by the restoring torsional couple set up in the suspension fibre. Now, if c be the torsional couple per unit twist, then for angular deflection θ, the total restoring couple is $c \,.\, \theta$.

$\therefore$ In equilibrium position, $\dfrac{GMm}{d^2} \times 2l = c\,\theta.$

From this, the value of G can be calculated. Using the arrangement of the quartz fibre and the mirror strip with gold balls as a torsion pendulum, the period T is found. Then $T = 2\pi \sqrt{I/c}$ where I = moment of inertia of the suspended system. From this c can be calculated.

The results obtained by him are very accurate. The value obtained for G by Boys is 6.6576×10^{-11} Nm^2 kg^{-2}.

Advantages : (1) The size of the apparatus is very much reduced. The disturbances due to convection currents are therefore almost negligible.

(2) By arranging the masses at different levels, the effect of the attraction of the heavier mass on the remote smaller mass is very much reduced.

(3) By the lamp and scale arrangement, very small deflections can be measured accurately.

(4) The use of a quartz fibre has made the apparatus very sensitive and accurate.

6.4. GRAVITATIONAL FIELD AND GRAVITATIONAL POTENTIAL

Gravitational Field : *The space around a body within which its gravitational force of attraction is perceptible is called its gravitational field.*

The gravitational field is an example of a vector field. Each point in this field has a vector associated with it. The intensity of the gravitational field at a point due to a body is the force experienced by a unit mass placed at that point.

Gravitational Potential : *The work done in moving a unit mass from infinity to a point in a gravitational field is called the gravitational potential at that point.*

Gravitational potential is always negative in sign, its highest value being zero at infinity. It is a scalar quantity.

Intensity of gravitational field at a point : *It is defined as the space rate of change of gravitational potential at the point. i.e.,* $F = -\,dV/dr$ where dV is the small change of gravitational potential for a small distance dr.

Gravitational potential due to a point mass. Consider a point A at a distance x from a particle of mass m (Fig. 6.2).

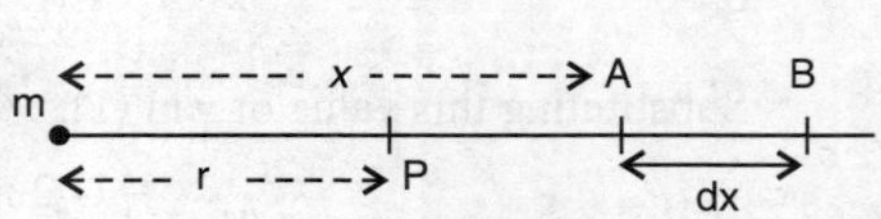

Fig. 6.2

Force of attraction experienced by a unit mass at A $\Big\} = \dfrac{G\,.\,m}{x^2}.$

Work done in displacing the unit mass from A to B through a distance dx $\Big\} = \dfrac{G\,.\,m}{x^2}\,dx.$

$\therefore$ The potential difference between A and $B = \delta V = \dfrac{G\,.\,m}{x^2}\,dx$

Hence the total work done in moving the unit mass from infinity to P or

The potential at $P = V = \displaystyle\int \delta V = \int_{\infty}^{r} \frac{G\,.\,m}{x^2}\,dx = -\frac{G\,.\,m}{r}$

Thus the gravitational potential has the maximum value of zero at infinity and decreases as the distance is decreased.

Equipotential Surface : *A surface at all the points of which the gravitational potential is the same is called an equipotential surface.*

For example, a spherical surface around a point mass with the mass as centre, is an equipotential surface. Since the potential on this surface is constant, no work is done against the gravitational force in moving a unit (or any other) mass along it.

6.5. GRAVITATIONAL POTENTIAL AND FIELD DUE TO A SPHERICAL SHELL

(*i*) **Point outside the shell :** Consider a point P outside the spherical shell at a distance r from its centre O (Fig. 6.3). Let a be the radius of the shell, ρ the mass per unit area of the surface of the shell, and M its total mass. Join OP and let $OP = r$. Consider a thin slice of the shell contained between two planes AB and CD drawn close to each other at right angles to OP. Join O and A, O and C and A and P.

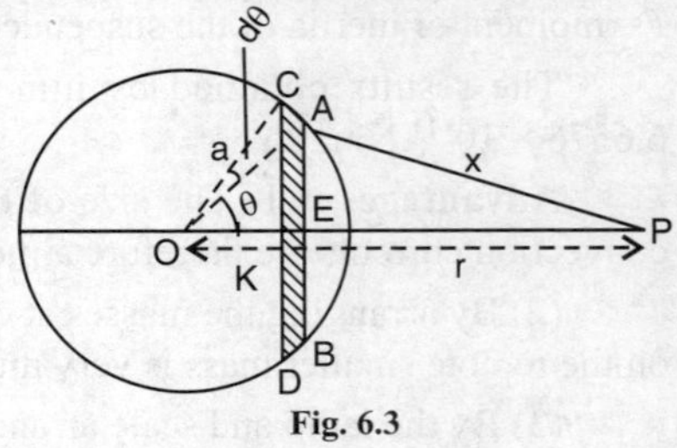

Fig. 6.3

Let $\angle AOP = \theta$, and $\angle AOC = d\theta$.

Now, AE = Radius of the slice = $a \sin\theta$

Circumference of the slice = $2\pi \times AE = 2\pi a \sin\theta$.

Width of the slice = $CA = a\,d\theta$.

Hence, surface area of the slice = $2\pi a \sin\theta \times a\,d\theta$

$= 2\pi a^2 \sin\theta\, d\theta$.

$\therefore$ Mass of the slice = $2\pi a^2 \rho \sin\theta\, d\theta$.

Let $PA = x$. Every point on the slice may be taken to be practically equidistant from P.

$$\therefore \left.\begin{array}{l}\text{Potential at } P\\ \text{due to the ring}\end{array}\right\} = dV = \frac{-G\,2\pi a^2 \rho \sin\theta\, d\theta}{x} \quad ...(1)$$

To find the value of x, consider the triangle OAP.

$$x^2 = a^2 + r^2 - 2ar\cos\theta$$

Differentiating, $2x\,dx = 2ar\sin\theta\,d\theta$ [$\because$ a and r are constants]

or
$$x = \frac{a\,.\,r\sin\theta\,d\theta}{dx}$$

Substituting this value of x in (1),

$$dV = \frac{-G\,2\pi a^2 \rho \sin\theta\,d\theta\,dx}{a\,.\,r\sin\theta\,d\theta} = \frac{-2\pi a\,.\,\rho G}{r}\,dx$$

If the entire shell is split up into slices of this kind, the value of PA will vary from $(r-a)$ to $(r+a)$. Hence,

$$\left.\begin{array}{l}\text{the potential at } P \text{ due}\\ \text{to the entire shell}\end{array}\right\} = V = \int_{r-a}^{r+a} \frac{-2\pi a \rho G}{r}\,dx.$$

$$= \frac{-2\pi a\,.\,\rho G}{r}[x]_{r-a}^{r+a} = \frac{-2\pi a \rho G}{r}\,.\,2a = -4\pi a^2 \rho \frac{G}{r}$$

Now $4\pi a^2 \rho$ = Mass of the whole shell.

$$V = -\frac{G\,.\,M}{r}.$$

This potential is the same as due to a mass M at O. Hence, the mass of the shell behaves as though it were concentrated at its centre.

(*ii*) **Point on the surface of the shell.** Let us consider a point which lies on the surface of the shell itself. The limits for the value of x will be 0 and $2a$. Hence

$$\left.\begin{array}{l}\text{Potential at a point on}\\ \text{the surface of the shell}\end{array}\right\} = V = \int_0^{2a} \frac{-2\pi\, a\, \rho\, G}{r}\, dx$$

$$= \frac{-2\pi \,.\, a \,.\, \rho \,.\, G}{r}[x]_0^{2a} = \frac{-4 \,.\, \pi \,.\, a^2\, \rho \,.\, G}{r} = \frac{-G \,.\, M}{r} = \frac{-G \,.\, M}{a}; (\because r = a)$$

$$\therefore \qquad V = \frac{-G \,.\, M}{a}$$

(*iii*) **Point inside the shell.** Let the point P be situated at K inside the shell, such that $OK = r$. The limits for the value of x will be $(a - r)$ and $(a + r)$.

$$\therefore \quad \left.\begin{array}{l}\text{Potential at a point } (K)\\ \text{inside the shell}\end{array}\right\} = V = \int_{a-r}^{a+r} \frac{-2\,\pi\, a\, \rho \,.\, G}{r}\, dx$$

$$= \frac{-2\pi\, a\, \rho \,.\, G}{r}[x]_{a-r}^{a+r} = -\,4\pi\, a\rho \,.\, G.$$

Multiplying and dividing by a, $V = \dfrac{-4\,\pi\, a^2\, \rho \,.\, G}{a} = \dfrac{-G \,.\, M}{a}$

$$\therefore \qquad V = -\frac{G \,.\, M}{a}$$

Hence the potential at all points inside a spherical shell is the same and is equal to the value of the gravitational potential on the surface.

GRAVITATIONAL FIELD. The intensity of the gravitational field F is given by $F = -\,dV/dr$.

(*i*) **At a point outside the shell :** $V = \dfrac{-GM}{r}$

$$\therefore \qquad F = \frac{-dV}{dr} = -\frac{d}{dr}\left[\frac{-G \,.\, M}{r}\right] = \frac{-G \,.\, M}{r^2} \qquad ...(i)$$

The negative sign indicates that the force is towards the centre O.

(*ii*) **At a point on the outer surface of the shell :** Putting $r = a$ in the expression (*i*), we get the intensity of the gravitational field at a point on the surface of the shell.

$$F = -\frac{GM}{a^2}$$

(*iii*) **At a point inside the shell.**

Potential $V = \dfrac{-GM}{a}$ = a constant. $\qquad \therefore F = \dfrac{-dV}{dr} = 0.$

Hence there is no gravitational field inside a spherical shell.

6.6. GRAVITATIONAL POTENTIAL AND FIELD DUE TO A SOLID SPHERE

(*i*) **Point outside the sphere. :** Let P be a point outside the sphere at a distance r from the centre O [Fig. 6.4(*a*)]. Let M be the mass of the sphere, a its radius and ρ its density. A solid sphere may be imagined to be made up of a large number of concentric shells. Each one of the shells produces a potential at the point P outside the shell, as if its entire mass is concentrated at the centre O.

Thus if m is the mass of one such shell,

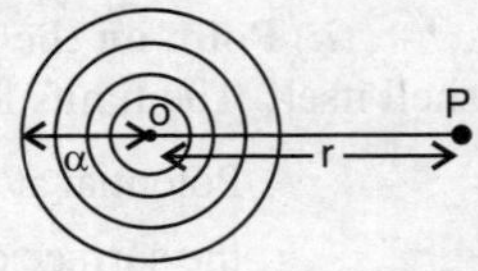

Fig. 6.4 (*a*)

The potential at P due to the shell $= \dfrac{-Gm}{r}$

Potential due to the whole sphere $V = -\Sigma \dfrac{Gm}{r} = -\dfrac{G}{r}\Sigma m$

Clearly, $\Sigma m = M$ = Mass of the solid sphere.

$$\therefore \qquad V = \frac{-GM}{r} \qquad ...(i)$$

(*ii*) **Point on the surface :** If the point P lies on the surface of the solid sphere, we have $r = a$. Putting $r = a$ in (*i*), we get,

The potential at a point on the surface $= \dfrac{-GM}{a}$

(*iii*) **Point inside the sphere :** Let the point now lie inside the solid sphere at a distance r from the centre O.

The solid sphere may be imagined to be made up of (*i*) an inner solid sphere of radius r and (*ii*) a hollow sphere of internal radius r and external radius a. The hollow sphere may be imagined to be made up of concentric shells with radii ranging from r to a.

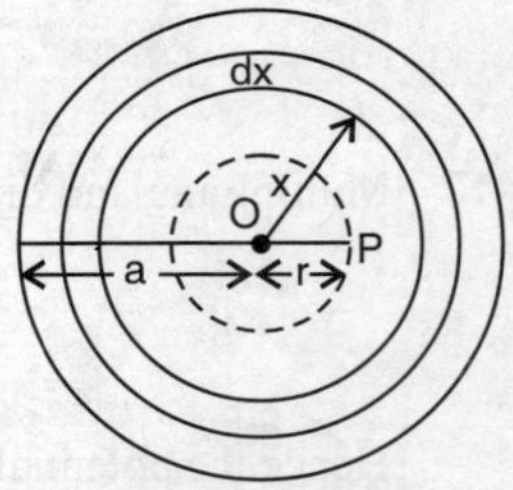

Fig. 6.4 (*b*)

Potential at P due to the whole solid sphere is

V = Potential at P due to the inner solid sphere (V_1) + Potential at P due to all the shells (V_2)

(*a*) **To determine the potential at P due to the inner solid sphere**

The point P lies on the surface of the inner solid sphere of radius r [Fig. 6.4 (*b*)].

Mass of the inner sphere $= \dfrac{4}{3}\pi r^3 \rho$.

Potential at P due to the inner sphere

$$= V_1 = \frac{-G\frac{4}{3}\pi r^3 \rho}{r} = -G\frac{4}{3}\pi r^2 \rho \qquad ...(i)$$

(*b*) **To determine the potential at P due to all the outer shells**

Consider one such shell of radius x and thickness dx. The point P lies inside the spherical shell.

Mass of the shell $= 4\pi x^2 dx\, \rho$.

$\therefore$ Potential at P due to this shell $= \dfrac{-G4\pi x^2 dx\rho}{x} = -G\,4\pi x\, dx\rho$

$$\therefore \quad \left.\begin{array}{l}\text{Potential at } P \text{ due}\\ \text{to all shells}\end{array}\right\} = V_2 = \int_r^a -G\,4\pi x\, dx\rho$$

$$= -G\,4\pi\rho \int_r^a x\,dx = -G\,4\pi\rho\left[\frac{x^2}{2}\right]_r^a = -G\,4\pi\rho\left(\frac{a^2 - r^2}{2}\right)$$

$\therefore$ Total potential at $P = V = V_1 + V_2$

$$= -G\frac{4}{3}\pi r^2\rho - G\,4\pi\rho\left(\frac{a^2-r^2}{2}\right) = -G\frac{4}{3}\pi\rho\left[r^2 + \frac{(3a^2-3r^2)}{2}\right]$$

$$= -G\frac{4}{3}\pi\rho\left[\frac{2r^2+3a^2-3r^2}{2}\right] = -G\frac{4}{3}\pi\rho\left[\frac{3a^2-r^2}{2}\right]$$

$$= -G\frac{4}{3}\pi a^3\rho\left[\frac{3a^2-r^2}{2a^3}\right] \qquad \text{(multiplying and dividing by } a^3\text{)}$$

$$\therefore \quad V = -GM\left[\frac{3a^2-r^2}{2a^3}\right] \qquad \left(\because \frac{4}{3}\pi a^3\rho = M\right)$$

GRAVITATIONAL FIELD

(*i*) **Point outside the sphere.** Potential $V = \dfrac{-GM}{r}$

$$\therefore \quad \text{intensity } F = \frac{-dV}{dr} = \frac{-d}{dr}\left[\frac{-G.M}{r}\right] = \frac{-G.M}{r^2} \qquad ...(i)$$

(*ii*) **Point on the surface of the sphere.** For a point on the surface of the solid sphere, $r = a$, and therefore

$$F = \frac{-G.M}{a^2} \qquad \text{[putting } r = a, \text{ in } (i)]$$

(*iii*) **Point inside the sphere.** Potential at a point inside the solid sphere at a distance r from the centre O,

$$V = -G.M\left[\frac{3a^2-r^2}{2a^3}\right]$$

$$\therefore \quad \left.\begin{array}{l}\text{Intensity of the}\\ \text{field at } P\end{array}\right\} = F = -\frac{dV}{dr} = -\frac{d}{dr}\left[-GM\left(\frac{3a^2-r^2}{2a^3}\right)\right]$$

$$= -\frac{GM}{a^3}r$$

Thus, the intensity of the gravitational field at a point inside a solid sphere is directly proportional to the distance of the point from the centre of the sphere.

Example 5 : *With what velocity should a body be projected vertically upwards from the surface of the earth so that it may first attain a height of R/2 where R is the radius of the earth ?* ($R = 6.4 \times 10^6$ *m*).

At the surface of the earth *i.e.*, at a distance R from its centre,

P.E. of the body $= -m\,MG/R = -mg\,R^2/R = -mgR$ $\qquad (\because MG = gR^2)$

At a distance $\left(\dfrac{R}{2}+R\right)$ or $\dfrac{3}{2}R$ from the centre of the earth

$$\text{P.E. of the body} = -MmG/\left(\frac{3}{2}R\right) = -\frac{2}{3}mg\,R$$

$$\therefore \quad \left.\begin{array}{l}\text{Increase in P.E.}\\ \text{of the body}\end{array}\right\} = -\frac{2}{3}mgR - (-mgR) = \frac{1}{3}mgR$$

If v be its velocity of projection, its K.E. $= \dfrac{1}{2}mv^2 = \dfrac{1}{3}mgR.$

$$\therefore \qquad v = \sqrt{\frac{2}{3} gR} = \sqrt{\frac{2}{3} (9.8)(6.4 \times 10^6)} = 6469 \text{ ms}^{-1}.$$

6.7. VARIATION OF g WITH LATITUDE OR ROTATION OF THE EARTH

Let us assume that the earth is a uniform sphere of radius R revolving about its polar diameter NS (Fig. 6.5). Consider a particle of mass m on the surface of the earth at a latitude λ. If the earth were at rest, a particle of mass m placed at P will experience a force mg along the radius PO towards O.

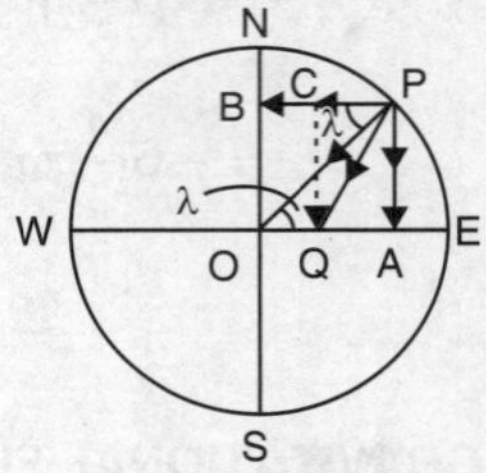

Fig. 6.5

Let ω be the angular velocity of the earth. As the earth revolves, the particle at P will execute circular motion with B as centre and BP as radius. A centrifugal force will develop and the centrifugal force acting on P along BP, away from $B = mBP.\ \omega^2$.

$$= m (R \cos \lambda)\, \omega^2 \qquad (\because BP = R \cos \lambda)$$

$$= m R\, \omega^2 \cos \lambda.$$

Force mg acts along PO. Resolve mg into two rectangular components (*i*) $mg \sin \lambda$ along PA and (*ii*) $mg \cos \lambda$ along PB. Out of the resolved component along PB, a portion $m R\, \omega^2 \cos \lambda$ is used in overcoming centrifugal force.

Let the net force be represented by PC. Then

$$PC = mg \cos \lambda - mR\, \omega^2 \cos \lambda \text{ and } PA = mg \sin \lambda.$$

The resultant force (mg') experienced by P is along PQ, such that

$$(PQ)^2 = (PC)^2 + (PA)^2 \text{ or } PQ = [(PC)^2 + (PA)^2]^{1/2}$$

i.e.,

$$mg' = [(mg \cos \lambda - mR\, \omega^2 \cos \lambda)^2 + (mg \sin \lambda)^2]^{1/2}$$

$$= mg \left[1 + \frac{R^2 \omega^4}{g^2} \cos^2 \lambda - \frac{2\, R\omega^2}{g} \cos^2 \lambda\right]^{\frac{1}{2}}$$

$$\therefore \qquad mg' = mg \left[1 - \frac{2R\omega^2}{g} \cos^2 \lambda\right]^{\frac{1}{2}} \qquad \left(\text{neglecting } \frac{R^2 \omega^4}{g^2} \cos^2 \lambda\right)$$

$$= mg \left[1 - \frac{R\omega^2 \cos^2 \lambda}{g}\right]$$

($\because R\omega^2/g$ is small, its higher powers can be neglected)

$$\therefore \qquad g' = g \left[1 - \frac{R\, \omega^2 \cos^2 \lambda}{g}\right]$$

Example 6: *How many times faster than the present speed would the earth have to rotate about its axis, in order that the apparent weight of bodies at equator be zero ? What should be the new period of rotation ?*

We have
$$g' = g\left(1 - \frac{R\omega^2 \cos^2 \lambda}{g}\right)$$

where g' = value of acceleration due to gravity at latitude λ.

At the equator, $\lambda = 0$ and $\therefore \cos^2 \lambda = 1$.

Hence,
$$g' = g\left(1 - \frac{R\omega^2}{g}\right)$$

Now, $$\frac{R\omega^2}{g} = \frac{(6.37\times10^6)}{9.78}\times\left(\frac{2\,\pi}{86164}\right)^2 = \frac{1}{289} \qquad ...(1)$$

In order that the weight of a body at equator may be zero, the value of g should be zero.

If the new angular speed of earth were ω', then $$\frac{R(\omega')^2}{g} = 1 \qquad ...(2)$$

Dividing (2) by (1), $$\left(\frac{\omega'}{\omega}\right)^2 = 289$$

or $$\left(\frac{\omega'}{\omega}\right) = \sqrt{289} = 17 \text{ or } \omega' = 17\omega.$$

Hence the earth should have about seventeen times the present angular velocity in order that apparent weight of bodies at equator be zero.

Now the earth makes one complete revolution in 86164 seconds. When the earth rotates 17 times faster, its new period will be 86164/17 = 5069 seconds = 1 h 24 m 29 s.

Example 7 : *If the earth were to cease rotating about its axis, what will be the change in the value of g at a place of latitude 45°, assuming the earth to be a sphere of radius 6.38 × 10⁶ metres.*

We have, $$g' = g\left(1-\frac{R\omega^2\cos^2\lambda}{g}\right)$$

[where g' = value of acceleration due to gravity at latitude λ] *i.e.,*

$$g' = g - R\omega^2\cos^2\lambda$$

[where g = value of acceleration due to gravity, if the earth were at rest].

Hence, $g - g'$ = change in the value of $g = R\omega^2\cos^2\lambda$.

Here, $$R = 6.38\times10^6\text{ m};\ \omega = \frac{2\pi}{24\times60\times60}\text{ rad s}^{-1};$$

$$\lambda = 45^\circ \text{ and } \therefore \cos^2\lambda = \frac{1}{2}.\ (g-g') = ?$$

$$\therefore\quad g-g' = (6.38\times10^6)\times\left(\frac{2\pi}{24\times60\times60}\right)^2\times\frac{1}{2} = 0.0169\text{ ms}^2.$$

6.8. VARIATION OF WITH *g* ALTITUDE

Let P be a point on the surface of the earth and Q another point at an altitude h (Fig. 6.6). Mass of the earth is M and radius of the earth is R. Let g be the acceleration due to gravity on the surface of the earth. Then

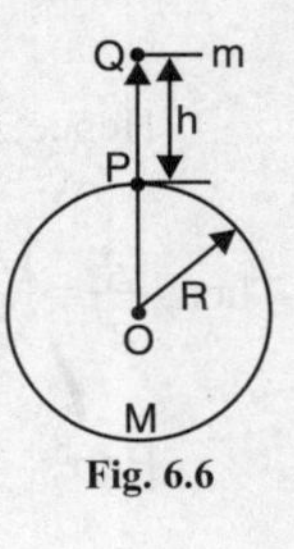

Fig. 6.6

$$\left.\begin{array}{l}\text{The force experienced by}\\ \text{a body of mass } m \text{ at } P\end{array}\right\} = mg = \frac{GMm}{R^2} \qquad ...(i)$$

$$\left.\begin{array}{l}\text{The force experienced by}\\ \text{a body of mass } m \text{ at } Q\end{array}\right\} = mg' = \frac{GMm}{(R+h)^2} \qquad ...(ii)$$

where g' is the acceleration due to gravity at an altitude h.

Dividing (*ii*) by (*i*),

$$\frac{g'}{g} = \frac{R^2}{(R+h)^2} = \frac{R^2}{R^2\,[1+(h/R)]^2} = \left(1+\frac{h}{R}\right)^{-2}$$

$$= \left(1-\frac{2h}{R}\right) \quad \text{[neglecting higher powers of } h/R\text{]}$$

or

$$g' = g\left(1-\frac{2h}{R}\right).$$

This shows that the acceleration due to gravity decreases with increase in altitude.

Example 8 : *How far away from earth does acceleration due to gravity become one percent of its value at the earth's surface ? Assume that the earth is a sphere of radius 6.38 × 10⁶ metres.*

Acceleration due to gravity on the earth's surface $\Big\}$ $= g = \dfrac{GM}{R^2}$

Acceleration due to gravity at a height $h = g' = \dfrac{GM}{(R+h)^2}$

Here, $R = 6.38 \times 10^6$ m ; $g' = (1/100)g$

or $\dfrac{g'}{g} = \dfrac{1}{100}$; $h = ?$

$$\frac{g'}{g} = \frac{R^2}{(R+h)^2} \text{ i.e., } \frac{1}{100} = \frac{R^2}{(R+h)^2} \text{ or } \frac{1}{10} = \frac{R}{R+h}$$

or $h = 9R = 9 \times 6.38 \times 10^6 = 5.742 \times 10^7$ m.

Example 9 : *A pendulum that beats seconds on the surface of the earth, is found to lose 10.8 seconds per day, when taken to the top of a hill 800 m high. What is the radius of the earth ?*

Let M and R be the mass and radius of the earth. Let g be the acceleration due to gravity on the surface of the earth. Then,

$$g = \frac{GM}{R^2}$$

Let g' be the acceleration due to gravity on the top of the hill. Then,

$$g' = \frac{GM}{(R+h)^2}.$$

Hence, $\dfrac{g}{g'} = \left(\dfrac{R+h}{R}\right)^2$. Let T and T' be the periods on the earth and on the top of the hill.

Then, $\dfrac{g}{g'} = \dfrac{(T')^2}{T^2}$

Here, $T = 2$ s ; $T' = \dfrac{86400}{86400-10.8} \times 2 = \dfrac{86400}{86389.2} \times 2.$

$$\therefore \quad \frac{(T')^2}{T^2} = \left(\frac{86400}{86389.2}\right)^2$$

Hence, $$\frac{R+h}{R} = \frac{86400}{86389.2} = 1.00012$$

or $$\frac{R+800}{R} = 1.00012 \text{ or } R = 6.666 \times 10^6 \text{ m.}$$

6.9. VARIATION OF g WITH DEPTH

Let g and g' be the values of acceleration due to gravity at P and Q respectively (Fig. 6.7). At P, the whole mass of the earth attracts the body.

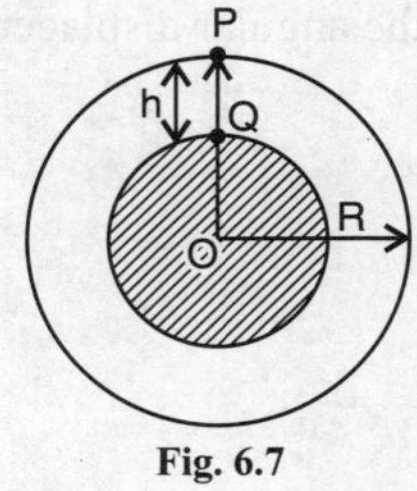

Fig. 6.7

$$\therefore \quad mg = \frac{GMm}{R^2} \quad ...(1)$$

where m = mass of the body,

M = mass of the earth and

R = Radius of the earth

At Q, the body is attracted by the mass of the earth of radius $(R - h)$.

$$\therefore \quad mg' = \frac{GM'm}{(R-h)^2} \quad ...(2)$$

Here, $$M = \frac{4}{3}\pi R^3 \rho \text{ and } M' = \frac{4}{3}\pi (R-h)^3 \rho$$

where ρ is the mean density of the earth.

Dividing (2) by (1), $$\frac{g'}{g} = \frac{M'}{M}\frac{R^2}{(R-h)^2}$$

$$= \frac{\frac{4}{3}\pi (R-h)^3 \rho}{\frac{4}{3}\pi R^3 \rho} \times \frac{R^2}{(R-h)^2} = \frac{(R-h)}{R} = \left(1 - \frac{h}{R}\right)$$

$$\therefore \quad g' = g\left(1 - \frac{h}{R}\right)$$

Therefore, the acceleration due to gravity decreases with increase of depth.

6.10. THE COMPOUND PENDULUM

Any rigid body capable of oscillating freely about a horizontal axis passing through it is a compound pendulum.

To find the period of oscillation of a compound pendulum :

Let O be the point of suspension and G the centre of mass (Fig. 6.8). In the equilibrium position, OG is vertical. $OG = h$. Suppose the body is given a small angular displacement about O and let go. The centre of mass G is displaced to G'. The body oscillates about the equilibrium position. It can be shown that the motion is simple harmonic. Let M be the mass of the pendulum. The restoring couple due to gravity = $Mgh \sin\theta$. The couple is also equal to $I\,(d^2\theta/dt^2)$ where I = M.I. of the body about the axis of rotation and $(d^2\theta/dt^2)$ = angular acceleration.

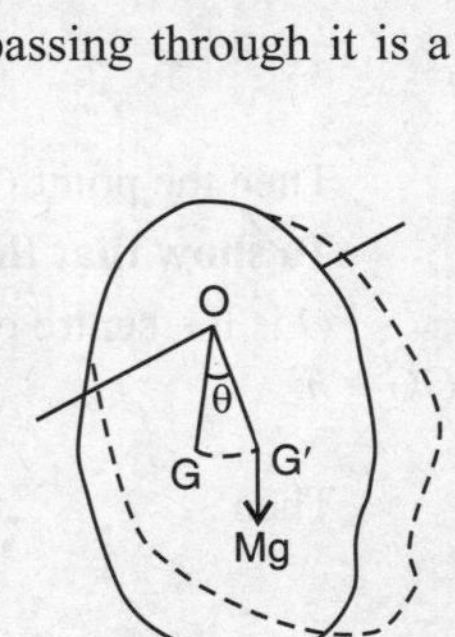

Fig. 6.8

The equation of motion of the body is

$$I \frac{d^2\theta}{dt^2} = - Mgh \sin \theta$$

or $$\frac{d^2\theta}{dt^2} = -\frac{Mgh}{I} \sin \theta = -\frac{Mgh}{I}\theta \qquad [\because \sin\theta = \theta \text{ when } \theta \text{ is small}]$$

Now Mgh/I is a constant quantity. Therefore, angular acceleration ($d^2\theta/dt^2$) is proportional to the angular displacement θ and the motion of the body is simple harmonic.

$$\text{Period } T = 2\pi \sqrt{\frac{\text{displacement}}{\text{acceleration}}} = 2\pi \sqrt{\frac{\theta}{\left(\frac{Mgh}{I}\theta\right)}}$$

$$= 2\pi \sqrt{\frac{I}{Mgh}}$$

$I = I_{cm} + Mh^2$ by the parallel axes theorem

$I = Mk^2 + Mh^2$ [where k = Radius of gyration of the body about an axis passing through cm]

or $I = M(k^2 + h^2)$.

$$\therefore \quad T = 2\pi \sqrt{\frac{M(k^2 + h^2)}{Mgh}} \text{ or } T = 2\pi \sqrt{\frac{k^2 + h^2}{gh}} \qquad ...(1)$$

Equivalent simple pendulum

A simple pendulum which oscillates with the same period as the compound pendulum is called the equivalent simple pendulum.

In the case of a simple pendulum of length L, the period

$$T = 2\pi \sqrt{\frac{L}{g}} \qquad ...(2)$$

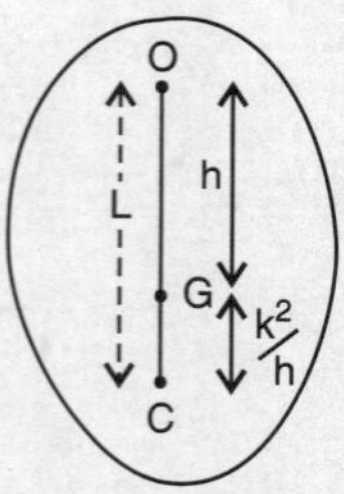

Fig. 6.9

Comparing expressions (1) and (2),

the length of the equivalent simple pendulum $= L = \frac{k^2 + h^2}{h}$

Centre of oscillation : Produce the line OG to C such that

$$OC = \frac{k^2 + h^2}{h} = h + \frac{k^2}{h}$$

Then the point C is called the centre of oscillation (Fig. 6.9).

To show that the centre of suspension and the centre of oscillation are interchangeable.

O is the centre of suspension and C is the centre of oscillation of a compound pendulum and $OG = h$.

Then $$OC = L = \frac{k^2 + h^2}{h}$$

$$\therefore \quad k^2 = hL - h^2 = h(L - h) = OG \times CG.$$

The symmetry of the expression $k^2 = OG \times CG$ shows that, if the body is suspended about a parallel axis through C, O would be the centre of oscillation. Hence the centre of oscillation and suspension are interchangeable.

Minimum time of oscillation of a compound pendulum

$$T = 2\pi\sqrt{\frac{k^2+h^2}{gh}}$$

T will be a minimum when, $\frac{d}{dh}\left[\frac{k^2+h^2}{h}\right] = 0$

i.e., $\frac{d}{dh}\left[h + \frac{k^2}{h}\right] = 0$, *i.e.,* $1 - \frac{k^2}{h^2} = 0$ or $k = h$

Therefore, the period is a minimum when the radius of gyration about a parallel axis through the centre of mass of the body = The depth of the cm below the point of suspension.

Determination of *g* with compound pendulum

A compound pendulum consists of a heavy uniform metal bar about a metre long. It has a number of holes drilled at regular intervals on either side of the centre of mass G [Fig. 6.10].

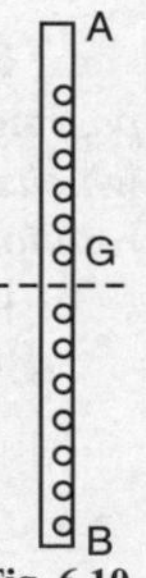

Fig. 6.10

The horizontal knife-edge is passed through the hole near the end A. The period of oscillation is determined and the distance of the knife-edge from the end A is measured. The experiment is repeated and the bar is made to oscillate about the knife-edge placed successively in the different holes from A to B. In each case the period of oscillation (T) and the distance of the position of the knife-edge from the same end A are noted.

A graph is plotted between the period, (on y-axis) and the distance from A (on the x-axis).

Two curves as shown in Fig. 6.11 are obtained. A horizontal line $PQRS$ is drawn cutting both the curves at points P, Q, R and S. P, Q, R and S are then the four points on the bar collinear with the centre of mass having the same period.

$PR = QS = L$, the length of the equivalent simple pendulum. Therefore, if t be its time period given by the ordinate of any one of the points P, Q, R or S we have,

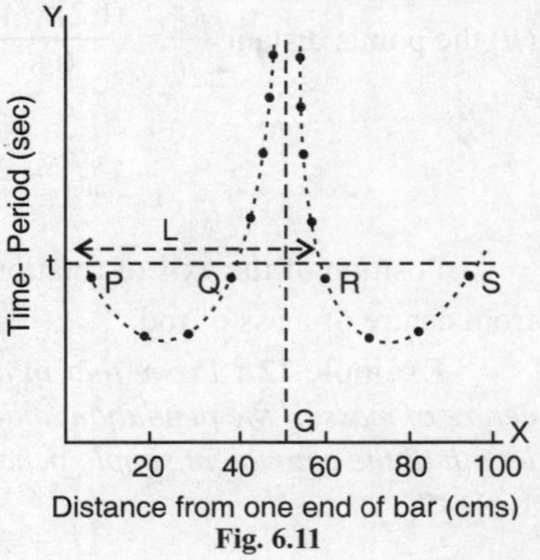

Fig. 6.11

$$t = 2\pi\sqrt{\frac{L}{g}} \text{ or } g = 4\pi^2\frac{L}{t^2}$$

Knowing L and t, we can calculate the value of g at the given place.

Example 10 : *A compound pendulum is formed by suspending a heavy ring of radius 4.905 metres. What is the length of the string for the minimum period ? Calculate this period.*

Since the ring is large and heavy, the weight of the string may be neglected so that the centre of mass of the compound pendulum will be the same as that of the ring, *i.e.*, the centre of the ring.

Let m and R be the mass and radius of the ring. If the ring oscillates in its own plane,

M.I. of the ring about an axis through its centre and $\perp r$ to its plane $\Big\} = I = mR^2 = mk^2$

where k is the radius of gyration of the ring about an axis through its centre and perpendicular to its plane.

$\therefore \quad k = R,$

The period is minimum when $k = h$, where h = the distance of the point of suspension from the centre of mass of the ring. This gives $h = R$.

i.e., the length of the string is zero and the ring is suspended at a point on its periphery.

We have, $T = 2\pi\sqrt{\dfrac{k^2+h^2}{gh}}$. T is minimum when $k = h$

$$\therefore \quad T_{min} = 2\pi\sqrt{\frac{k^2+k^2}{kg}} = 2\pi\sqrt{\frac{2k}{g}} = 2\pi\sqrt{\frac{2R}{g}} \qquad (\because k = R)$$

Here, $R = 4.905$ m, $g = 9.81$ ms^{-2}.

Hence $T_{min} = 2\pi\sqrt{\dfrac{2\times 4.905}{9.81}} = 6.282$ s.

Example 11 : *A uniform rod of length 1.2 m oscillates about a horizontal axis of rotation passing through one end. Find the period of oscillation. Find the positions of the other points about which the period is same. Also calculate the minimum period possible and the position of the axis of rotation for obtaining the minimum period.*

Here, $k^2 = l^2/12 = (1.2)^2/12$; $h = l/2 = 0.6$ m.

$$\therefore \quad T = 2\pi\sqrt{\frac{k^2+h^2}{gh}} = 2\pi\sqrt{\frac{(1.2)^2/12+(0.6)^2}{9.8\times 0.6}} = 1.794 \text{ s}$$

The position of other points, about which the period is same : (*i*) the other end of the rod ; (*ii*) the points distant $\dfrac{k^2}{h} = \dfrac{(1.2)^2/12}{0.6} = 0.2$ m from the centre and on either side.

$$T_{min} = 2\pi\sqrt{\frac{2k}{g}} = 2\pi\sqrt{\frac{2\times(1.2/\sqrt{12})}{9.8}} = 1.671 \text{ s}$$

Position of the axis of rotation for obtaining the minimum period = $k = 1.2/\sqrt{12} = 0.346$ m from centre of mass of rod.

Example 12 : *Prove that in a compound pendulum, there are four points collinear with the centre of mass of the pendulum about which it has the same period of oscillation. Hence obtain the length of the equivalent simple pendulum.*

$$T = 2\pi\sqrt{\frac{k^2+h^2}{g.h}}; \text{ Squaring, } T^2 = \frac{4\pi^2(k^2+h^2)}{g.h}$$

Rearranging, $h^2 - \dfrac{g.T^2}{4\pi^2}h + k^2 = 0$

This is a quadratic equation in h and gives two values of h.

$$h_1 = \frac{g.T^2}{8\pi^2} + \sqrt{\frac{g^2T^4}{64\pi^4} - k^2}$$

and
$$h_2 = \frac{g.T^2}{8\pi^2} - \sqrt{\frac{g^2T^4}{64\pi^4} - k^2}$$

Thus there are two points on either side of *cm* lying at distances h_1 and h_2 respectively about which the time period of the compound pendulum has the same value. Hence in all there are four points collinear with the *cm* about which the periods are equal.

$$h_1 + h_2 = g \cdot T^2/(4\pi^2) = \text{Length of the equivalent simple pendulum.}$$

Example 13 : *A uniform circular disc of radius R oscillates in a vertical plane about a horizontal axis. Find the distance of the axis of rotation from the centre for which the period is minimum. What is the value of this period ?*

We have, $T = 2\pi\sqrt{(k^2 + h^2)/(g \cdot h)}$

T is minimum when $h = k$.

$$\therefore \quad T_{min} = 2\pi\sqrt{(k^2 + k^2)/(kg)} = 2\pi\sqrt{2k/g}$$

M.I. of a disc about an axis perpendicular to its plane and passing through its centre $= I = Mk^2 = \frac{1}{2}MR^2$

where M is the mass of the disc and R, its radius.

$$\therefore \quad k = R/\sqrt{2}.$$

Thus, T is minimum when the distance of the axis of rotation from the centre is $R/\sqrt{2}$.

$$T_{min} = 2\pi\sqrt{\frac{2R/\sqrt{2}}{g}} = 2\pi\sqrt{\frac{R\sqrt{2}}{g}} = 2\pi\sqrt{\frac{1.414\,R}{g}}$$

Example 14 : *A uniform circular disc of 0.2 m radius oscillates in its own plane about a point on its circumference. Calculate the period of oscillation.*

We have, $T = 2\pi\sqrt{(k^2 + h^2)/(g \cdot h)}$

Here, $I = Mk^2 = MR^2/2$ or $k^2 = R^2/2$; $h = R$

$$\therefore \quad T = 2\pi\sqrt{\frac{(R^2/2) + R^2}{g \cdot R}} = 2\pi\sqrt{\frac{3R}{2g}} = 2\pi\sqrt{\frac{3 \times 0.2}{2 \times 9.8}}$$

$$= 1.099 \text{ s}$$

EXERCISE VI

1. State and explain Newton's law of Gravitation.
2. Describe Boys method for measuring the gravitational constant.
3. Explain the terms 'gravitational potential', 'gravitational field', 'gravitational constant' and 'equipotential surface'.
4. Determine the gravitational potential and the gravitational field at a point (*i*) inside and (*ii*) outside a hollow spherical shell.
5. Derive an expression for the gravitational potential due to a uniform sphere at a point (*i*) inside the sphere and (*ii*) outside the sphere.
6. Explain how g varies with altitude, latitude, and depth. Obtain a general relation for the variation of g in each case.
7. The moon describes a circular orbit of radius 3.8×10^8 m about the earth in 27 days and the earth describes a circular orbit of radius 1.5×10^{11} m round the sun in 365 days. Determine the mass of the sun in terms of that of the earth. [**Ans.** $M_s = 3.366 \times 10^5\, M_e$]
8. Given the mass of the earth is 81.5 times the mass of the moon and the diameter of the moon is 0.27 times that of the earth. Calculate the value of gravity at the surface of the moon. [**Ans.** $0.1683\, g$]
9. Calculate the period of revolution of the moon around the earth from the following data : Radius of the earth = 6.4×10^6 m ; Distance of the moon from the centre of the earth = 3.84×10^8 m ; Acceleration due to gravity on the surface of the earth = 9.8 ms^{-2}. [**Ans.** 2.41×10^6 s]

10. Assuming that Mars is a sphere of radius 3400 km and density 3900 kg m^{-3} and that the Earth is a sphere of radius 6370 km and density 5500 kg m^{-3}, calculate the value of acceleration due to gravity on the surface of Mars. [**Ans.** 3.71 ms^{-2}]

11. The radius of earth is 6.637×10^6 m, its mean density = 5570 kg m^{-3} and the gravitational constant = 6.67×10^{-11} Nm^2 kg^{-2}. Calculate the earth's surface potential.

[**Hint :**
$$V = -\frac{GM}{a} = \frac{-G\frac{4}{3}\pi . a^3\rho}{a} = -G\frac{4}{3}\pi . a^2\rho$$
$$= -(6.67 \times 10^{-11})\ \frac{4}{3}\pi\ (6.637 \times 10^6)^2 \times 5570$$
$$= -6.853 \times 10^7 \text{ J kg}^{-1}.$$

12. Deduce an expression for the gravitational potential due to a sphere at an external point. Hence calculate the amount of work required to send a body of mass *m* from the earth's surface to a height (*i*) *R*/2, (*ii*) 10 *R* and (*iii*) 1000 *R*, where *R* is the radius of the earth. Express the result in *m*, *R* and *g*.

$$\left[\textbf{Ans. } (i)\ \frac{1}{3}mgR,\ (ii)\ \frac{10}{11}mgR,\ (iii)\ \frac{1000}{1001}mgR\right]$$

13. Find the gravitational field intensity 3000 km above the surface of the earth.

[**Hint :**
$$F = -\frac{GM}{r^2}.$$

Here, $M = 5.977 \times 10^{24}$ kg ; $r = 6371 + 3000 = 9371$ km $= 9.371 \times 10^6$ m

$$\therefore \quad F = -\frac{6.67 \times 10^{-11} \times 5.977 \times 10^{24}}{(9.371 \times 10^6)^2}$$
$$= 4.542 \text{ ms}^{-2}$$

14. Assuming the earth to be a sphere of uniform density, calculate the acceleration due to gravity at a point (*a*) 1600 km above the surface, (*b*) 1600 km below the surface and the rate of variation of acceleration (*c*) above and (*d*) below the earth's surface. Radius of the earth = 6400 km and $g = 9.8$ ms^{-2}.

[**Ans.** (*a*) 6.392 ms^{-2} (*b*) 7.35 ms^{-2} (*c*) $\frac{dg}{g} = -2\frac{dr}{r}$ (*d*) $\frac{dg}{g} = -\frac{dr}{R-r}$

15. Assuming that the whole variation of the weight of a body with its position on the earth's surface is due to rotation of the earth, find the difference in the weight of a kilogram as measured at the equator and at the poles. [**Ans.** 3.395×10^{-2} N]

16. The earth is rotating fast enough such that the bodies at latitude of 30° just begin to fly away from the surface of the earth. Find the number of revolutions of earth per day. [**Ans.** 12.72 per day]

17. Compare the apparent loss of weight of a body to its real weight when it is kept on the surface of the earth (*i*) on its equator ; (*ii*) on latitude 60°, because of rotation of the earth. [**Ans.** 0.0347 ; 0.0174]

7 Moment of Inertia

CHAPTER

7.1. INTRODUCTION

Definition : If a rigid body consists of a finite number of particles of masses m_1, m_2, m_3 etc., at distances r_1, r_2, r_3, etc., from a given straight line XY (Fig. 7.1), the moment of inertia of the body about the given line is given by

$$I = m_1 r^2{}_1 + m_2 r_2{}^2 + m_3 r^2{}_3 + ... = \Sigma m r^2$$

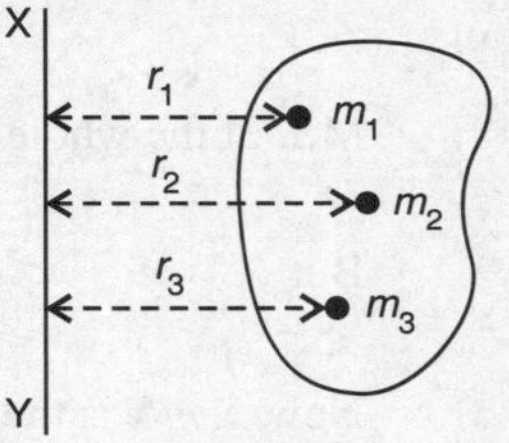

Fig. 7.1

Unit : *kgm*^{2};

Dimensional formula: $[I] = [ML^2]$

Radius of Gyration : Suppose the whole mass of the body (M) is concentrated at a point distant k from the axis such that $M.k^2 = \Sigma m r^2 = I$.

Then k is called the radius of gyration of the body about the given axis. $k = \sqrt{I/M}$.

Physical Significance of M.I. In translational motion $F = m a$.

In rotational motion $\vec{\tau} = I \alpha$. This suggests that just as we associate a force with the linear acceleration of a body, so we may associate a torque with the angular acceleration of a body about a given axis. Mass M is a measure of the resistance a body offers to having its translational motion changed by a given force. Similarly, moment of inertia I is a measure of the resistance a body offers to having its rotational motion changed by a given torque. Thus *M.I. plays the same role in rotational motion as mass does in translational motion.*

7.2. PERPENDICULAR AXES THEOREM

Statement : *If I_x and I_y are the moments of inertia of a lamina about two rectangular axes OX and OY in its plane, its moment of inertia about an axis OZ, perpendicular to its plane, is $I_z = I_x + I_y$.*

Proof : Let OX, OY be the two perpendicular axes in the plane of the lamina and OZ an axis perpendicular to the lamina (Fig. 7.2). Consider a particle P, of mass m in the plane of the lamina. x, y and r are the distances of the particle from OY, OX and OZ respectively.

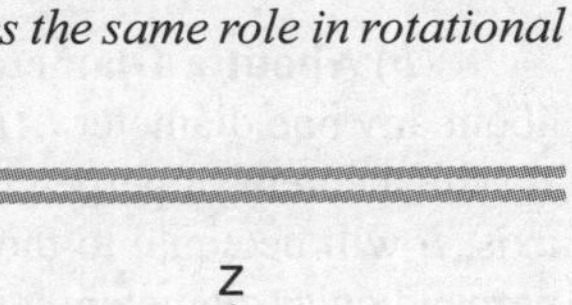

Fig. 7.2

Moment of inertia of the particle about $OZ = m r^2$

$\therefore$ M.I. of the lamina about $OZ = \Sigma m r^2$

Similarly, M.I. of the lamina about $OX = \Sigma m y^2$

and M.I. of the lamina about $OY = \Sigma m x^2$

But $r^2 = x^2 + y^2$

$$\therefore \quad \Sigma m r^2 = \Sigma m (x^2 + y^2) = \Sigma m x^2 + \Sigma m y^2$$

$$\Sigma m r^2 = I_z; \Sigma m x^2 = I_y; \Sigma m y^2 = I_x.$$

$$\therefore \quad I_z = I_x + l_y$$

7.3. THEOREM OF PARALLEL AXES

Statement : *If I is the moment of inertia of a body about an axis through its centre of mass and I' its moment of inertia about a parallel axis at a perpendicular distance h from the first axis, then $I' = I + Mh^2$, where M is the mass of the body.*

Proof : AB is an axis passing through the centre of mass of the body G (Fig. 7.3.) CD is a parallel axis at a perpendicular distance h from AB. M is the mass of the body.

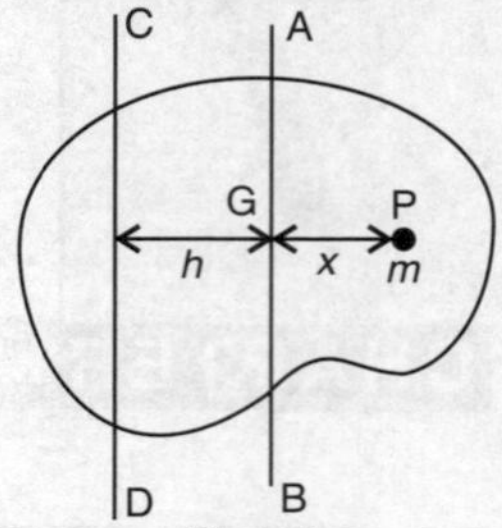

Fig. 7.3

Consider a particle P of mass m at a distance x from AB.

M.I. of the particle P about $AB = mx^2$

M.I. of the whole body about $AB = I = \Sigma\, m\, x^2$

M.I. of the particle P about $CD = m(x + h)^2$

$$= m\,(x^2 + h^2 + 2\,x\,h)$$

$$= mx^2 + m\,h^2 + 2\,m\,x\,h.$$

M.I. of the whole body about CD

$$= I' = \Sigma\, m\, x^2 + \Sigma\, m\, h^2 + \Sigma\, 2\, m\, x\, h.$$

But $\quad \Sigma\, m\, x^2 = I;\ \Sigma\, m\, h^2 = M\, h^2.$

$\therefore \quad I' = I + M\, h^2 + 2\, h\, \Sigma\, m\, x.$

Now, $\Sigma\, m\, x$ is the algebraic sum of the moments of all the particles about G.

Since the body is balanced about the centre of mass G, $\Sigma\, m\, x = 0$.

$\therefore \quad I' = I + M\, h^2.$

7.4. MOMENT OF INERTIA OF A THIN CIRCULAR RING

(*a*) **About an axis through its centre and perpendicular to its plane.** Let M be the mass and R the radius of the ring (Fig. 7.4). Consider a particle of mass m of the ring. Its moment of inertia about an axis through O and perpendicular to the plane of the ring is $m.R^2$.

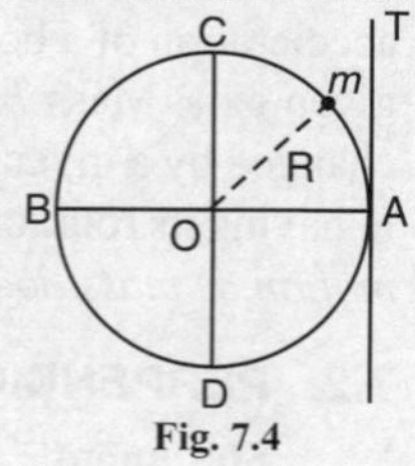

Fig. 7.4

Therefore, the moment of inertia I of the entire ring about the given axis is

$$I = \Sigma\, m\, R^2 = R^2\, \Sigma\, m \text{ or } \mathbf{I = MR^2}$$

(*b*) **About a Diameter.** By symmetry, the moment of inertia of the ring about any one diameter (AB) is the same as about any other (CD). Let it be I_d. By the theorem of perpendicular axes, the moment of inertia about the central axis, I, will be equal to the sum of its moments of inertia about two mutually perpendicular diameters lying in its plane. Thus

$$I = I_d + I_d = MR^2$$

$$I_d = \frac{1}{2}\, MR^2$$

(*c*) **About a Tangent.** By the theorem of parallel axes, the moment of inertia about a tangent is $I_t = I_d + MR^2 = \dfrac{3}{2} MR^2$.

7.5. MOMENT OF INERTIA OF A CIRCULAR DISC ABOUT AN AXIS THROUGH ITS CENTRE AND PERPENDICULAR TO ITS PLANE

Consider a uniform circular disc of mass M. R is the radius of the disc, O the centre and ρ the mass per unit area. Consider an elementary ring of radius x and width dx (Fig. 7.5).

Area of the ring = circumference × width

$$= 2\,\pi\, x\, dx.$$

Mass of the ring = Area × Mass per unit area = $2\,\pi\, x\, dx\,.\,\rho$

$= 2\,\pi\,\rho\, x\, dx$

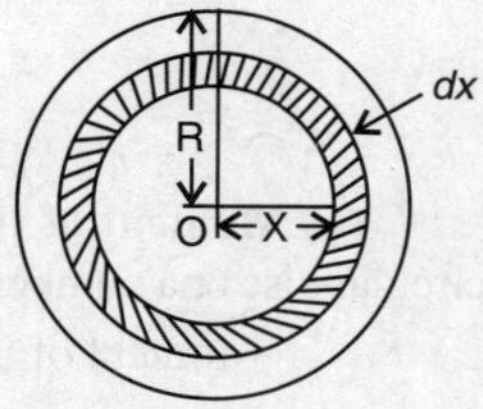

Fig. 7.5

(*i*) M.I. of this ring about an axis through its centre perpendicular to its plane = $2\,\pi\,\rho\, x\, dx\,.\, x^2$

∴ M.I. of the whole disc about this axis is

$$I = \int_0^R 2\,\pi\,\rho\, x^3\, dx = 2\,\pi\,\rho\left[\frac{x^4}{4}\right]_0^R = 2\,\pi\,\rho\,.\frac{R^4}{4} = \frac{\pi\,\rho\, R^4}{2} = \pi\, R^2\,\rho\,\frac{R^2}{2}$$

$$\therefore \qquad I = \frac{1}{2} M R^2. \qquad (\because\ \pi R^2 \rho = M)$$

(*ii*) *M.I. of the disc about a diameter* = $I_d = \frac{1}{4} M R^2$ by perpendicular axes theorem.

(*iii*) *Moment of Inertia about a Tangent in its Plane*

By the application of the theorem of parallel axes,

$$I_t = I_d + M R^2 = \frac{1}{4}\, M R^2 + M R^2 = \frac{5}{4}\, M R^2.$$

7.6. (i) MOMENT OF INERTIA OF A CYLINDER ABOUT AN AXIS PERPENDICULAR TO ITS LENGTH AND PASSING THROUGH ITS CENTRE OF MASS

M is the mass of the cylinder, *R* its radius, *l* its length, ρ its mass per unit volume and *O* the centre of gravity of the cylinder. Let *YOY′* be the axis passing through the centre of mass and perpendicular to the length of the cylinder. Divide the cylinder into thin discs and consider one such disc of thickness *dx* at a distance *x* from *O* (Fig. 7.6).

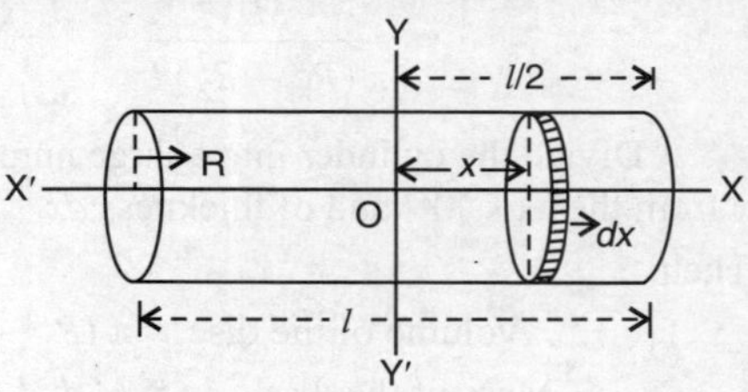

Fig. 7.6

Volume of the disc = $\pi\, R^2\, dx$

Mass of the disc = $\pi\, R^2\, \rho\, dx$

Moment of Inertia of the disc about its diameter

$$= \pi\, R^2\,\rho\, dx\,\frac{R^2}{4} = \frac{1}{4}\,\pi\, R^4\,\rho\,(dx).$$

By the theorem of parallel axes, the M.I. of the disc about *YOY′*

$$= \frac{1}{4}\,\pi\, R^4\,\rho\,(dx) + \pi\, R^2\,\rho\,(dx)\, x^2 = \pi\, R^2\,\rho\left(\frac{R^2}{4} + x^2\right) dx.$$

Moment of Inertia of the whole cylinder about *YOY′* is

$$I = \pi\, R^2\,\rho \int_{-l/2}^{+l/2}\left(\frac{R^2}{4} + x^2\right) dx = 2\,\pi\, R^2 \rho \int_0^{l/2}\left(\frac{R^2}{4} + x^2\right) dx$$

$$= 2\,\pi\, R^2\,\rho\left[\frac{R^2}{4}x + \frac{x^3}{3}\right]_0^{l/2} = 2\,\pi\, R^2\,\rho\left[\frac{R^2}{4}\left(\frac{l}{2}\right) + \frac{l^3}{24}\right]$$

$$= \pi\, R^2\,\rho l\left[\frac{R^2}{4} + \frac{l^2}{12}\right]$$

$$\therefore \qquad I = M\left[\frac{l^2}{12} + \frac{R^2}{4}\right] \qquad (\because M = \pi R^2 l \rho)$$

(*ii*) **Moment of Inertia of a Solid Cylinder about its Own Axis.** A solid cylinder is a thick circular disc or a number of thin circular discs each of mass m and radius R placed one above the other.

Moment of inertia of one disc $= m R^2/2$.

$\therefore$ Moment of inertia of the cylinder $= I = \Sigma \dfrac{m R^2}{2} = \dfrac{M R^2}{2}$

7.7. M.I. OF A HOLLOW CYLINDER ABOUT AN AXIS PASSING THROUGH THE CENTRE AND PERPENDICULAR TO THE LENGTH OF THE CYLINDER

Let M be the mass of the cylinder, l, its length, and R_1, and R_2, its external and internal radii respectively. Let YOY' be the axis though its centre O and perpendicular to its own axis XX' (Fig. 7.7). Then,

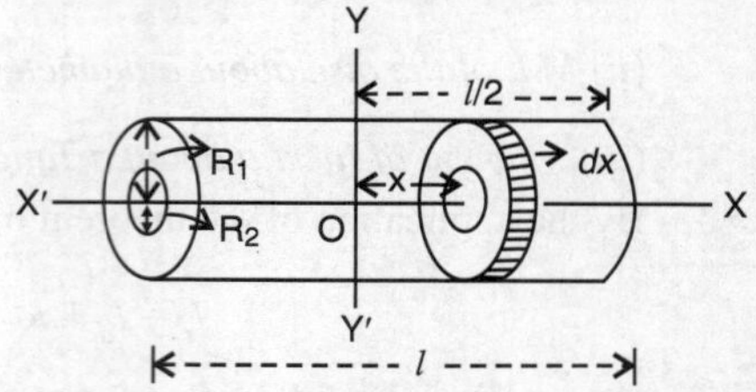

Fig. 7.7

face area of the cylinder

$$= \pi (R_1^2 - R_2^2)$$

and its volume $= \pi (R_1^2 - R_2^2)\, l$

$\therefore$ its mass per unit volume

$$= \frac{M}{\pi (R_1^2 - R_2^2)\, l}$$

Divide the cylinder into a large number of annular discs. Consider one such disc at a distance x from the axis YY' and of thickness dx.

Then,

Volume of the disc $= \pi (R_1^2 - R_2^2)\, dx$

$\therefore$ Mass of the disc $= M\, dx / l$

M.I. of the disc about its diameter $= \dfrac{M}{l} dx \times \left(\dfrac{R_1^2 + R_2^2}{4}\right)$

M.I. of the disc about the parallel axis

$$YY' = \frac{M}{l} dx \frac{R_2^2 + R_2^2}{4} + \frac{M}{l} dx \cdot x^2$$

M.I. of the whole cylinder about the axis YY' is

$$I = 2\int_0^{l/2} \left[\frac{M\,(M_1^2 + R_2^2)}{4\,l} dx + \frac{M}{l} x^2\, dx\right]$$

$$\therefore \qquad I = M\left[\left(\frac{R_1^2 + R_2^2}{4}\right) + \frac{l^2}{12}\right]$$

7.8. M.I. OF A HOLLOW CYLINDER ABOUT ITS OWN AXIS

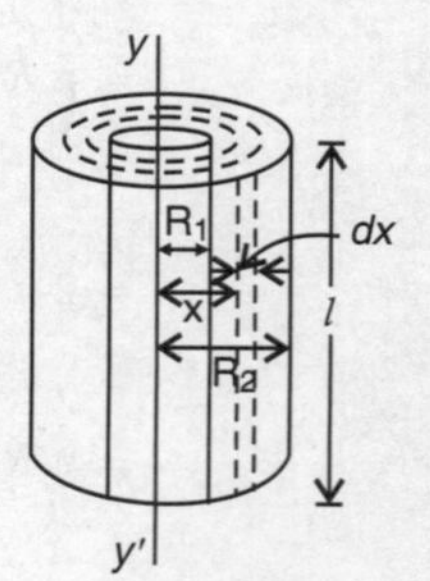

Fig. 7.8

Consider an annular cylinder of mass M, density ρ, length l, inner radius R_1 and outer radius R_2 (Fig. 7.8). Consider an element of length l, radius x and thickness dx.

Mass of the element $= dm = (2\,\pi x l \cdot dx)\, \rho$

M.I. of the element about the axis

$$YY' = x^2 dm$$

$\therefore$ M.I. of the whole cylinder about the cylinder axis

$$I = \int_{R_1}^{R_2} x^2\, dm = 2\,\pi\, l\, \rho \int_{R_1}^{R_2} x^3\, dx.$$

$$= 2\,\pi\, l\, \rho \left(\frac{R_2^4 - R_1^4}{4} \right) = 2\,\pi\, l\, \rho \left(\frac{(R_2^2 + R_1^2)\,(R_2^2 - R_1^2)}{4} \right)$$

But $\quad M = \rho\, \pi\, (R_2{}^2 - R_1{}^2)\, l$

$\therefore \quad I = \frac{1}{2} M\,(R_1^2 + R_2^2)$

For a solid cylinder of radius R, $I = \frac{1}{2} MR^2$ $\quad (\because R_1 = 0)$.

7.9. MOMENT OF INERTIA OF A SOLID SPHERE

(*a*) **About a diameter.** Consider a solid sphere of radius R and mass M (Fig. 7.9). Let ρ be the density of the sphere. Then its mass $M = \frac{4}{3}\,\pi\, R^3\, \rho$.

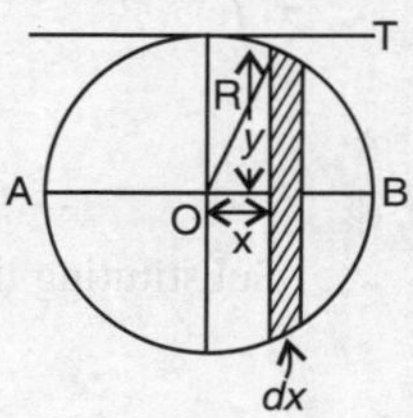

Fig. 7.9

Divide the sphere into a number of thin discs by planes perpendicular to the diameter AB. Consider a disc of thickness dx at a distance x from the centre O.

The radius of the disc $= y = \sqrt{R^2 - x^2}$

Volume of the disc $= \pi\,(R^2 - x^2)\, dx.$

Mass of the disc $= \pi\,(R^2 - x^2)\, dx \times \rho$

M.I. of this disc about $AB = \dfrac{\pi\,\rho\,(R^2 - x^2)\, dx \times (R^2 - x^2)}{2}$

$$= \frac{\pi\,\rho}{2}\,(R^4 - 2R^2\, x^2 + x^4)\, dx.$$

M.I. of the whole sphere about the diameter AB is

$$I = \int_{-R}^{R} \frac{\pi\,\rho}{2}\,(R^4 - 2\,R^2 x^2 + x^4)\, dx = 2 \times \frac{\pi\,\rho}{2} \int_0^R (R^4 - 2\,R^2 x^2 + x^4)\, dx$$

$$= \pi\,\rho \left[R^4 x - 2R^2 \cdot \frac{x^3}{3} + \frac{x^5}{5} \right]_0^R = \pi\,\rho \left[R^5 - \frac{2\,R^5}{3} + \frac{R^5}{5} \right] = \pi\,\rho \cdot \frac{8\,R^5}{15}$$

$$= \frac{2}{5} \times \frac{4}{3}\,\pi\, R^3\, \rho \times R^2 \qquad \left[\because M = \frac{4}{3}\,\pi\, R^3\, \rho \right]$$

or $\quad \boldsymbol{I = \frac{2}{5}\, MR^2}$

(*b*) **About a Tangent.** The tangent at a point on the sphere is at a distance R from the diameter. Now by the application of the theorem of parallel axes, the M.I. of the sphere about the tangent T is

$$I_t = I + MR^2 = \frac{2}{5}\, MR^2 + MR^2 = \frac{7}{5}\, \mathbf{MR^2}.$$

7.10. M.I. OF A HOLLOW SPHERE ABOUT ITS DIAMETER

Let M be the mass of the hollow sphere and R_1 and R_2 be its inner and outer radii respectively (Fig. 7.10).

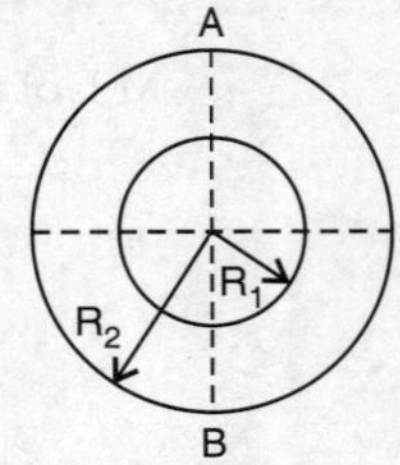

Fig. 7.10

Volume of the hollow sphere

$$= \frac{4}{3}\pi\left(R_2^3 - R_1^3\right)$$

∴ mass per unit volume of the sphere

$$\rho = \frac{3M}{4\pi(R_2^3 - R_1^3)} \quad ...(1)$$

M.I. of the hollow sphere = [M.I. of the solid sphere of radius R_2] – [M.I. of the removed solid sphere of radius R_1].

$$I = \frac{2}{5}(M_2 R_2^2 - M_1 R_1^2)$$

$$= \frac{2}{5}\left[\left(\frac{4}{3}\pi R_2^3 \rho\right) R_2^2 - \left(\frac{4}{3}\pi R_1^3 \rho\right) R_1^2\right]$$

$$I = \frac{2}{5} \times \frac{4}{3}\pi\rho(R_2^5 - R_1^5) \quad ...(2)$$

Substituting the value of ρ in Eq. (2), we have,

$$I = \frac{2}{5} \times \frac{4}{3}\pi\left[\frac{3M}{4\pi(R_2^3 - R_1^3)}\right](R_2^5 - R_1^5)$$

$$I = \frac{2}{5}M\left(\frac{R_2^5 - R_1^5}{R_2^3 - R_1^3}\right)$$

Note : M.I. of a hollow sphere about a tangent

$$I_T = \frac{2}{5}\mathbf{M}\left[\frac{\mathbf{R}_2^5 - \mathbf{R}_1^5}{\mathbf{R}_2^3 - \mathbf{R}_1^3}\right] + \mathbf{MR}_2^2$$

7.11. DETERMINATION OF THE MOMENT OF INERTIA OF A FLY-WHEEL

A flywheel is a large heavy wheel with a long cylindrical axis passing through its centre (Fig. 7.11). Its shape is such that its most of the mass is concentrated at the rim. Its centre of gravity lies on its axis of rotation so that it can come to rest in any desired position.

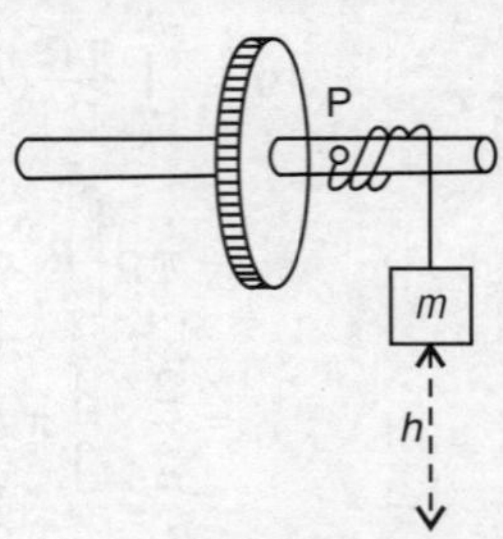

Fig. 7.11

Experimental Procedure : The fly-wheel, mounted on ball bearings, is placed at a height h above the ground on a suitable support with its axle horizontal. A small peg P is attached to the axle. A loop made at one end of a fine cord is fastened to the peg. The cord is wrapped several times round the axle and a mass m is attached to the other end of the cord. The number of turns n_1, of the cord round the axle is noted.

The mass m is allowed to fall vertically down. As the mass falls, the fly-wheel begins to rotate. After sometime, the cord is completely unwound and the loop comes off the peg. The wheel is allowed to rotate till it comes to rest. The number of rotations n_2 made by the fly-wheel in coming to rest and the time t taken are noted. The height h through which the mass has fallen is also noted.

The radius r of the axle is obtained by measuring its diameter at several places by means of a vernier calipers.

I is then calculated from the formula derived below.

Moment of inertia of the fly-wheel	$= I$
Mass suspended from the cord	$= m$
Radius of the axle	$= r$
Angular velocity of the fly-wheel when the mass is just detached	$= \omega$
Velocity gained by the mass m when the cord leaves the axle	$= v$
Average work done against friction per rotation	$= f$
Energy lost in doing work against friction in n_1 rotations	$= n_1 . f$
Potential energy lost by the mass m	$= mgh$
Kinetic energy of translation gained by the mass m	$= \frac{1}{2} m v^2$
Kinetic energy of rotation gained by the fly-wheel	$= \frac{1}{2} I \omega^2$

Now, loss in **P.E.** = [gain in **K.E.** by mass] + [gain in **K.E.** by flywheel] + [energy used against friction].

$$\therefore \quad mgh = \frac{1}{2} mv^2 + \frac{1}{2} I \omega^2 + n_1 f.$$

But $v = r\,\omega$, where r is the radius of the axle.

$$mgh = \frac{1}{2} mr^2 \omega^2 + \frac{1}{2} I \omega^2 + n_1 f. \quad ...(1)$$

Determination of ω : When the cord leaves the axle, the wheel has an angular velocity ω. After making n_2 rotations in a time t, its velocity is reduced to zero owing to the friction force. Assuming that the frictional forces remains constant during this time, the average angular velocity of the wheel is $(\omega + 0)/2 = \omega/2$. Thus,

$$\frac{\omega}{2} = \frac{2\pi n_2}{t} \text{ or } \omega = \frac{4\pi n_2}{t} \quad ...(2)$$

Determination of f : The kinetic energy $\frac{1}{2} I\omega^2$ gained by the wheel is dissipated in time t in doing work against the frictional forces.

$$\therefore \quad \frac{1}{2} I\omega^2 = n_2 f$$

$$\text{or} \quad f = \frac{1}{2} \frac{I\omega^2}{n_2} \quad ...(3)$$

Putting the value of f in Eq. (1), we get

$$mgh = \frac{1}{2} m r^2 \omega^2 + \frac{1}{2} I\omega^2 \left(1 + \frac{n_1}{n_2}\right) \quad ...(4)$$

$$I . \frac{n_2 + n_1}{n_2} = \frac{2\, mgh}{\omega^2} - mr^2 .$$

Substituting for ω from Eq. (2) and rearranging, we get

$$I = \frac{mn_2}{n_2 + n_1}\left[\frac{ght^2}{8\pi^2 n_2^2} - r^2\right] \quad ...(5)$$

7.12. MASS OF A FLY-WHEEL IS USUALLY CONCENTRATED AT THE RIM

A fly-wheel of 'large' moment of inertia gains 'large' amount of kinetic energy of rotation when mounted on the shaft of an engine. Now the moment of inertia of a mass varies as the square of its distance from the axis of rotation. Hence the mass must be situated at a greater distance from the axis to have its moment of inertia more. Thus to get the large moment of inertia of the fly-wheel, its mass is usually made to concentrate at its rim.

Example 1 : *A diatomic molecule consists of two atoms of masses m_1 and m_2 kg. r is the distance between the two atoms or the internuclear distance. Calculate the moment of inertia of the system about an axis passing through the centre of gravity of the system and perpendicular to the line joining the atoms.*

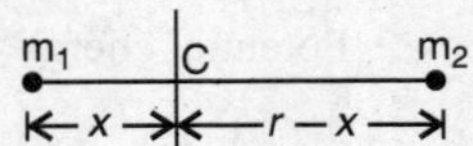

Fig. 7.12

Let m_1 be at the origin of coordinates, and the centre of mass C at a distance x from m_1 along the x-axis (Fig. 7.12). Then,

$$x = \frac{(m_1)(0) + (m_2)(r)}{m_1 + m_2} = \frac{m_2 r}{m_1 + m_2}$$

Now, the distance of C from m_2 is

$$(r - x) = r - \frac{m_2 r}{m_1 + m_2} = \frac{m_1 r}{m_1 + m_2}.$$

The moment of inertia I of the system about the axis through C and perpendicular to the line joining them, will simply be the sum of the moments of inertia of m_1 and m_2 about this axis, *i.e.*,

$$I = m_1 x^2 + m_2 (r - x)^2$$

$$= m_1\left(\frac{m_2 r}{m_1 + m_2}\right)^2 + m_2\left(\frac{m_1 r}{m_1 + m_2}\right)^2$$

$$= \frac{m_1 m_2^2 r^2 + m_2 m_1^2 r^2}{(m_1 + m_2)^2} = \frac{m_1 m_2}{m_1 + m_2} r^2$$

$\frac{m_1 m_2}{m_1 + m_2}$ is called the *'reduced mass'* μ of the system m_1, m_2.

$$\therefore \quad I = \mu r^2$$

Example 2 : *If the moment of inertia of a cylinder about an equatorial axis is to be a minimum, determine the ratio l/R, when the mass of the cylinder is kept constant.*

Here, $$I = M\left(\frac{R^2}{4} + \frac{l^2}{12}\right) \quad ...(1)$$

For **M.I.** to be minimum, $dI/dR = 0$

$$M\left(\frac{2R}{4} + \frac{2l}{12}\cdot\frac{dl}{dR}\right) = 0 \text{ i.e., } \frac{dl}{dR} = -\frac{3R}{l}. \quad ...(2)$$

Let ρ be the density of the material of the cylinder.
Mass of the cylinder $M = \pi R^2 l \times \rho$ ...(3)

$$\frac{dM}{dR} = \pi\rho\left(2Rl + R^2\frac{dl}{dR}\right) = 0 \qquad [\because M = \text{constant}]$$

$$\therefore \quad \frac{dl}{dR} = -\frac{2l}{R} \qquad ...(4)$$

Equating equations (2) and (4), we get

$$\frac{2l}{R} = \frac{3R}{l} \text{ or } \frac{\mathbf{1}}{\boldsymbol{R}} = \sqrt{\frac{3}{2}}.$$

EXERCISE VII

1. Explain the terms : moment of inertia and radius of gyration. State and prove the theorem of parallel and perpendicular axes.
2. Define moment of inertia. Derive an expression for the moment of inertia of a thin circular ring about an axis perpendicular to its plane and passing through the centre. Hence, find the moment of inertia of a thin circular ring about its diameter.
3. Find the moment of inertia of a uniform circular disc about an axis passing through its centre and perpendicular to its plane. Hence, find the moment of inertia about a tangent to the disc and about a diameter.
4. (*a*) Derive an expression for the moment of inertia of a solid cylinder of radius R and length l about an axis passing through the centre of gravity and perpendicular to the axis of the cylinder.

   (*b*) Find the moment of inertia of a solid cylinder about its axis.
5. Derive an expression for the moment of inertia of a hollow cylinder (*i*) about its own axis, and (*ii*) about an axis passing through its centre and perpendicular to its own axis.
6. Derive an expression for the moment of inertia of a sphere (*i*) about its diameter; (*a*) about a tangent.
7. Find an expression for the moment of inertia of a hollow sphere about (*i*) a diameter, and (*ii*) a tangent.
8. Describe, with necessary theory, a method of determining the moment of inertia of a fly-wheel in the laboratory.
9. What must be the relation between length l and radius R of a cylinder, if the moment of inertia of the cylinder about its axis is to be the same as its moment of inertia about the equatorial axis ?

   $$\left[\textbf{Hint } \frac{MR^2}{2} = M\left(\frac{R^2}{4} + \frac{l^2}{12}\right) \therefore \frac{l}{R} = \sqrt{3}\right]$$
10. Show that the radii of gyration of a circular disc and circular ring of the same radius about a tangential axis are in the ratio $\sqrt{\frac{5}{6}}$.
11. Four spheres, each of diameter 2 a and mass m, are placed with their centres on the four corners of a square of side b. Calculate the moment of inertia of the system about any side of the square.

    [**Ans.** (8/5) $m\,a^2 + 2\,m\,b^2$]
12. The two atoms of sodium iodide molecule, NaI, have masses 3.8×10^{-26} kgm. and 21.0×10^{-26} kgm respectively. If the distance between their centres is 2.9×10^{-10} m and they rotate around their centre of gravity, find the moment of inertia.

    [**Ans.** 27×10^{-46} kg m^2]

8 Impact of Elastic Bodies

CHAPTER

8.1. IMPULSE OF A FORCE

The impulse I of a constant force F acting for a time t is defined as $F \times t$.

$$I = F \times t.$$

By Newton's second law, $F = ma$.

If u and v are the initial and final velocities of the particle,

$$a = (v - u)/t$$

$$\therefore \qquad I = Ft = mat = m\left(\frac{v-u}{t}\right)t = m\,(v - u)$$

Thus *the impulse of a force is equal to the change in momentum produced.*

Impulsive Force : Definition. *An impulsive force is an infinitely great force acting for a very short interval of time, such that their product is finite.*

The force and the time cannot be measured because one is too great and the other is too small. Nevertheless, their product, which is definite, is capable of measurement. This we have seen, is the impulse of the impulsive force and is equal to the change in momentum produced. Hence an impulsive force is always measured by the change in momentum produced. In practice, the conditions of an impulsive force are never realized. Some approximate examples of impulsive force are : (1) the blow of a hammer on a pile and (2) the force exerted by the bat on a cricket ball.

Example 1. *A pile driver of mass 3000 kg falls through a height of 5m on to a pile of mass 1000 kg. If the pile is driven 0.24 m into the ground, find the resistance of the ground (supposed uniform).*

Sol. Velocity with which the pile driver impinges on the pile

$$v = \sqrt{2gh} = \sqrt{2 \times 9.8 \times 5} = 9.9 \text{ ms}^{-1}$$

Momentum of the pile driver = $mv = 3000 \times 9.9$ kg ms^{-1}.

After the impact the total mass in motion is 4000 kg.

Let V be the common velocity of the driver and pile immediately after the impact.

By the principle of conservation of momentum,

$$4000 \times V = 3000 \times 9.9 \text{ or } V = 7.4 \text{ ms}^{-1}.$$

Now the velocity is destroyed in 0.24 m. If a is the retardation, we get

$$0 = (7.4)^2 - 2a\,(0.24) \text{ or } a = 114.1 \text{ ms}^{-2}$$

The retarding force is $4000 \times 114.1 = 4.56 \times 10^5$ N

The resistance of the ground R must equal the retarding force plus the weight of the driver and pile.

$$\therefore \qquad R = 4.56 \times 10^5 + 4000 \times 9.8 = 4.95 \times 10^5 \text{ N}$$

8.2. WHAT IS A COLLISION ?

We learn much about atomic, nuclear and elementary particles, experimentally by observing collisions between them. The study of collisions of molecules in gases has developed into kinetic theory of gases. The study of collisions is based on the principles of conservation of momentum and conservation of energy.

In a collision, a relatively large force acts on each colliding particle for a relatively short time. The force is called an *impulsive force*. When a bat hits a ball, the bat exerts a large force on the ball. Both the ball and the bat are deformed during the collision.

The force of interaction may be due to different causes in different cases.

Thus in the collision between two billiard balls, the force of interaction is due to elasticity. It comes into existence only when the two billiard balls come into physical contact. In the case of scattering of α particles by the nuclei, it is the electrostatic force that causes the interaction.

An alpha particle projected towards the nucleus of an atom will be repelled by the electrostatic forces due to the nucleus. In this case, the particles will not touch each other. Even then this process is called a collision because a relatively strong force acts between the particles and this force has a marked effect on the motion of the α particle.

Elastic and Inelastic collisions : There are two types of collision :

(*i*) elastic and (*ii*) inelastic.

(*i*) Elastic collisions are those in which the total kinetic energy before and after the collision remains unchanged. Collisions between atomic, nuclear and fundamental particles are the true elastic collisions. Collisions between ivory or glass balls can be treated as approximately elastic collisions. In such a collision between two particles, we have

$$m_1u_1 + m_2u_2 = m_1v_1 + m_2v_2$$

and $$\frac{1}{2}m_1u_1^2 + \frac{1}{2}m_2u_2^2 = \frac{1}{2}m_1v_1^2 + \frac{1}{2}m_2v_2^2$$

where m_1 and m_2 are the respective masses of the two particles and u_1, u_2 and v_1, v_2 their velocities *before* and *after* the collision.

(*ii*) If the K. E. is not conserved, the collision is said to be inelastic. When two bodies stick together after collision, the collision is said to be completely inelastic. For example, the collision between a bullet and its target is completely inelastic when the bullet remains embedded in the target.

Completely Inelastic Collision : Suppose a body of mass m_1 moving with a velocity u_1 collides with a body of mass m_2 moving with velocity u_2 in the same direction. The two bodies stick together after collision and they move with a final common velocity V in the same direction as the original. It is not necessary to restrict the discussion to one dimensional motion. Using only the conservation of momentum principle,

$$m_1u_1 + m_2u_2 = (m_1 + m_2)\,V.$$

From this the value of V can be determined if u_1 and u_2 are known.

FUNDAMENTAL PRINCIPLES OF IMPACT

1. Newton's law of impact-coefficient of restitution. When two bodies impinge directly, their relative velocity after impact is in a constant ratio to their relative velocity before impact and is in the opposite direction. This constant ratio depends only on the material of the bodies and not on their masses or velocities. It is called the coefficient of restitution and is denoted by the letter *e*. If u_1, u_2 be the velocities of two bodies before the impact and v_1, v_2 the velocities after impact,

$$\frac{v_1 - v_2}{u_1 - u_2} = -e \text{ or } v_1 - v_2 = -e\,(u_1 - u_2)$$

Here, $(u_1 - u_2)$ and $(v_1 - v_2)$ are their relative velocities, before and after the impact. e lies between 0 and 1. If $e = 0$, the bodies are called perfectly plastic bodies. If $e = 1$, the bodies are called perfectly elastic bodies. For two glass balls, $e = 0.94$; For two lead balls, $e = 0.2$.

Definition of coefficient of restitution. *The ratio, with a negative sign, of the relative velocity of two bodies after impact to their relative velocity before impact is called the coefficient of restitution.*

2. Motion of two smooth bodies perpendicular to the line of impact. When two smooth bodies impinge, there is no tangential action between them. Hence there is no change of momentum along the common tangent. Hence, there is no change of velocity for either body along the tangent. In other words, there is no change in the velocity of a body in a direction perpendicular to the common normal due to impact.

3. Principle of conservation of momentum. The total momentum of two bodies after impact along the *common normal* should be equal to the total momentum before the impact along the same direction.

The above three principles are sufficient to determine the change in motion of two impinging smooth bodies.

Definitions. (*i*) Two bodies are said to impinge *directly* when the direction of motion of each is along the common normal at the point where they touch.

(*ii*) Two bodies are said to impinge *obliquely* if the direction of motion of either or both is not along the common normal at the point of contact.

(*iii*) The common normal at the point of contact is called the line of impact. Thus, in the case of two spheres the line of impact is the line joining their centres.

8.3. OBLIQUE IMPACT OF A SMOOTH SPHERE ON A FIXED SMOOTH PLANE

Let XY be the fixed plane. Let the sphere strike the fixed plane at point P. Then if C is the centre of the sphere, CP is the common normal at the point of contact of the plane and the sphere. Let u and v be the velocities of the sphere before and after impact making angles α and θ respectively with the common normal CP [Fig. 8.1]. By Newton's experimental law, the relative velocity of the sphere along the common normal after impact is $-e$ times its relative velocity along the common normal before impact.

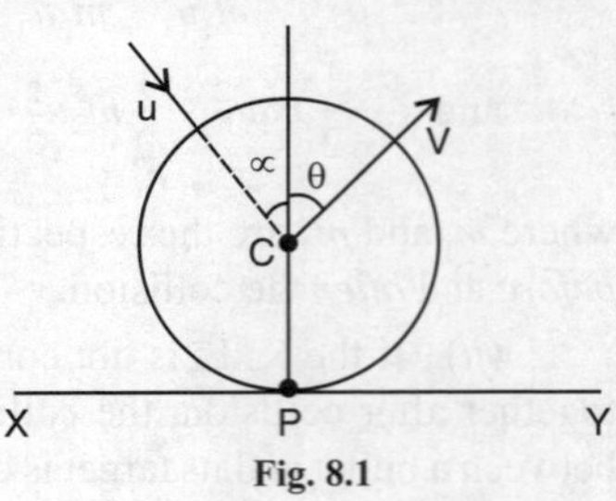

Fig. 8.1

$$v \cos \theta - 0 = -e(-u \cos \alpha - 0)$$

or $$v \cos \theta = eu \cos \alpha \quad ...(1)$$

Since both the sphere and the plane are smooth, there is no force in a direction parallel to the plane. Hence the velocity of the sphere resolved parallel to the plane is unaltered by the impact.

$$\therefore \quad v \sin \theta = u \sin \alpha \quad ...(2)$$

Dividing (1) by (2), $\cot \theta = e \cot \alpha \quad ...(3)$

Squaring and adding (1) and (2),

$$v^2 = u^2 (\sin^2 \alpha + e^2 \cos^2 \alpha) \quad ...(4)$$

Equations (4) and (3) give the velocity and direction of the sphere after impact.

Cor. 1. The impulse of the plane on the sphere is measured by the change of momentum of the sphere measured along the normal.

$$I = m[v \cos \theta - (-u \cos \alpha)] = m[v \cos \theta + u \cos \alpha]$$
$$= m[eu \cos \alpha + u \cos \alpha]$$
$$\therefore \quad I = mu(1 + e) \cos \alpha$$

Cor. 2. If $e = 1$, then $\theta = \alpha$ and $v = u$ *i.e.*, when a perfectly elastic sphere impinges on a fixed smooth plane its velocity is unaltered in magnitude by the impact and the angle of reflection is equal to the angle of incidence.

Cor.3. If $e = 0$, then $\cot\theta = 0$ or $\theta = 90°$ [from (3)] and $v = u \sin\alpha$ [from (4)]. Thus if the sphere and the plane are both inelastic, the sphere moves along the plane with velocity $u \sin\alpha$.

Cor. 4. If $\alpha = 0$ then $\theta = 0$ and $v = eu$ from (3) and (4) *i.e.*, if a sphere impinges normally on a horizontal plane, it rebounds vertically with velocity eu.

Cor. 5. The change in K.E. of the sphere due to impact on the plane is given by

$$\frac{1}{2}m(v^2 - u^2) = \frac{1}{2}m(v+u)(v-u) = \frac{1}{2}I(v+u)$$

Here, $m(v - u) = I$ = Impulse of the force of the sphere on the plane.

Example 1. *A steel ball is let fall through a height of 0.64 m on a plate of steel. The height through which it rebounds is 0.36 m. Calculate the coefficient of restitution.*

Sol. Let u be the velocity of the ball when it strikes the plane. Then $u^2 = 2gh$. The ball rebounds vertically with velocity eu after the first impact. The height to which the particle rebounds after the first impact $= \dfrac{(eu)^2}{2g} = h_1$; *i.e.*, $h_1 = \dfrac{e^2u^2}{2g}$

But $$u^2 = 2gh \; \therefore \; h_1 = e^2h \text{ or } e^2 = \frac{h_1}{h} \text{ or } e = \sqrt{\frac{h_1}{h}}$$

Here, $$h_1 = 0.36\ m;\ h = 0.64\ m;\ \therefore\ e = \sqrt{\frac{0.36}{0.64}} = 0.75$$

8.4. DIRECT IMPACT OF TWO SMOOTH SPHERES

A smooth sphere of mass m_1 moving with a velocity u_1 impinges on another smooth sphere of mass m_2 moving in the same direction with velocity u_2. If e is the coefficient of restitution between them, find the velocities of the spheres after impact.

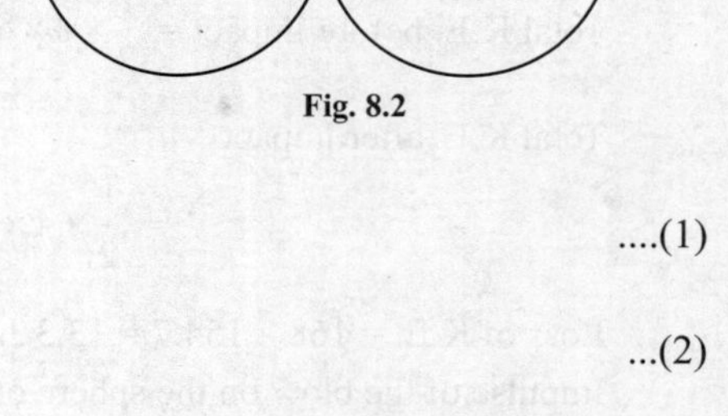

Fig. 8.2

Since the spheres are smooth, there is no impulsive force on either along the common tangent. Hence in this direction their velocities after impact are the same as their original velocities *i.e.*, zeroes.

Let v_1 and v_2 be the velocities of the two spheres along the common normal after impact [Fig. 8.2].

By the principle of conservation of momentum,

$$m_1v_1 + m_2v_2 = m_1u_1 + m_2u_2 \qquad \text{....(1)}$$

By Newton's experimental low,

$$v_1 - v_2 = -e(u_1 - u_2) \qquad \text{...(2)}$$

Multiplying (2) by m_2 and adding to (1),

$$v_1(m_1 + m_2) = m_2u_2(1+e) + u_1(m_1 - em_2)$$

$$\therefore \quad v_1 = \frac{m_2u_2(1+e) + u_1(m_1 - em_2)}{(m_1 + m_2)} \qquad \text{...(3)}$$

Multiplying (2) by m_1 and subtracting from (1),

$$v_2(m_1 + m_2) = m_1u_1(1+e) + u_2(m_2 - em_1)$$

$$\therefore \qquad v_2 = \frac{m_1u_1 (1+e) + u_2 (m_2 - em_1)}{(m_1 + m_2)} \qquad \text{....(4)}$$

Equations (3) and (4) give the velocities of the two spheres after impact.

Cor. 1. The impulse of the blow on the sphere of mass m_1 = change of momentum produced in it $= m_1 (v_1 - u_1) = \dfrac{m_1m_2 (1+e)(u_2 - u_1)}{m_1 + m_2}$.

This is equal and opposite to the impulse of the blow on the sphere of mass m_2.

Cor. 2. If $e = 1$ and $m_1 = m_2$ then, $v_1 = u_2$ and $v_2 = u_1$. Thus, if two equal perfectly, elastic spheres impinge directly, they interchange their velocities.

Example 1. *A ball of mass 8 kg, moving with a velocity of 10 ms^{-1} impinges directly on another mass 24 kg, moving at 2 ms^{-1} in the opposite direction. If e = 0.5, find the velocities of the balls after impact.*

Sol. Let v_1 and v_2 be the velocities of the balls after impact (Fig. 8.3).

By the principle of conservation of momentum,

$$8v_1 + 24v_2 = 8 \times 10 - 24 \times 2$$

By Newton's experimental law,

$$v_1 - v_2 = -0.5\,[10 - (-2)]$$

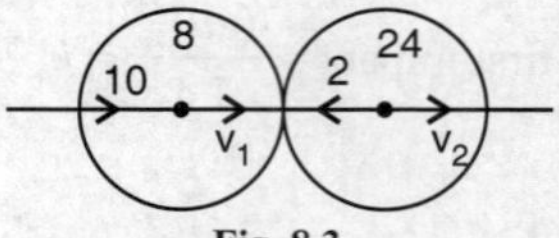

Fig. 8.3

Solving $v_1 = -3.5$ ms^{-1} and $v_2 = 2.5$ ms^{-1}

Example 2. *A smooth sphere of mass 4 kg moving with a velocity of 8 ms^{-1} impinges directly on a smooth sphere of mass 5kg moving in the same direction with a velocity of 4 ms^{-1}. Find the velocities of the spheres after impact. Calculate also the loss of K. E. due to the impact and the impulse of the blow on the sphere of smaller mass, (e = 0.5).*

Sol. Let v_1 and v_2 be the velocities of the smaller and bigger spheres after impact.

By the principle of conservation of momentum,

$$4v_1 + 5v_2 = 4 \times 8 + 5 \times 4 \qquad \text{...(1)}$$

By Newton's experimental law,

$$v_1 - v_2 = -0.5\,[8 - 4] \qquad \text{...(2)}$$

Solving, $v_1 = 4.667$ ms^{-1} and $v_2 = 6.667$ ms^{-1}

Total K.E. before impact $= \frac{1}{2} \times 4 \times 8^2 + \frac{1}{2} \times 5 \times 4^2 = 168$ J

Total K.E. after impact

$$= \frac{1}{2} \times 4 \times (4.667)^2 + \frac{1}{2} \times 5 \times (6.667)^2 = 154.7 \text{ J}$$

Loss of K.E. = 168 – 154.7 = 13.3 J.

Impulse of the blow on the sphere of mass 4 kg

$$= m(v_1 - u_1) = 4(4.667 - 8) = -13.33 \text{ J.}$$

Example 3. *A ball of mass m_1 moving with velocity u_1 strikes another ball of mass m_2 which is stationary. If the collision is elastic, calculate the fraction of the kinetic energy transferred to the second ball.*

Sol. From Eq. (4), $v_2 = \dfrac{2m_1u_1}{(m_1 + m_2)}$ $\quad (\because u_2 = 0 \text{ and } e = 1)$

Initial K. E. of the first ball $= \frac{1}{2} m_1 u_1^2$.

K.E. transferred to the second ball $= \frac{1}{2} m_2 v_2^2 = \frac{1}{2} m_2 \left(\frac{2m_1 u_1}{m_1 + m_2} \right)^2$

Fraction of K.E. transferred to second ball

$$= \frac{\frac{1}{2} m_2 \left(\frac{2m_1 u_1}{m_1 + m_2} \right)^2}{\frac{1}{2} m_1 u_1^2} = \frac{4m_1 m_2}{(m_1 + m_2)^2}.$$

Example 4. *A ball impinges on another equal ball moving with same speed in a direction perpendicular to its own, the line joining the centres of the balls at the instant of impact being perpendicular to the direction of motion of the second ball; if e be the coefficient of restitution, show that the direction of motion of the second ball is turned through* $\tan^{-1}\left(\frac{1+e}{2}\right)$.

Sol. Let C_1 and C_2 be the centres of the balls A and B. Let u be the velocity of the ball A before impact along the line of centres and u the velocity of B, perpendicular to the line of centres. Since the velocity of the ball A is u along the line of centres, after impact it will move only along the line of centres. Let it move with velocity v. Let v_1 be the velocity of the second ball B at an angle θ with the line of centres [Fig. 8.4].

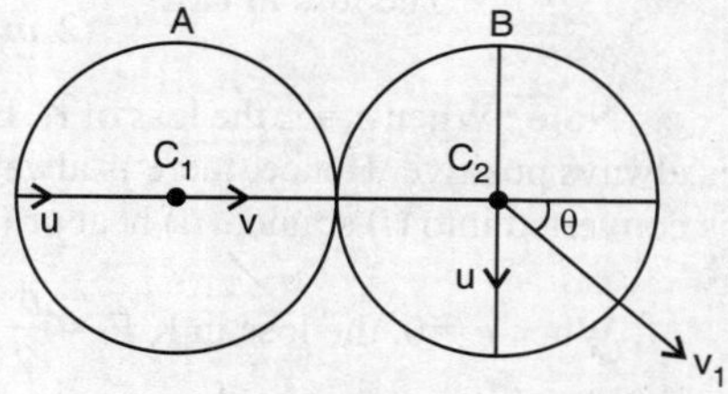

Fig. 8.4

By the principle of conservation of momentum,

$$mv_1 \cos\theta + mv = mu \qquad \text{where } m \text{ is the mass of each ball}$$

or $$v_1 \cos\theta + v = u \qquad ...(1)$$

By Newton's experimental law, $v_1 \cos\theta - v = -e(0 - u) = eu$

or $$v_1 \cos\theta - v = eu \qquad ...(2)$$

The velocity of the second ball B perpendicular to the line of centres is unaltered by the impact.

Hence, $$v_1 \sin\theta = u \qquad ...(3)$$

Adding (1) and (2), $2v_1 \cos\theta = u(1 + e)$ ...(4)

Dividing (4) by (3), $2 \cot\theta = (1 + e)$ or $\cot\theta = \frac{(1+e)}{2}$.

The angle through which the direction of motion of the second ball is turned through is $90° - \theta$, *i.e.*, $\tan^{-1} \frac{1+e}{2}$.

8.5. LOSS OF K.E. DUE TO DIRECT IMPACT OF TWO SMOOTH SPHERES

Let m_1, m_2 be the masses, u_1 and u_2, v_1 and v_2 their velocities before and after impact and e the coefficient of restitution. Then, by the principle of conservation of linear momentum,

$$m_1 v_1 + m_2 v_2 = m_1 u_1 + m_2 u_2 \qquad ...(1)$$

By Newton's experimental law,

$$v_1 - v_2 = -e(u_1 - u_2) \qquad ...(2)$$

Square both equations, multiply the square of the second by m_1m_2 and add the results. Then,

$$\left.\begin{matrix}(m_1^2 + m_1m_2)\, v_1^2 + \\ (m_2^2 + m_1\, m_2)\, v_2^2\end{matrix}\right\} = (m_1u_1 + m_2u_2)^2 + e^2m_1m_2(u_1 - u_2)^2$$

$$\therefore \quad m_1\,(m_1 + m_2)v_1^2 + m_2(m_1 + m_2)v_2^2 = (m_1u_1 + m_2u_2)^2 + m_1m_2(u_1 - u_2)^2 + e^2m_1m_2(u_1 - u_2)^2 - m_1m_2(u_1 - u_2)^2$$

$$\therefore \quad (m_1 + m_2)\,(m_1v_1{}^2 + m_2v_2^2) = (m_1 + m_2)\,(m_1u_1^2 + m_2u_2^2) - m_1m_2\,(u_1 - u_2)^2\,(1 - e^2)$$

$$\therefore \quad \frac{1}{2}m_1v_1^2 + \frac{1}{2}m_2v_2^2 = \frac{1}{2}m_1u_1^2 + \frac{1}{2}m_2u_2^2 - \frac{1}{2}\frac{m_1m_2}{m_1 + m_2}(u_1 - u_2)^2\,(1 - e^2)$$

Now, $\frac{1}{2}m_1v_1^2 + \frac{1}{2}m_2v_2^2$ = K.E. after impact.

$\frac{1}{2}m_1u_1^2 + \frac{1}{2}m_2u_2^2$ = K.E. before impact.

$\therefore$ The loss in K.E. $= \frac{1}{2}\frac{m_1m_2}{m_1 + m_2}(u_1 - u_2)^2\,(1 - e^2)$

Note : When $e = 1$, the loss of K. E. is zero. In general $e < 1$ so that $(1 - e^2)$ is positive. $(u_1 - u_2)^2$ is always positive. Hence, there is always a loss of K.E. due to impact. The K.E. lost during impact is converted into (*i*) sound, (*ii*) heat or (*iii*) vibration or rotation of the colliding bodies.

When $e = 0$, the loss in K.E. $= \frac{1}{2}\frac{m_1m_2\,(u_1 - u_2)^2}{(m_1 - m_2)}$

i.e., there is maximum loss of K.E. on impact of plastic bodies.

8.6. OBLIQUE IMPACT OF TWO SMOOTH SPHERES

A smooth sphere of mass m_1 moving with velocity u_1 impinges obliquely on a smooth sphere of mass m_2 moving with velocity u_2. If the directions of motion before impact make angles α and β with the common normal, find the velocities and direction of the spheres after impact.

Let *AB* be the common normal (Fig.8.5). Let v_1 and v_2 be the velocities of the two spheres after impact making angles θ and ϕ with the common normal *AB*. *Before impact* velocities along the common normal *AB* are $u_1 \cos \alpha$ and $u_2 \cos \beta$ and velocities perpendicular to *AB* are $u_1 \sin \alpha$ and $u_2 \sin \beta$. After impact velocities along *AB* are $v_1 \cos \theta$ and $v_2 \cos \phi$ and perpendicular to *AB* are $v_1 \sin \theta$ and $v_2 \sin \phi$.

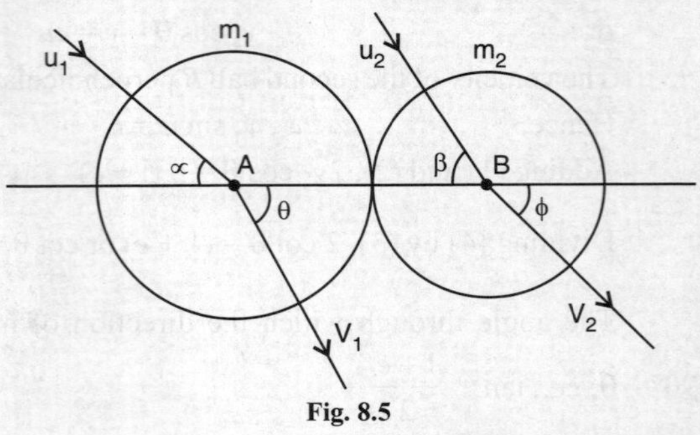

Fig. 8.5

By the principle of conservation of momentum, the total momentum of the two spheres along the common normal is unaltered by the impact.

$$\therefore \quad m_1v_1 \cos \theta + m_2v_2 \cos \phi = m_1u_1 \cos \alpha + m_2u_2 \cos \beta \qquad ...(1)$$

By Newton's experimental law (on relative velocities along the common normal),

$$v_1 \cos \theta - v_2 \cos \phi = -\,e\,(u_1 \cos \alpha - u_2 \cos \beta) \qquad ...(2)$$

Since there is no force perpendicular to the common normal *AB*, the velocities of the spheres perpendicular to the common normal *AB* remain unaltered due to impact. Hence

$$v_1 \sin\theta = u_1 \sin\alpha \qquad ...(3)$$

and $$v_2 \sin\phi = u_2 \sin\beta \qquad ...(4)$$

Multiplying (2) by m_2 and adding to (1),

$$v_1 \cos\theta = \frac{(m_1 - em_2)\, u_1 \cos\alpha + m_2\,(1+e)\, u_2 \cos\beta}{(m_1 + m_2)} \qquad ...(5)$$

Multiplying (2) by m_1 and subtracting from (1),

$$v_2 \cos\phi = \frac{(m_2 - em_1)\, u_2 \cos\beta + m_1\,(1+e)\, u_1 \cos\alpha}{(m_1 + m_2)} \qquad ...(6)$$

Squaring (3) and (5) and adding we get v_1^2 and hence we can find v_1. Dividing (3) by (5) we get $\tan\theta$. Similarly, from (4) and (6) we can get v_2 and $\tan\phi$. Therefore, v_1, v_2, θ and ϕ are determined uniquely.

Cor. 1. The impulse of the blow on the sphere of mass m_1 = its change of momentum measured along the common normal

$$= m_1 v_1 \cos\theta - m_1 u_1 \cos\alpha$$

$$= m_1\,(v_1 \cos\theta - u_1 \cos\alpha) = \frac{m_1\, m_2\,(1+e)}{(m_1 + m_2)}\,(u_2 \cos\beta - u_1 \cos\alpha)$$

This is equal and opposite to the impulse on the sphere of mass m_2.

8.7. LOSS OF K.E. DUE TO OBLIQUE IMPACT

The velocities of the spheres perpendicular to the common normal are unaltered. Therefore, the loss of K.E. is the same as in the case of direct impact if we substitute $u_1 \cos\alpha$ and $u_2 \cos\beta$ for u_1 and u_2 respectively.

$$\therefore \qquad \text{The loss in K.E.} = \frac{m_1 m_2\,(1-e^2)}{2\,(m_1 + m_2)}\,(u_1 \cos\alpha - u_2 \cos\beta)^2$$

Example 1. *A ball of mass m impinges obliquely on a ball of mass M at rest. If m = eM, prove that the directions of motion of the balls are at right angles after impact.*

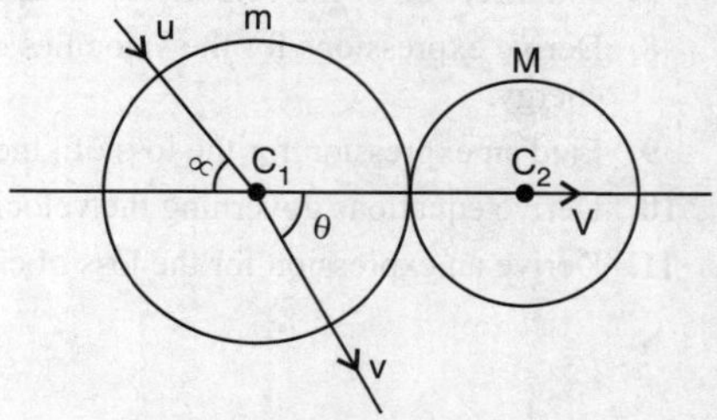

Fig. 8.6

Sol. Before impact, let u be the velocity of the ball of mass m and α its direction of motion with the common normal (Fig. 8.6). Before impact, the second ball of mass M is at rest. After impact, the second ball of mass M will move along the common normal, because the force on it during the period of impact is only along the common normal. Let the velocity of the second ball of mass M be V. After impact, let v be the velocity of the ball of mass m making an angle θ with the common normal.

By the principle of conservation of momentum,

$$mv \cos\theta + MV = mu \cos\alpha + 0$$

i.e., $$eMv \cos\theta + MV = eMu \cos\alpha$$

or, $$ev \cos\theta + V = eu \cos\alpha \qquad ...(1)$$

By Newton's experimental law,

$$v \cos\theta - V = -\,e\,(u \cos\alpha - 0) \text{ or } v \cos\theta - V = -\,eu \cos\alpha \qquad ...(2)$$

Adding (1) and (2), $(1 + e)\, v \cos\theta = 0$ or $\cos\theta = 0$ or $\theta = 90°$.

Hence, the directions of motion of the balls are at right angles after impact.

Example 2. *If an oblique collision occurs between two equal smooth perfectly elastic spheres one of which is initially at rest, show that their paths after impact are at right angles to one another.*

Sol. Let u_1 be the velocity and α_1 inclination to the line of impact of A before impact (Fig. 8.7). After impact, let these be v_1 and β_1. B is initially at rest. So after impact it moves along the line of impact with velocity v_2. By the principle of conservation of momentum,

$$mv_1 \cos \beta_1 + mv_2 = mu_1 \cos \alpha_1$$

or $$v_1 \cos \beta_1 + v_2 = u_1 \cos \alpha_1 \quad ...(1)$$

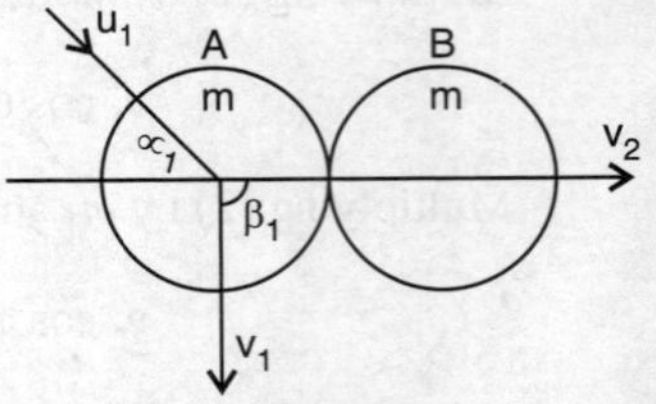

Fig. 8.7

By Newton's experimental law,

$$v_1 \cos \beta_1 - v_2 = -e\,(u_1 \cos \alpha_1 - 0) \quad ...(2)$$

Adding (1) and (2),

$$2v_1 \cos \beta_1 = (1 - e)\,(u_1 \cos \alpha_1) = 0 \; (\because e = 1)$$

v_1 can not be zero.

$\therefore$ $$\cos \beta_1 = 0 \text{ or } \beta_1 = 90°.$$

EXERCISE VIII

Section A

1. Mention the three principles that hold good when an impact takes place between two smooth spheres. Define "coefficient of restitution".
2. Define impulse of a force. How is it measured? What is an impulsive force?
3. Define : 'Line of impact". Distinguish between direct and oblique impacts.
4. Explain "Elastic collision"; "Inelastic collision".
5. Explain the momentum and impulse.
6. State and explain the difference between elastic and inelastic collision.

Section B

7. A smooth sphere has an oblique impact on a fixed smooth plane. Derive expressions for the magnitude and direction of the velocity of the sphere after impact.
8. Derive expressions for the velocities of two smooth spheres after direct impact. Find the loss of kinetic energy.
9. Find an expression for the loss of kinetic energy due to direct collision between two spheres.
10. Derive equations governing the velocities and directions of two smooth spheres after oblique impact.
11. Derive an expression for the loss of energy during the oblique impact of two solid spheres.

9 CHAPTER
Projectile on Inclined Plane, Motion of Two Interacting Bodies and Bifilar Pendulum

9.1. RANGE ON AN INCLINED PLANE

A particle is projected with a velocity u at an angle α to the horizontal from a point O on an inclined plane, inclined at an angle β to the horizontal. The direction of projection lies in the vertical plane through OA the line of greatest slope, of the plane. Let the particle strike the inclined plane at A. Then OA $(= R)$ is the range on the inclined plane (Fig. 9.1).

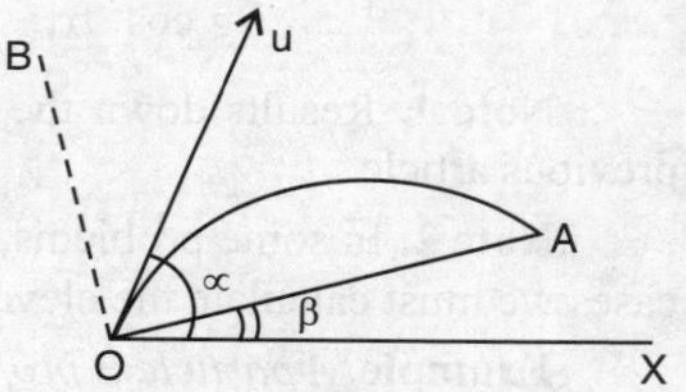

Fig. 9.1

Let OX and OA be respectively the horizontal and inclined plane through the point of projection O. OB is a line perpendicular to OA.

Component of initial velocity u along $OA = u\cos(\alpha - \beta)$

Component of initial velocity u along $OB = u\sin(\alpha - \beta)$

The projectile moves with a vertical retardation g.

Acceleration along $OA = -g\sin\beta$

Acceleration along $OB = -g\cos\beta$

Now, let T be the time taken by the particle to go from O to A. When the particle reaches A after time T, the distance moved perpendicular to the plane is zero.

Hence, $$0 = u\sin(\alpha - \beta).T - \frac{1}{2}g\cos\beta.T^2 \qquad \left(\because s = ut + \frac{1}{2}at^2\right)$$

$$T = \frac{2u\sin(\alpha - \beta)}{g\cos\beta} \qquad ...(1)$$

When the particle strikes A after time T, the distance OA $(= R)$ moved is the range on the inclined plane.

$$\therefore \quad R = u\cos(\alpha - \beta).T - \frac{1}{2}g\sin\beta.T^2$$

$$= u\cos(\alpha - \beta)\frac{2u\sin(\alpha - \beta)}{g\cos\beta} - \frac{1}{2}g\sin\beta\frac{4u^2\sin^2(\alpha - \beta)}{g^2\cos^2\beta}$$

$$= \frac{2u^2\sin(\alpha - \beta)}{g\cos^2\beta}[\cos(\alpha - \beta)\cos\beta - \sin(\alpha - \beta)\sin\beta]$$

$$= \frac{2u^2\sin(\alpha - \beta)\cos\alpha}{g\cos^2\beta}$$

9.2. RANGE AND TIME OF FLIGHT DOWN AN INCLINED PLANE

The particle is projected down the plane from O at an elevation α [Fig. 9.2]. Initial velocities along and perpendicular to OA are $u\cos(\alpha+\beta)$ and $u\sin(\alpha+\beta)$. Acceleration along and perpendicular to OA are $g\sin\beta$ and $-g\cos\beta$. When the particle reaches A after time T_1, the distance moved perpendicular to the inclined plane is zero. Therefore,

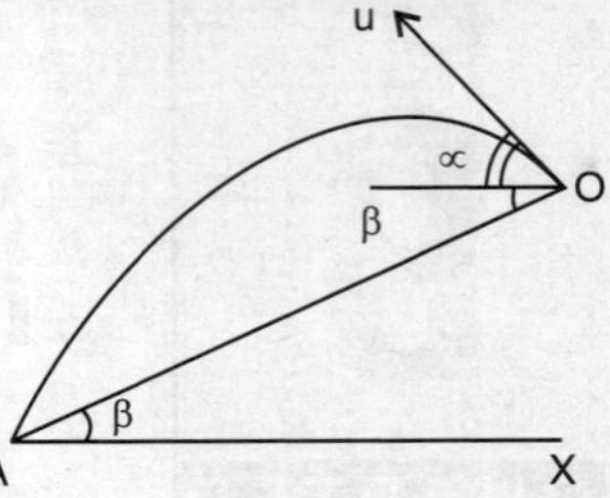

Fig. 9.2

$$0 = u\sin(\alpha+\beta).T_1 - \frac{1}{2}g\cos\beta.T_1^2 \text{ or } T_1 = \frac{2u\sin(\alpha+\beta)}{g\cos\beta}$$

$$\text{Range} = OA = R_1 = u\cos(\alpha+\beta).T_1 + \frac{1}{2}g\sin\beta.T_1^2$$

$$= \frac{2u^2\sin(\alpha+\beta)\cos\alpha}{g\cos^2\beta}$$

Note 1. Results down the plane can be obtained by putting $-\beta$ for β in the results of the previous article.

Note 2. In some problems, the elevation relative to the inclined plane may be given. In such cases we must calculate the elevation relative to the horizontal.

Example. *A particle is projected with a velocity of 32 ms^{-1} at an angle of* 60° *to the horizontal. Find the range on a plane inclined at* 30° *to the horizontal when projected (i) up the plane and (ii) down the plane.*

(*i*) When the particle is projected up the plane, the range on inclined plane is given by

$$R = \frac{2u^2\sin(\alpha-\beta)\cos\alpha}{g\cos^2\beta}$$

Here, $u = 32$ ms^{-1} ; $\alpha = 60°$; $\beta = 30°$; $g = 9.8$ ms^{-2}

$$\therefore \quad R = \frac{2\times 32\times 32\times\sin 30°\times\cos 60°}{9.8\cos^2 30°} = 69.66 \text{ m.}$$

(*ii*) When the particle is projected down the plane at an angle α with the horizontal, the range down the plane is given by

$$R_1 = \frac{2u^2\sin(\alpha+\beta)\cos\alpha}{g\cos^2\beta}$$

$$= \frac{2\times 32\times 32\times\sin 90°\times\cos 60°}{9.8\cos^2 30°} = 139.3 \text{ m}$$

Maximum Range : *To find the direction of projection for the maximum range on the inclined plane.*

$$R = \frac{2u^2\sin(\alpha-\beta)\cos\alpha}{g\cos^2\beta}$$

$$= \frac{u^2}{g\cos^2\beta}[\sin(2\alpha-\beta)-\sin\beta]$$

The value of R depends on α, for given values of u and β. Hence R is a maximum when $\sin(2\alpha-\beta) = 1$; *i.e.*, when $2\alpha - 2\beta = 90°$ or $\alpha = (45 + \beta/2)°$.

$\therefore$ The maximum range on the inclined plane

$$R_m = \frac{u^2}{g\cos^2\beta}(1-\sin\beta)$$

$$\therefore \quad R_m = \frac{u^2(1-\sin\beta)}{g(1-\sin^2\beta)} = \frac{u^2(1-\sin\beta)}{g(1+\sin\beta)(1-\sin\beta)}$$

$$= \frac{u^2}{g(1+\sin\beta)} \quad \text{...(3)}$$

Note : $\alpha = 45° + \beta/2$. Then $\alpha - \beta = 45 - \beta/2$ and $90° - \alpha = 45° - \beta/2$.

This shows that the direction giving the maximum range bisects the angle between the vertical and the inclined plane.

9.3. TO SHOW THAT FOR A GIVEN VELOCITY OF PROJECTION AND A GIVEN RANGE ON THE INCLINED PLANE THERE ARE TWO DIRECTIONS OF PROJECTION WHICH ARE EQUALLY INCLINED TO THE DIRECTION FOR THE MAXIMUM RANGE

Now, $$R = \frac{u^2}{g\cos^2\beta}[\sin(2\alpha-\beta) - \sin\beta]$$

The range R and the values of u and β are given.

Hence $\sin(2\alpha - \beta)$ is constant. There are two values of $(2\alpha - \beta)$, each less than 180° which satisfy the above equation. Let the corresponding values of α be α_1 and α_2. Then

$$2\alpha_1 - \beta = 180 - (2\alpha_2 - \beta) \text{ or } \alpha_1 - \beta/2 = 90 - (\alpha_2 - \beta/2)$$

$$\alpha_1 - (45 + \beta/2) = (45 + \beta/2) - \alpha_2.$$

$(45 + \beta/2)$ is the angle of projection giving the maximum range. Therefore, it follows that the direction giving maximum range bisects the angle between the two angles of projection that can give a particular range (Fig. 9.3).

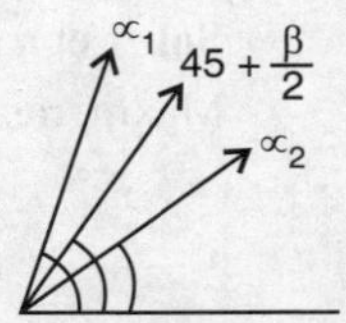

Fig. 9.3

Example 1. *Prove that for a given velocity of projection the ratio between the maximum ranges up and down an inclined plane inclined at an angle β to the horizon is $\frac{1-\sin\beta}{1+\sin\beta}$.*

We have already proved that,

$$\text{Maximum range up the inclined plane} = R_m = \frac{u^2}{g(1+\sin\beta)}$$

$$\text{Range down an inclined plane} = R_1 = \frac{2u^2\sin(\alpha+\beta)\cos\alpha}{g\cos^2\beta}$$

$$= \frac{u^2}{g\cos^2\beta}[\sin(2\alpha+\beta) + \sin\beta]$$

This range is maximum when $\sin(2\alpha + \beta) = 1$.

$$\text{Maximum range down the inclined plane} = R_{m1} = \frac{u^2(1+\sin\beta)}{g\cos^2\beta} = \frac{u^2}{g(1-\sin\beta)}$$

Hence $$\frac{R_m}{R_{m1}} = \frac{u^2}{g\,(1+\sin\beta)} \times \frac{g\,(1-\sin\beta)}{u^2} = \frac{1-\sin\beta}{1+\sin\beta}$$

Example 2. *A particle is projected at an angle α with the horizontal from the foot of the plane, whose inclination to the horizontal is β. Show that it will strike the plane at right angles if* $\cot\beta = 2\tan(\alpha-\beta)$.

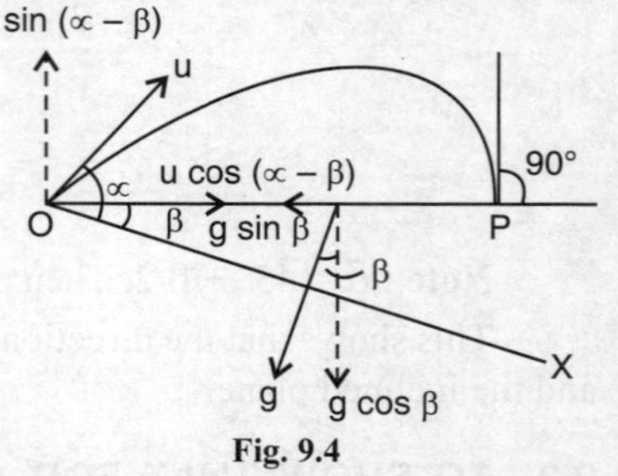

Fig. 9.4

Sol. Let u be the velocity of projection. The components of u parallel and perpendicular to the plane are $u\cos(\alpha-\beta)$ and $u\sin(\alpha-\beta)$ respectively. Accelerations in these directions are $g\sin\beta$ and $g\cos\beta$ respectively (Fig. 9.4).

Time of flight on the inclined plane

$$= T = \frac{2u\sin(\alpha-\beta)}{g\cos\beta} \quad ...(1)$$

Since the particle hits the plane at right angles to the plane, the velocity of the particle parallel to the plane at that instant is zero.

$$0 = u\cos(\alpha-\beta) - g\sin\beta \, . \, T \text{ or } T = \frac{u\cos(\alpha-\beta)}{g\sin\beta} \quad ...(2)$$

Equating (1) and (2), $$\frac{2u\sin(\alpha-\beta)}{g\cos\beta} = \frac{u\cos(\alpha-\beta)}{g\sin\beta}$$

or $$\cot\beta = 2\tan(\alpha-\beta).$$

Example 3. *The angular elevation of an enemy's position on a hill h metres high is β. Show that in order to shell it, the initial velocity of the projectile must not be less than* $\sqrt{gh(1+\operatorname{cosec}\beta)}$.

Sol. Let u be the velocity of projection.

Maxm. range up a plane inclined at an angle α with the horizontal

$$= R_m = \frac{u^2}{g\,(1+\sin\beta)}$$

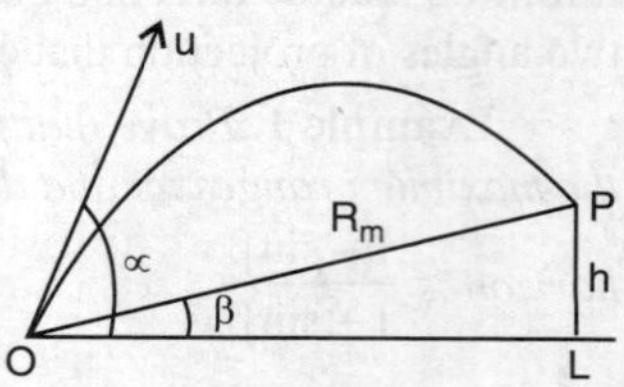

Fig. 9.5

If the shell is to hit enemy's position $R_m = \dfrac{h}{\sin\beta}$ (Fig. 9.5)

$$\therefore \quad \frac{u^2}{g\,(1+\sin\beta)} = \frac{h}{\sin\beta}$$

$$\therefore \quad u^2 \geq \frac{gh(1+\sin\beta)}{\sin\beta}$$

or $$u \geq \sqrt{gh(1+\operatorname{cosec}\beta)}$$

Example 4. *The greatest range on the horizontal plane of a projectile starting with a certain velocity is 150 m. What is the greatest range when projected with the same velocity up an inclined plane inclined 30° to the horizon ?*

Sol. Let u be the velocity of projection.

Max. range on the horizontal plane $= \dfrac{u^2}{g} = 150\ m.$

Max. range on the inclined plane

$$= \frac{u^2}{g\,(1+\sin\beta)} = \frac{150}{1+\sin 30^\circ} = 100 \text{ m}$$

9.4. TWO-BODY PROBLEM AND THE REDUCED MASS

A two-body system can be effectively reduced to one-body system by introducing the concept of *reduced mass*. Let us consider two particles of masses m_1 and m_2 whose instantaneous position vectors with respect to an origin O in an inertial reference frame are $\boldsymbol{r}_1$ and $\boldsymbol{r}_2$ (Fig. 9.6).

The vector distance of m_1 from m_2 is $\boldsymbol{r} = \boldsymbol{r}_1 - \boldsymbol{r}_2$.

The particles exert gravitational forces of attraction on each other which act along the vector $\boldsymbol{r}$, and are thus "central" forces. Let $\boldsymbol{F}_{12}$ be the force experienced by m_1 due to m_2 and $\boldsymbol{F}_{21}$ be the force on m_2 due to m_1 due to mutual interaction. Then the equations of motion for m_1 and m_2 with respect to O are

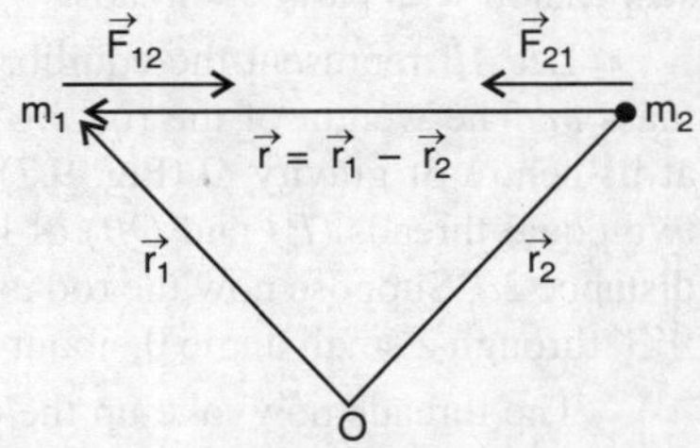

Fig. 9.6

$$m_1 \frac{d^2\boldsymbol{r}_1}{dt^2} = \boldsymbol{F}_{12} \text{ and } m_2 \frac{d^2\boldsymbol{r}_2}{dt^2} = \boldsymbol{F}_{21}.$$

By Newton's third law, $\boldsymbol{F}_{21} = -\boldsymbol{F}_{12} = \boldsymbol{F}$ (say). Then

$$\frac{d^2\boldsymbol{r}_1}{dt^2} = -\frac{\boldsymbol{F}}{m_1} \text{ and } \frac{d^2\vec{\boldsymbol{r}}_2}{dt^2} = \frac{\boldsymbol{F}}{m_2}.$$

Subtracting these equations, we get

$$\frac{d^2(\boldsymbol{r}_1 - \boldsymbol{r}_2)}{dt^2} = -\left(\frac{1}{m_1} + \frac{1}{m_2}\right)\boldsymbol{F}$$

But from fig. $\boldsymbol{r}_1 - \boldsymbol{r}_2 = \boldsymbol{r}$

$$\therefore \quad \frac{d^2\boldsymbol{r}}{dt^2} = -\left(\frac{1}{m_1} + \frac{1}{m_2}\right)F\,\hat{r}$$

where F is the magnitude of the force and is any function of $\boldsymbol{r}$, and $\hat{r}$, is the unit vector along $\boldsymbol{r}$.

Put $$\frac{1}{\mu} = \frac{1}{m_1} + \frac{1}{m_2}$$

$$\mu = \frac{m_1 m_2}{m_1 + m_2}$$ is called the *reduced mass of* two particles.

Then, $$\frac{d^2\boldsymbol{r}}{dt^2} = -\frac{1}{\mu}F\,\hat{r}$$

or $$\mu\frac{d^2\boldsymbol{r}}{dt^2} = -F\,\hat{r}.$$

This equation represents a one-body problem because it is similar to equation of motion of a single particle of mass μ at a distance $\boldsymbol{r}$ from mass m_1, considered as fixed origin of inertial frame. Thus the two-body problem is reduced to one-body problem and the relative motion is represented by the motion of a fictitious particle of mass μ (placed at m_2), acted on by the same force $\boldsymbol{F}$.

Let m_1 and m_2 be the masses of the electron and proton of the hydrogen atom. Their *reduced mass* is given by

$$\mu = \frac{m_1 m_2}{m_1 + m_2} = \frac{m_1}{1 + (m_1/m_2)} \approx m_1\left(1 - \frac{m_1}{m_2}\right)$$

$\left(\frac{m_1}{m_2}\right)$ is very small in comparison with 1. $\therefore \mu \approx m_1$.

9.5. BIFILAR PENDULUM-PARALLEL THREADS

When a heavy uniform rod is suspended horizontally by means of two equal vertical threads at equal distances from the centre of gravity of the rod, the arrangement is called a bifilar suspension. The suspension may be made by means of two parallel or two nonparallel threads. We consider here suspension with parallel threads.

Let *AB* represent the equilibrium position of the rod of mass *m*. The weight of the rod *mg* acts vertically downwards at its centre of gravity *O* (Fig. 9.7). The rod is suspended by two equal threads (*PA* and *QB*) of length *l* and separated by a distance 2*d*. Suppose now the rod is displaced into the position *A′B′* through a small angle θ, about a vertical axis through *O*.

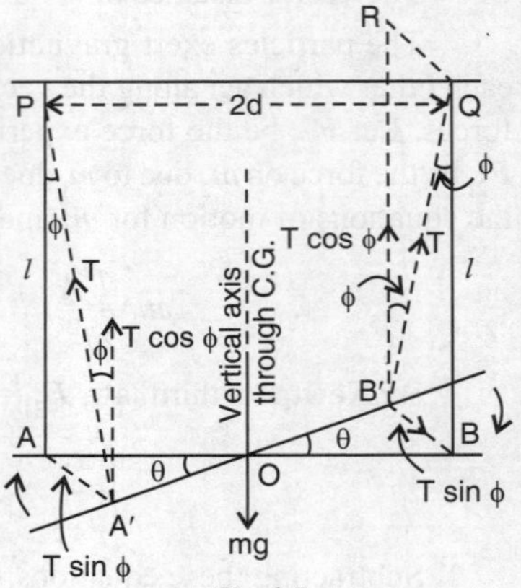

Fig. 9.7

The threads now take up the position *PA′* and *QB′* at an angle θ with their original positions.

Let *T* be the tension in each thread, acting upwards along the thread. *T* may be resolved into two components, one vertically and the other horizontally. The vertical components are each of magnitude $T\cos\phi$, which balance the weight of the rod. Hence, $2T\cos\phi = mg$ or $T\cos\phi = mg/2$.

Now from the figure, $d\theta = l\phi$ or $\phi = \dfrac{d\theta}{l}$

$$\cos\phi = \cos\left(\frac{d\theta}{l}\right) = 1 \qquad \text{since } \theta \text{ is small.}$$

$$\therefore \qquad T = mg/2.$$

Now the horizontal component of the tension ($T\sin\phi$) is along $B'B$ and $A'A$. The two horizontal components of the tension are equal, opposite and parallel. Hence they constitute a couple, tending to bring the rod back into its original position. The two forces act at right angles to the axis of the rod. Hence moment of the restoring couple = $(T\sin\phi)\,2d = T\phi\,2d$. $(\because \sin\phi = \phi)$

$$= T\left(\frac{\theta d}{l}\right)2d = \frac{mg}{2}\frac{\theta d}{l}2d = \frac{mgd^2}{l}\theta.$$

Let *I* be the M.I., of the rod about its axis of rotation and $d^2\theta/dt^2$, its angular acceleration. Hence the equation of motion of the rod is given by

$$I\frac{d^2\theta}{dt^2} = \frac{-mgd^2}{l}\theta \quad \text{or} \quad \frac{d^2\theta}{dt^2} = \frac{-mgd^2}{Il}\theta$$

Clearly, $\dfrac{d^2\theta}{dt^2} \propto \theta$ since $\dfrac{mgd^2}{Il}$ is a constant quantity.

Hence the rod executes S.H.M., and its time period is

$$t = \frac{2\pi}{\sqrt{mgd^2/(Il)}} = 2\pi\sqrt{\frac{Il}{mgd^2}}$$

Now $I = mk^2$ where *k* is the radius of gyration of the rod about the axis of rotation. Hence

$$t = 2\pi\frac{k}{d}\sqrt{\frac{l}{g}}$$

or

$$g = 4\pi^2k^2l/(d^2t^2).$$

Thus *g* at the given place is determined.

9.6. TO DETERMINE EXPERIMENTALLY THE M.I. OF A RECTANGULAR BLOCK

The bifilar pendulum consists of a heavy rectangular metal block *ABCDEFGH* provided with symmetrical suspension pins on each face (Fig. 9.8). The block is suspended by two equal parallel torsionless strings from a horizontal scale.

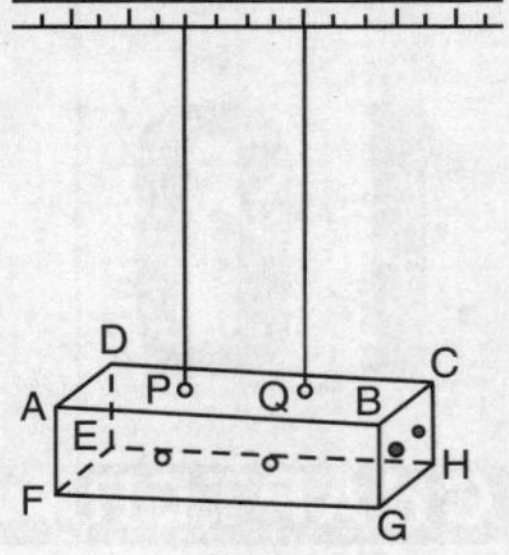

Fig. 9.8

The pendulum is suspended symmetrically by the two equal strings with the face *ABCD* horizontal and the distance between them $(2a_1)$ is made the same both at the scale and on the block. The block is turned slightly about a vertical axis through *ABCD* and released. The period of oscillation T_1 is found. The length (y) of each string is measured. Observations are taken when the block is suspended with the faces *ABGF* and *BCHG* respectively horizontal. Let T_2 and T_3 be the periods. Mass of the block (M) is found. Its length *AB* (l), breadth *AD* (b) and thickness *AF* (d) are also found.

M.I. of the block about an axis passing through the centre of gravity, $\perp r$ to the face *ABCD* is given by

$$I_1 = \frac{mga_1^2\, T_1^2}{4\pi^2 y}$$

When the axis is $\perp r$ to *ABGF*, $I_2 = \dfrac{Mga_2^2\, T_2^2}{4\pi^2 y}$

When the axis is $\perp r$ to *BCHG*, $I_3 = \dfrac{Mga_3^2\, T_3^2}{4\pi^2 y}$

M. I. may also be calculated using the formula

$$I_1 = M\left(\frac{l^2+b^2}{12}\right); I_2 = M\left(\frac{l^2+d^2}{12}\right); I_3 = M\left(\frac{b^2+d^2}{12}\right)$$

EXERCISE IX

1. A particle is projected with velocity u at an angle α to the horizontal. Find the range of particle on the inclined plane through its point of projection inclined at an angle ϕ to the horizontal. ***(Madras 1995)***
2. Show that the motion of two interacting bodies may be reduced to the motion of a single body. Hence explain the concept of reduced mass.
3. The internuclear distance between the two protons in a hydrogen molecule is 0.074 nm. Calculate the moment of inertia of the hydrogen molecule about the axis of rotation passing through the centre of mass. (Given, the mass of proton = 1.673×10^{-27} kg.)

$$\left[\textbf{Hint. } \mu = \frac{m_1 m_2}{m_1+m_2} = \frac{(1.673\times10^{-27})(1.673\times10^{-27})}{(1.673\times10^{-27}+1.673\times10^{-27})} = 8.365\times10^{-28}\text{kg.}\right.$$

$$I = \mu r^2 = (8.365 \times 10^{-28})(0.074 \times 10^{-9})^2 = 4.58 \times 10^{-48} \text{ kg m}^2].$$

10 Mechanics of a Rigid Body

CHAPTER

10.1. MOMENT OF INERTIA OF A UNIFORM ROD

(*a*) About an axis through its centre and perpendicular to its length

Let *AB* be a thin uniform rod of length l and mass M (Fig. 10.1). Its mass per unit length = M/l. YY' is the axis of rotation passing through the centre of gravity G of the rod and perpendicular to its length.

Consider a small element of thickness dx, at a distance x from G. Mass of the element = $(M/l)dx$.

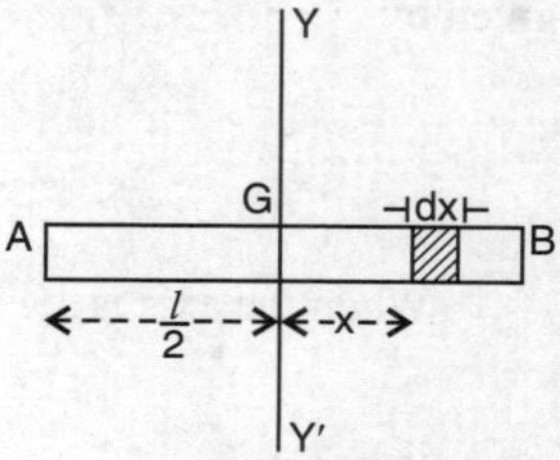

Fig. 10.1

$$\text{Its M.I. about the axis through } G = \frac{M}{l}\,.\,dx\,.\,x^2$$

The moment of inertia I of the whole rod about YY' axis will be the sum of the moments of inertia of all such elements lying between $x = -l/2$ (at A) and $x = l/2$ (at B), *i.e.,*

$$I = \int_{-l/2}^{l/2} \frac{M}{l}\,.\,x^2 dx = \frac{2M}{l}\int_0^{l/2} x^2\, dx$$

or $$I = \frac{Ml^2}{12}$$

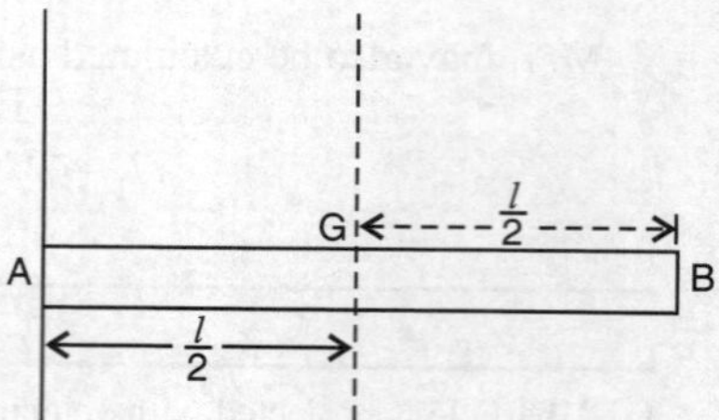

Fig. 10.2

(*b*) The M.I. of rod about the parallel axis through one end A and perpendicular to the rod can be calculated by theorem of parallel axes as (Fig. 10.2)

$$I' = \frac{M\,l^2}{12} + M\left(\frac{l}{2}\right)^2 = \frac{Ml^2}{3}$$

10.2. MOMENT OF INERTIA OF A RECTANGULAR LAMINA

(*a*) About an axis through its centre and parallel to one side

Let *ABCD* represent a plane rectangular lamina of mass M, length l and breadth b (Fig. 10.3). Let YOY' be an axis passing through the centre of gravity O and parallel to the side b of the lamina.

Area of the lamina = lb

$$\left.\begin{array}{l}\text{Mass per unit area}\\ \text{of the lamina}\end{array}\right\} = \frac{M}{lb}$$

Consider a small rectangular strip of the lamina of width dx parallel to the axis YOY' at a distance x from the axis.

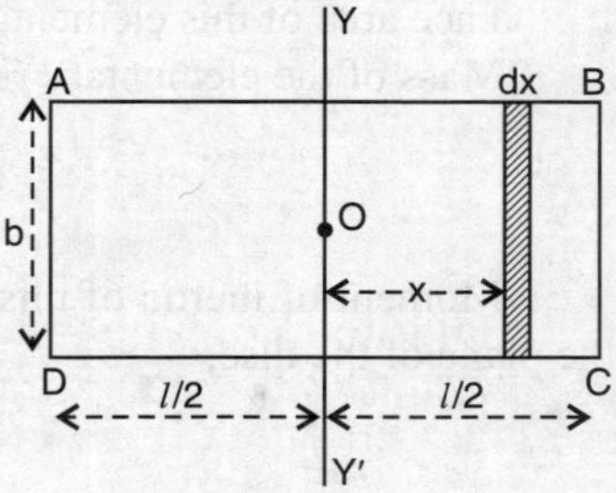

Fig. 10.3

Area of the strip $= b\,dx$

$$\text{Mass of the strip} = \frac{M}{lb} b\,dx = \frac{M}{l} dx$$

Moment of inertia of the strip about the axis YOY'

$$= \frac{M}{l} dx \,.\, x^2 = \frac{M}{l} x^2 dx$$

The moment of inertia I_b of the whole rectangular lamina about the axis YOY' is the sum of the moments of inertia of all such strips lying between $x = -\frac{l}{2}$ and $x = \frac{l}{2}$

$$I_b = \int_{-l/2}^{l/2} \frac{M}{l} x^2\, dx = \frac{2M}{l} \int_0^{l/2} x^2\, dx$$

$$\therefore \qquad I_b = \frac{Ml^2}{12}$$

Similarly, it can be proved that the moment of inertia I_l, of the lamina about an axis passing through the centre of gravity and parallel to the side l is given by,

$$I_l = \frac{Mb^2}{12}$$

(*b*) About an axis perpendicular to its plane and passing through its centre

Let this moment of inertia be I. By the perpendicular axes theorem,

$$I = I_l + I_b = \frac{Mb^2}{12} + \frac{Ml^2}{12}$$

$$\therefore \qquad I = M\frac{(l^2 + b^2)}{12}$$

(*c*) About one side. The moment of inertia of the rectangular lamina about the side AD or BC is given by the parallel axes theorem.

$$I_{AD} = \frac{Ml^2}{12} + M\left(\frac{l}{2}\right)^2 = \frac{Ml^2}{3}$$

10.3. MOMENT OF INERTIA OF AN ANNULAR RING OR DISC

(*a*) About an axis passing through its centre and perpendicular to its plane.

Let M be the mass and r and R the inner and outer radii of the annular disc (Fig. 10.4).

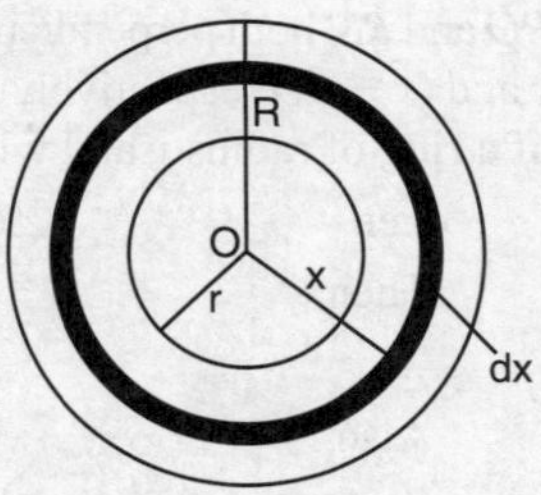

Fig. 10.4

$$\text{Mass per unit face area of the disc} = \frac{M}{\pi (R^2 - r^2)}.$$

The annular disc may be imagined to be made up of a number of thin circular rings. Consider one such ring of radius x and of width dx.

Face area of this elementary ring = $2\pi\, x\, dx$

Mass of the elementary ring

$$= \frac{M}{\pi(R^2 - r^2)} 2\pi\, x\, dx = \frac{2Mx\, dx}{(R^2 - r^2)}$$

Moment of inertia of this ring about an axis through the centre *O*, and perpendicular to the plane of the disc,

$$= \frac{2Mx\, dx}{(R^2 - r^2)} . x^2 = \frac{2Mx^3\, dx}{(R^2 - r^2)}$$

The moment of inertia *I*, of the annular disc may be obtained by integrating the above expression between limits $x = r$ and $x = R$.

$$I = \int_r^R \frac{2Mx^3}{(R^2 - r^2)} dx$$

$$= \frac{2M}{(R^2 - r^2)} \int_r^R x^3\, dx = \frac{M}{2(R^2 - r^2)} (R^4 - r^4) = \frac{M}{2(R^2 - r^2)} (R^2 + r^2)(R^2 - r^2)$$

$$\therefore \qquad I = \frac{M(R^2 + r^2)}{2}$$

***(b)* About its diameter.** From the symmetrical shape of the annular ring, it is obvious that its moment of inertia about one diameter is equal to the moment of inertia about another diameter. If *I* be the moment of inertia about any diameter, then according to the theorem of perpendicular axes, we have sum of moments of inertia about two perpendicular diameters (axes) = moment of inertia about an axis through *O* and perpendicular to its plane.

$$\therefore \qquad I + I = \frac{M(R^2 + r^2)}{2}$$

or,
$$I = \frac{M(R^2 + r^2)}{4}$$

10.4. MOMENT OF INERTIA OF A SPHERICAL SHELL

***(a)* About a diameter.** Consider a thin spherical shell of mass *M* and radius *R* (Fig. 10.5). Let *O* be its centre. *AB* and *CD* are two mutually perpendicular diameters. The moment of inertia about *AB* is to be determined.

Mass per unit area of the shell = $M/4\pi R^2$.

Consider a thin slice of the shell between two parallel planes *PQ* and *RS*, both perpendicular to *AB*. Let these planes be at distances *x* and $(x + dx)$ respectively from the centre. This slice is in the form of a ring of radius *y* and width *PR*.

Let $\angle COP = \theta$ and $\angle POR = d\theta$

Then $y = R\cos\theta$, and $x = R\sin\theta$

$\therefore \qquad dx = R\cos\theta\, d\theta = y\, d\theta$

Also $\qquad PR = R\, d\theta$

Surface area of the slice = $2\pi y . PR = 2\pi y R\, d\theta = 2\pi R\, dx$

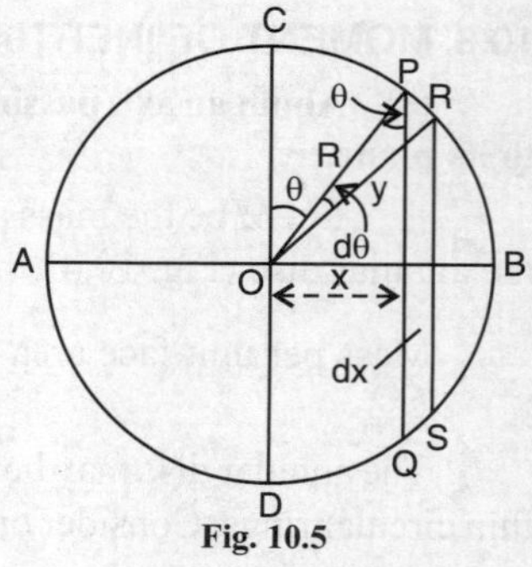

Fig. 10.5

$$\text{Mass of the slice} = 2\pi R\, dx. \times \frac{M}{4\pi R^2} = \frac{M}{2R} dx$$

$$\text{Moment of inertia of the slice about } AB = \frac{M}{2R} dx \,.\, y^2 = \frac{M}{2R} dx\,(R^2 - x^2)$$

Moment of inertia I, of the whole shell about the diameter can be obtained by integrating the above expression between $x = -R$ and $x = R$.

$$I = \int_{-R}^{R} \frac{M}{2R}(R^2 - x^2)\, dx = \frac{2M}{2R}\int_{0}^{R}(R^2 - x^2)\, dx$$

$$\therefore \qquad I = \frac{2}{3} MR^2$$

(*b*) About a tangent. Any tangent to the spherical shell is at a distance R from a parallel diameter. The moment of inertia of the shell about any diameter is the same. Hence applying the theorem of parallel axes, the moment of inertia of the shell about a tangent is given by

$$I = \frac{2}{3} M\,R^2 + MR^2 \quad \text{or} \quad I = \frac{5}{3} MR^2.$$

10.5. KINETIC ENERGY OF A ROTATING BODY

Consider a rigid body of mass M rotating with an angular velocity ω about an axis through O (Fig. 10.6). The body is composed of a large number of particles of masses $m_1, m_2, m_3, \ldots$ situated at distances $r_1, r_2, r_3, \ldots$ from the axis of rotation. Every particle revolves with the same angular velocity ω.

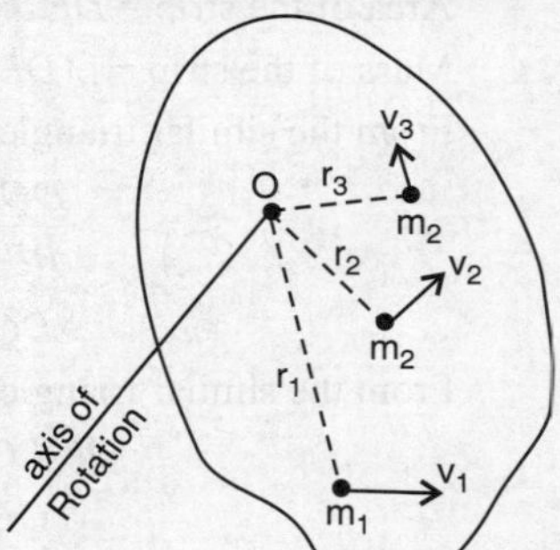

Fig. 10.6

The linear velocity (v_1) of the particle of mass m_1 is $r_1\omega$. Then

$$\text{K.E of the particle of mass } m_1 = \frac{1}{2} m_1 v_1^2 = \frac{1}{2} m_1\, r_1^2\, \omega^2.$$

Similarly, K.E. of particle of mass $m_2 = \frac{1}{2} m_2 r_2^2 \omega^2$ and so on.

$$\therefore \quad \text{The K.E. of the whole body } E = \frac{1}{2} m_1 r_1^2\, \omega^2 + \frac{1}{2} m_2 r_2^2\, \omega^2 + \ldots..$$

$$= \frac{1}{2}(\Sigma m r^2)\, \omega^2$$

$$= \frac{1}{2} I\omega^2$$

Here, $I = \Sigma mr^2$, is the moment of inertia of the body about the axis of rotation.

$$\therefore \qquad E = \frac{1}{2} I\omega^2$$

If $\qquad \omega = 1;\ I = 2 \times E$

Hence the moment of inertia of a body may also be defined as twice its kinetic energy of rotation, the angular velocity of the body being unity.

Angular momentum, $L = I\omega$

$\therefore$ Kinetic energy of rotation $E = \dfrac{L^2}{2I}$

This is the relation between rotational kinetic energy and angular momentum of a body about the same axis.

10.6. MOMENT OF INERTIA OF A TRIANGULAR LAMINA ABOUT AN AXIS PASSING THROUGH ITS ONE SIDE

Consider a triangular lamina *ABC* (10.7).

M is the mass of the lamina.

Altitude of the lamina $AP = H$.

Area of the lamina $= \dfrac{1}{2}\ base \times altitude = \dfrac{1}{2}.a.H$

Mass per unit area, $\rho = \left(\dfrac{2M}{aH}\right)$

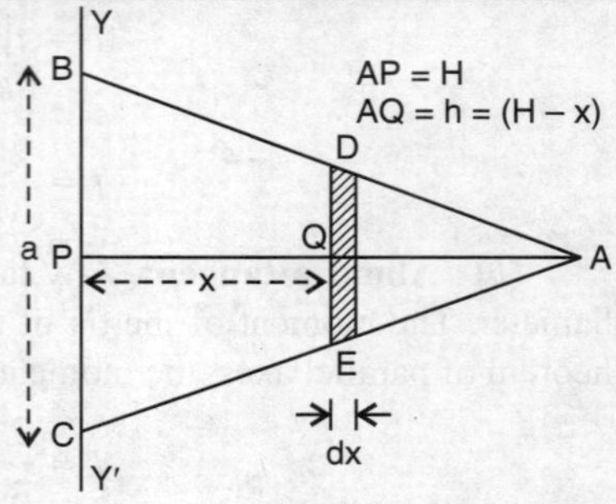

Fig. 10.7

The triangular lamina is made up of a number of thin *strips*, parallel to *BC*.

Consider one such strip *DE*, of width *dx*, at a distance *x* from the base *BC* (axis *YY'*).

Area of the strip = *DE.dx*.

Mass of the strip = $\rho(DE).dx$

From the similar triangles *AQD* and *APB*, we have

$$\frac{DQ}{BP} = \frac{AQ}{AP} = \frac{h}{H} \qquad (\because AQ = h)$$

$$\therefore \quad DQ = BP.h/H.$$

From the similar triangles *AQE* and *APC*, we have

$$QE/PC = AQ/AP = h/H.$$

$$\therefore \quad QE = PC.h/H.$$

$$DQ + QE = (BP.h/H) + (PC.h/H).$$

$$DE = (BP + PC).h/H = BC.h/H = a.h/H.$$

$$\therefore \quad \text{mass of the strip} = \rho.a.\frac{h}{H}.dx = \rho\, a\left(\frac{H-x}{H}\right)dx$$

M.I. of the strip about the axis *YY'*

$$= \left[\rho\, a\left(\frac{H-x}{H}\right)dx\right] \times x^2$$

M.I. of the whole lamina about the axis *YY'*,

$$I = \int_0^H \rho\, a.\left(\frac{H-x}{H}\right)x^2.dx.$$

$$I = \frac{\rho\, a}{H}\int_0^H (Hx^2 - x^3).dx = \frac{\rho\, aH^3}{12} = \left(\frac{2M}{aH}\right)\left(\frac{aH^3}{12}\right)$$

$$\therefore \quad I = \frac{MH^2}{6}$$

10.7. TORQUE

Consider a particle at a point P (Fig. 10.8). Let the position vector of P relative to an origin O be $\boldsymbol{r}$. Let $\boldsymbol{F}$ be the force acting on the particle. Then the torque $\vec{\tau}$ acting on the particle with respect to the origin O is defined as

$$\vec{\tau} = \vec{r} \times \vec{F} \qquad ...(1)$$

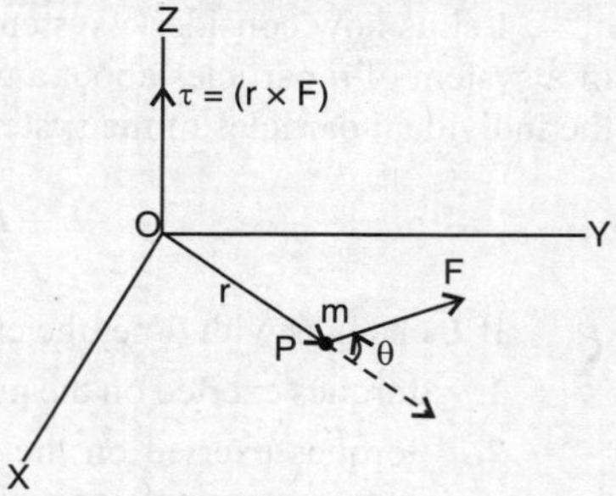

Fig. 10.8

Torque is a vector quantity. Its magnitude is given by

$$\tau = rF \sin \theta \qquad ...(2)$$

Here, θ is the angle between $\boldsymbol{r}$ and $\boldsymbol{F}$. Its direction is normal to the plane formed by $\boldsymbol{r}$ and $\boldsymbol{F}$. If one swings $\boldsymbol{r}$ into $\boldsymbol{F}$ through the smaller angle between them with the curled fingers of the right hand, the direction of the extended thumb gives the direction of $\vec{\tau}$.

Torque has the dimensions of force times distance, *i.e.*, ML^2T^{-2}.

The unit of torque is newton metre (Nm).

10.8. ANGULAR MOMENTUM

Consider a particle of mass m and linear momentum $\boldsymbol{p}$ at a position $\boldsymbol{r}$ relative to the origin O (Fig. 10.9). The angular momentum $\boldsymbol{l}$ of the particle with respect to the origin O is defined as

$$\boldsymbol{l} = \boldsymbol{r} \times \boldsymbol{p} = m(\boldsymbol{r} \times \boldsymbol{v})$$

Angular momentum is a vector. Its magnitude is given by

$$l = r\,p \sin \theta$$

Here, θ is the angle between $\boldsymbol{r}$ and $\boldsymbol{p}$. Its direction is normal to the plane formed by $\boldsymbol{r}$ and $\boldsymbol{p}$. The sense is given by the right-hand rule.

The unit of angular momentum is kg m^2s^{-1} or Js.

For circular motion, $v = r\,\omega$. The magnitude of $\boldsymbol{I}$ is $mr^2\omega = I\omega$.

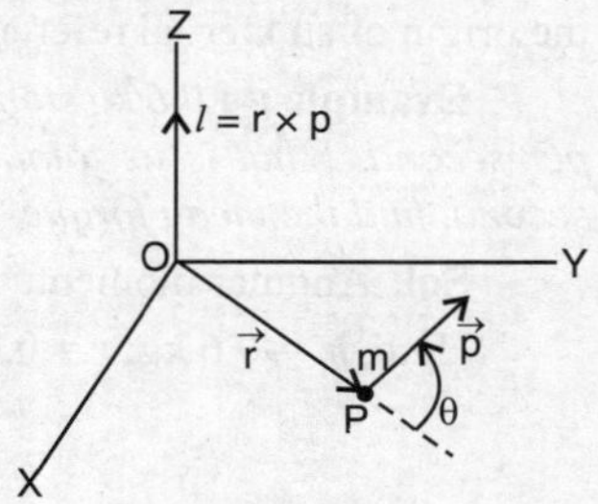

Fig. 10.9

10.9. RELATION BETWEEN TORQUE AND ANGULAR MOMENTUM

We have the relation $\quad \boldsymbol{l} = \boldsymbol{r} \times \boldsymbol{p}$

Differentiating this relation with respect to t,

$$\frac{d\boldsymbol{l}}{dt} = \frac{d\boldsymbol{r}}{dt} \times \boldsymbol{p} + \boldsymbol{r} \times \frac{d\boldsymbol{p}}{dt}$$

$$d\boldsymbol{r}/dt = \boldsymbol{v} = \text{instantaneous velocity.}$$

Now, $\qquad (d\boldsymbol{r}/dt) \times \boldsymbol{p} = \boldsymbol{v} \times (m\boldsymbol{v}) = 0.$

and $d\boldsymbol{p}/dt$ = Force $\boldsymbol{F}$ acting on the particle.

$$\therefore \qquad \frac{d\boldsymbol{l}}{dt} = \boldsymbol{r} \times \boldsymbol{F}$$

But, by definition, $\boldsymbol{r} \times \boldsymbol{F} = \vec{\tau}$ = torque of $\boldsymbol{F}$ about O.

$$\therefore \qquad \frac{d\boldsymbol{l}}{dt} = \vec{\tau}\,.$$

Thus, the time rate of change of the angular momentum of a particle is equal to the torque acting on it.

10.10. ANGULAR MOMENTUM OF A SYSTEM OF PARTICLES

Let us now consider a system of many particles. To calculate the total angular momentum $\boldsymbol{L}$ of a system of n particles about a given point, we must add vectorially the angular momenta of all the individual particles of the system about this same point. So we have

$$\boldsymbol{L} = \boldsymbol{l}_1 + \boldsymbol{l}_2 + \ldots + \boldsymbol{l}_n = \sum_{i=1}^{i=n} \boldsymbol{l}_i$$

If $\boldsymbol{L}$ changes with time, the change $d\boldsymbol{L}/dt$ can arise from two sources:

1. torques exerted on the particles of the system by internal forces between the particles, and
2. torques exerted on the particles of the system by external forces. The internal forces between any two particles are equal and opposite and are directed along the line joining the two particles. So the torque resulting from each internal action-reaction force pair is zero. Thus the total internal torque is zero.

Hence the contribution from the first source is zero. For our reference point, therefore, only the second source remains. Thus

$$\vec{\tau}_{ext} = d\boldsymbol{L}/dt$$

where $\vec{\tau}_{ext}$ = sum of all the external torques acting on the system.

In words the time rate of change of the total angular momentum of a system of particles about the origin of an inertial reference frame is equal to the sum of external torques acting on the system.

Example : *A 0,6 kg stone is revolved at the end of 0.6 m long string at the rate of 3 revolutions per second. What is its total angular momentum? If after 30s it is making only 1 revolution per second, find the mean torque.*

Sol. Angular momentum $L = mrv = mr^2\omega$ ($\because\ v = r\omega$)

Here, $m = 0.6$ kg; $r = 0.6$ m and $\omega = 3 \times 2\pi = 6\pi$ rad s^{-1}; $L = ?$

$$L = (0.6) \times (0.6)^2 \times 6\pi = 4.072 \text{ joule-second.}$$

$$\vec{\tau} = \frac{dL}{dt} \text{ (Torque is the rate of change of angular momentum)}$$

Angular velocity after $30s = \omega' = 1 \times 2\pi = 2\pi$ rad s^{-1}.

$\therefore$ Angular momentum after $30s = L' = mr^2\omega'$

$$= (0.6) \times (0.6)^2 \times 2\pi = 1.357 \text{ joule-second.}$$

$$\vec{\tau} = \frac{\text{change in angular momentum}}{\text{Time}} = \frac{4.072 - 1.357}{30} = 0.0905 \text{ Nm}$$

10.11. CONSERVATION OF ANGULAR MOMENTUM

$$\vec{\tau}_{ext} = \frac{d\vec{L}}{dt}.$$

Suppose there is no external torque acting on a body. $\vec{\tau}_{ext} = 0$. Then $d\boldsymbol{L}/dt = 0$ or $\boldsymbol{L}$ = a constant.

This is the principle of conservation of angular momentum and may be stated thus:

When the resultant external torque acting on a system is zero, the total vector angular momentum of the system remains constant.

Some examples of conservation of angular momentum

1. Dumbbell Experiment :

Consider a person, holding a pair of dumbbells in his two hands stretched outwards, standing on a turn table (Fig. 10.10 a).

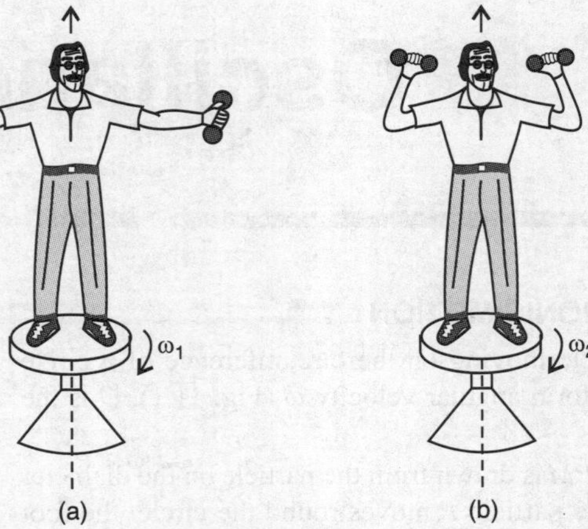

Fig. 10.10

Initially, the turn table is rotating with an angular velocity ω_1.

The M.I. of the person (including the weights) = I_1.

Initial angular momentum = $I_1\omega_1$.

When the person pulls his hands inwards (Fig. 10.10 b), his angular velocity is increased.

The turn table begins to rotate with an angular velocity ω_2.

There is no external torque acting on the system.

Since, the angular momentum is conserved,

$$I_1\omega_1 = I_2\omega_2$$

But $\quad I_2 < I_1$

$\therefore \quad \omega_2 > \omega_1$

2. A ballet dancer can increase or decrease angular speed of a spin about a vertical axis. Since $I\omega$ is a constant, when he pulls his arms and legs in 'I' decreases and ω increases, and hence he can spin faster. Similarly, when he extends his arms and legs as far out as possible, I increases and ω decreases and hence he can spin slower. The principle is similar to that of the dumbbell experiment.

3. Each planet revolves about the sun in elliptical orbit, with sun at one focus of ellipse (Fig. 10.11).

When the planet is at farthest position A, its moment of inertia (I) is *maximum*. When the planet is at nearest position B, its moment of inertia (I) is *minimum*. Therefore according to relation $L = I\omega$ = constant, the angular velocity of planet at farthest position A is minimum, while at nearest position B, it is maximum.

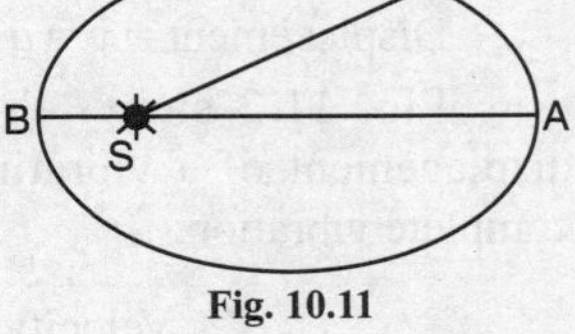

Fig. 10.11

11

CHAPTER

Waves and Oscillations

11.1. SIMPLE HARMONIC MOTION

Let P be a particle moving on the circumference of a circle of radius a with a uniform angular velocity ω (Fig. 11.1). O is the centre of the circle.

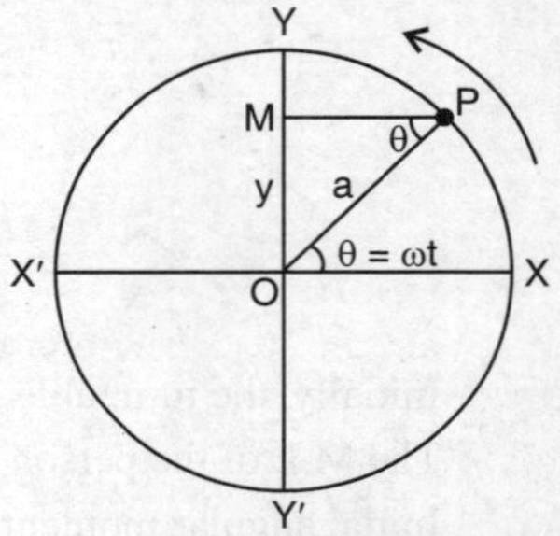

Fig. 11.1

A perpendicular PM is drawn from the particle on the diameter YY' of the circle. As the particle P moves round the circle, the foot of the perpendicular M vibrates along the diameter YY'. Since the motion of P is uniform, the motion of M is periodic. As the particle P completes one revolution, the foot of the perpendicular M completes one *vertical oscillation*. The distance OM is called the displacement and is denoted by y.

The particle moves from X to P in time t.

$$\angle POX = \angle MPO = \theta = \omega t$$

From the ΔMPO,

$$\sin\theta = \sin\omega t = \frac{OM}{a}$$

or

$$OM = y = a\sin\omega t$$

OM is called the displacement of the vibrating particle.

The displacement of a vibrating particle at any instant can be defined as its distance from the mean position of rest.

The maximum displacement of a vibrating particle is called its *amplitude*.

Displacement $= y = a\sin\omega t$ ...(1)

Fig. 11.2 shows the changes in the displacement of a vibrating particle in one complete vibration.

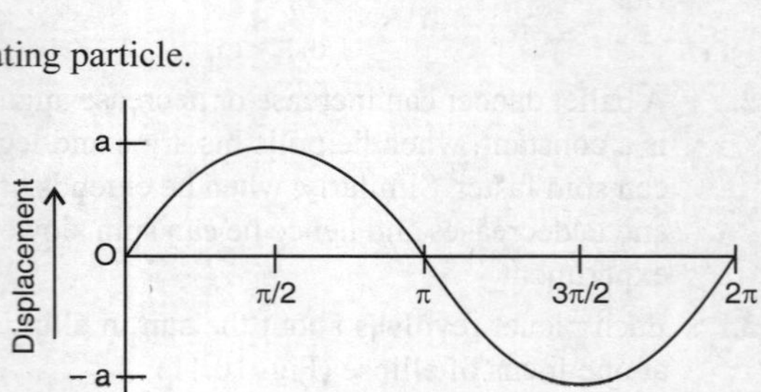

Fig. 11.2

$$\text{Velocity} = v = \frac{dy}{dt} = a\omega\cos\omega t = \omega\sqrt{(a^2 - y^2)} \quad ...(2)$$

$$\text{Acceleration} = \frac{d^2y}{dt^2} = -a\,\omega^2\sin\,\omega t = -\,\omega^2 y \quad ...(3)$$

Thus, acceleration is directly proportional to displacement and directed towards a fixed point. This type of motion is called *simple harmonic motion*.

Following are the characteristics of simple harmonic motion:

(*i*) The motion is periodic.

(*ii*) The motion is along a straight line about the mean or equilibrium position.

(*iii*) The acceleration is proportional to displacement.

(*iv*) Acceleration is directed towards the mean or equilibrium position.

Definition : *If a particle moves in a straight line, so that its acceleration is always directed towards a fixed point on the line, and is proportional to its displacement from the fixed point, the particle is said to move with simple harmonic motion.*

Eq. (3) can be written as

$$\frac{d^2y}{dt^2} + \omega^2 y = 0 \quad ...(4)$$

This is the differential equation of S.H.M.

$$\text{The period} = T = \frac{2\pi}{\omega}$$

$$\text{The frequency} = n = \frac{1}{T} = \frac{\omega}{2\pi}.$$

Examples of S.H.M.

1. The vertical oscillations of a spiral spring suspended from a rigid support, and loaded at the lower end. This is linear type of S.H.M.

2. The vibrations of a simple pendulum. This is angular type of S.H.M.

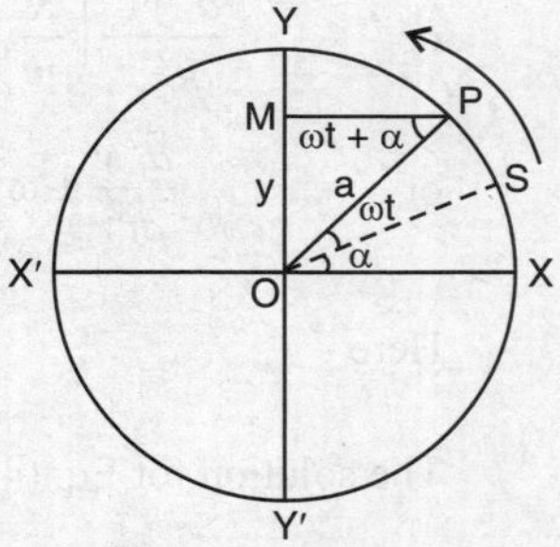

Fig. 11.3

Phase : Consider a particle starting from S and moving on the circumference of a circle (Fig. 11.3). $\angle SOX = \alpha$. The particle moves from S to P in time t.

$$\angle OPM = (\omega t + \alpha).$$

$$\text{Displacement } y = a \sin(\omega t + \alpha).$$

α is called *initial phase or epoch of the S.H.M.*

The angle $(\omega t + \alpha)$ is called the *phase* of the S.H.M.

11.2. FREE VIBRATIONS OF A BODY

When a body, free to oscillate, is displaced from its equilibrium position and no external driving or resisting force is acting on it, it continues to oscillate with a *constant amplitude and its own natural frequency*. Such vibrations of a body are called free vibrations.

Example. A simple pendulum oscillating in vacuum. When the bob of simple pendulum (in vacuum) is displaced from its mean position and left, it executes simple harmonic motion.

The simple pendulum vibrates with a time period T given by

$$T = 2\pi\sqrt{\frac{l}{g}}.$$

The time period (T) depends only on the length of the pendulum (l) and the acceleration due to gravity (g) at the place. The pendulum will continue to oscillate with the same time period and amplitude for any length of time. The amplitude of swing remains constant. In such cases, there is no loss of energy by friction or otherwise. In all similar cases, the vibrations will be undamped free vibrations.

Differential Equation of an Undamped Vibration

Let m be the mass of a particle executing S.H.M.

Kinetic energy of the particle for displacement $y = \frac{1}{2}m\left(\frac{dy}{dt}\right)^2$.

At the same instant, the potential energy of the particle $= \frac{1}{2}Ky^2$.

Here, K is the restoring force per unit displacement.

The total energy at any instant $= \frac{1}{2}m\left(\frac{dy}{dt}\right)^2 + \frac{1}{2}Ky^2$.

For an undamped harmonic oscillator, this total energy remains constant.

$$\therefore \quad \frac{1}{2}m\left(\frac{dy}{dt}\right)^2 + \frac{1}{2}Ky^2 = \text{constant} \qquad ...(1)$$

Differentiating Eq. (1) with respect to time,

$$m\frac{d^2y}{dt^2} + Ky = 0 \qquad ...(2)$$

$$\text{or} \quad \frac{d^2y}{dt^2} + \left(\frac{K}{m}\right)y = 0 \qquad ...(3)$$

$$\text{or} \quad \frac{d^2y}{dt^2} + \omega^2 y = 0 \qquad ...(4)$$

$$\text{Here} \quad \omega^2 = \frac{K}{m}$$

The solution for Eq. (4) is

$$y = a\sin(\omega t - \alpha)$$

$$y = a\sin\left[\left(\sqrt{\frac{K}{m}}\right)t - \alpha\right]$$

The period of vibration is $T = \frac{2\pi}{\omega} = \frac{2\pi}{\sqrt{\left(\frac{K}{m}\right)}}$

Thus, in the case of undamped free vibratitons, the differential equation is

$$\frac{d^2y}{dt^2} + \left(\frac{K}{m}\right)y = 0 \qquad ...(5)$$

11.3. DAMPED VIBRATIONS

When a body executing simple harmonic vibrations is left to itself, the amplitude of vibrations gradually decreases. After some time, the vibrations completely die out. This is because the motion of the body is resisted by various frictional effects. Retarding forces are called into play due to the viscosity or internal friction of the body and the resistance of the air. These forces reduce the amplitude and thus damp the oscillations. Such oscillations of a body are called damped vibrations.

Example. In actual practice, when a simple pendulum vibrates in air medium, there are frictional forces (resistance of air). So energy is dissipated in each vibration. The amplitude of swing decreases continuously with time. Finally the oscillations die out. Such vibrations are called **free damped** vibrations. The dissipated energy appears as heat either within the system itself or in the surrounding medium.

Differential Equation of a Damped Vibration

Consider a body of mass m. Let y be the displacement of the body from the equilibrium position at any instant.

The instantaneous velocity is $\frac{dy}{dt}$.

When a body is oscillating in a resisting medium, two forces are acting on the body.

1. A restoring force directly proportional to the displacement y, but acting in the opposite direction. It may be written as

$$-Ky.$$

Here, K is the restoring force per unit displacement.

2. A frictional (or damping) force proportional to the velocity, but opposite to the direction of motion. It may be written as

$$-\mu\frac{dy}{dt}.$$

Here, μ is a positive constant depending upon the force of resistance.

Therefore, the differential equation in the case of free-damped vibrations is,

$$m\frac{d^2y}{dt^2} = -Ky - \mu\frac{dy}{dt}$$

$$m\frac{d^2y}{dt^2} + Ky + \mu\frac{dy}{dt} = 0$$

or $$\frac{d^2y}{dt^2} + \left(\frac{\mu}{m}\right)\frac{dy}{dt} + \left(\frac{K}{m}\right)y = 0 \qquad ...(1)$$

Put $$\frac{\mu}{m} = 2b \text{ and } \frac{K}{m} = k^2.$$

Then, $$\frac{d^2y}{dt^2} + 2b\frac{dy}{dt} + k^2y = 0 \qquad ...(2)$$

General Solution. Let us try the solution,

$$y = Ae^{pt}$$

Then, $$\frac{dy}{dt} = Ape^{pt} \text{ and } \frac{d^2y}{dt^2} = Ap^2e^{pt}$$

Substituting these values in Eq. (2),

$$Ap^2e^{pt} + 2bApe^{pt} + k^2Ae^{pt} = 0$$

or $$p^2 + 2bp + k^2 = 0$$

or $$p = -b \pm \sqrt{(b^2 - k^2)}$$

The general solution is

$$y = Ae^{\{-b+\sqrt{(b^2-k^2)}\}t} + Be^{\{-b-\sqrt{(b^2-k^2)}\}t}$$

$$y = e^{-bt}[Ae^{\{\sqrt{(b^2-k^2)}\}t} + Be^{-\{\sqrt{(b^2-k^2)}\}t}]$$

If $b^2 < k^2$, $\sqrt{(b^2-k^2)}$ becomes imaginary.

$\therefore$ $$\sqrt{(b^2-k^2)} = \sqrt{-1(k^2-b^2)} = i\sqrt{(k^2-b^2)} = i\beta$$

Here $$i = \sqrt{-1} \text{ and } \beta = \sqrt{(k^2-b^2)}$$

The solution becomes,

$$y = e^{-bt}[Ae^{i\beta t} + Be^{-i\beta t}]$$

$$y = e^{-bt}[A(\cos\beta t + i\sin\beta t) + B(\cos\beta t - i\sin\beta t)]$$

$$y = e^{-bt}[(A+B)\cos\beta t + i(A-B)\sin\beta t]$$

Put $(A + B) = C \sin\delta$ and $i(A - B) = C\cos\delta$.

$$y = Ce^{-bt}\sin(\beta t + \delta) \qquad ...(3)$$

Here C and δ are constants.

Eq. (3) shows that the motion is oscillatory.

The amplitude of oscillation is Ce^{-bt}.

The amplitude is not constant.

The amplitude decreases exponentially with time.

Finally, the amplitude becomes zero after a long time (Fig. 11.4).

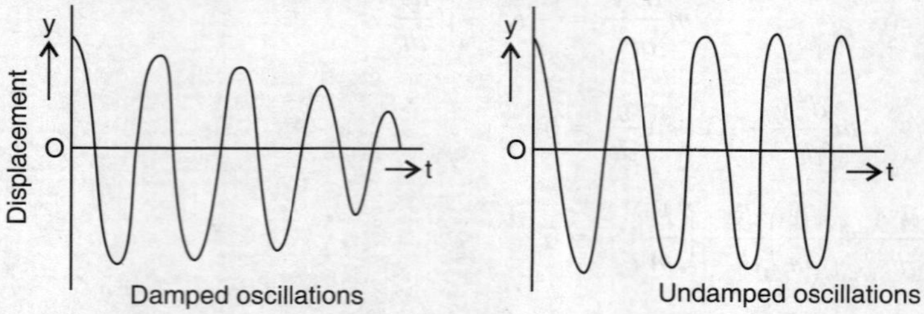

Fig. 11.4

The period of damped vibration is

$$T = \frac{2\pi}{\beta} = \frac{2\pi}{\sqrt{k^2-b^2}} = \frac{2\pi}{\sqrt{\frac{K}{m} - \frac{\mu^2}{4m^2}}}$$

The period of damped vibration is greater than the period of undamped vibration.

11.4. FORCED VIBRATIONS

A body set in vibration gradually loses its amplitude due to retarding forces and thus loses energy. If some external periodic force is constantly applied on the body, the body continues to oscillate under the influence of such external forces. Such vibrations of the body are called *forced vibrations*.

When an external periodic force is applied to a body executing damped vibrations, the body shows a tendency to vibrate with its natural frequency. But the external periodic force applied tries to impress on the body its own frequency of vibration. As time advances, the natural vibrations die out due to frictional forces. The body finally vibrates with the frequency of vibration of the applied external periodic force.

The amplitude of the forced vibration of the body depends on the difference between the natural frequency and the frequency of the applied force. The amplitude will be large if difference in frequencies is small.

Resonance : The particular case of forced vibration in which the frequency of the applied periodic force is equal to the natural frequency of the body itself is called *resonance*. When the frequencies are equal, the applied force helps to increase the amplitude of the body at each step. So when resonance occurs, the body vibrates with a large amplitude.

Examples of Resonance : (1) A vibrating tuning fork is held just above the open end of a tube containing water. If the level of water in the tube is gradually lowered, the length of the air column increases. The air column is thrown into *forced vibrations*. When the natural frequency of the air column is equal to the frequency of the tuning fork, a very loud sound is produced. This is due to resonance.

(2) Another phenomenon of resonance is the tuning of a radio receiver to a desired broadcasting station. By turning the tuning knob, the local oscillator in the radio continuously changes its frequency of oscillation. At the point when the frequency of the oscillator is equal to the frequency of the station, the sound is maximum.

(3) The shank of an excited tuning fork is pressed on the sounding board of the sonometer. The length of the sonometer wire is adjusted so that its frequency is equal to the frequency of the tuning fork. Now, the wire begins to vibrate with a large amplitude and a paper rider placed on the wire is violently thrown off.

Differential Equation of Forced Vibrations

Consider a system oscillating about an equilibrium position under an external *periodic force*. Let y be its displacement from the equilibrium position at an instant during the oscillation. Its instantaneous velocity is $\frac{dy}{dt}$.

Three forces are acting upon the system at this instant.

1. A restoring force proportional to the displacement y, but acting in the opposite direction. This may be written as

$$-Ky$$

Here, K is the restoring force per unit displacement.

2. A frictional force proportional to the velocity but acting in the opposite direction. This may be written as

$$-\mu\frac{dy}{dt}.$$

Here, μ is a positive constant depending upon the force of resistance.

3. An external periodic force represented by

$$F \sin pt.$$

Here p is the angular frequency of the applied periodic force.

Let m be the mass of the system and $\frac{d^2y}{dt^2}$ the instantaneous acceleration.

The equation of motion of the system is

$$m\frac{d^2y}{dt^2} = -Ky - \mu\frac{dy}{dt} + F \sin pt$$

$$\therefore \quad m\frac{d^2y}{dt^2} + \mu\frac{dy}{dt} + Ky = F \sin pt \quad \text{... (1)}$$

General Solution. The particular solution of Eq. (1) representing the forced vibrations is

$$y = a \sin (pt - \alpha) \quad \text{...(2)}$$

$$\frac{dy}{dt} = ap \cos (pt - \alpha) \quad \text{...(3)}$$

$$\frac{d^2y}{dt^2} = -ap^2 \sin (pt - \alpha) = -p^2y \quad \text{...(4)}$$

Substituting these values in Eq. (1),

$$-mp^2 a \sin (pt - \alpha) + \mu\, ap \cos (pt - \alpha) + Ka \sin (pt - \alpha) = F \sin pt$$

$$-mp^2 a[\sin pt \cos \alpha - \cos pt \sin \alpha] + \mu ap[\cos pt \cos \alpha + \sin pt \sin \alpha]$$
$$+ Ka [\sin pt \cos \alpha - \cos pt \sin \alpha] - F \sin pt = 0 \quad \text{... (5)}$$

When $\sin pt = 1$; $\cos pt = 0$

$$-mp^2 a \cos \alpha + \mu ap \sin \alpha + Ka \cos \alpha - F = 0 \quad \text{...(6)}$$

When $\cos pt = 1$; $\sin pt = 0$

$$+mp^2 a \sin \alpha + \mu\, ap \cos \alpha - Ka \sin \alpha = 0. \quad \text{...(7)}$$

Dividing Eq. (7) by $\cos \alpha$ and simplifying

$$\tan \alpha = \frac{\mu p}{(K - mp^2)} = \frac{A}{B} \quad \text{...(8)}$$

From Eq. (8),
$$\sin \alpha = \frac{A}{\sqrt{A^2 + B^2}} \quad \text{...(9)}$$

$$\cos \alpha = \frac{B}{\sqrt{A^2 + B^2}} \quad \text{...(10)}$$

Dividing Eq. (6) by $\cos \alpha$,

$$-mp^2 a + \mu\, ap \tan \alpha + Ka - \frac{F}{\cos \alpha} = 0 \quad \text{...(11)}$$

or $\quad a [(K - mp^2) + \mu p \tan \alpha] = \dfrac{F}{\cos \alpha}$

But $\quad (K - mp^2) = B$, and $\mu p = A$.

Substituting the values of $\tan \alpha$ and $\cos \alpha$ in Eq. (11),

$$a\left[B + \frac{A^2}{B}\right] = \frac{F\sqrt{A^2 + B^2}}{B}$$

$$a = \frac{F}{\sqrt{A^2 + B^2}}$$

Substituting the values of A and B

$$a = \frac{F}{\sqrt{\mu^2 p^2 + (K - mp^2)^2}} \quad \text{...(12)}$$

$$y = a \sin (pt - \alpha)$$

or $$y = \frac{F}{\sqrt{\mu^2 p^2 + (K - mp^2)^2}} \sin(pt - \alpha) \qquad ...(13)$$

Another solution is obtained when $F = 0$.

$$y = Ce^{-bt} \sin(\beta t + \delta) \qquad ...(14)$$

Here, $$\beta = \sqrt{\frac{K}{m} - \frac{\mu^2}{4m^2}}$$

The general solution is

$$y = Ce^{-bt} \sin(\beta t + \delta) + \frac{F}{\sqrt{\mu^2 p^2 + (K - mp^2)^2}} \sin(Pt - \alpha) \qquad ...(15)$$

The first term is the transient term and dies away with time.

The second term is called the steady state term. During the steady state, the oscillator performs forced oscillations with the impressed force frequency.

Resonance : The amplitude of the system executing forced vibration is

$$\frac{F}{\sqrt{\mu^2 p^2 + (K - mp^2)^2}} \qquad ...(16)$$

The amplitude is maximum when the denominator is minimum.

This is possible if $K - mp^2 = 0$ or $K = mp^2$

or $$p = \sqrt{\frac{K}{m}}. \qquad ...(17)$$

or $$p = \omega$$

i.e., when frequency of the external force = natural frequency of the system.

If friction is present, the amplitude at resonance $= \frac{F}{\mu p} = \frac{F}{\mu\sqrt{K/m}}$

or amplitude at resonance $= \frac{F}{\mu}\sqrt{\frac{m}{K}}$

Further, the amplitude will be infinite if μ is also zero.

In practice some damping is always present and μ is never zero. Hence the amplitude at resonance becomes very large but not infinite.

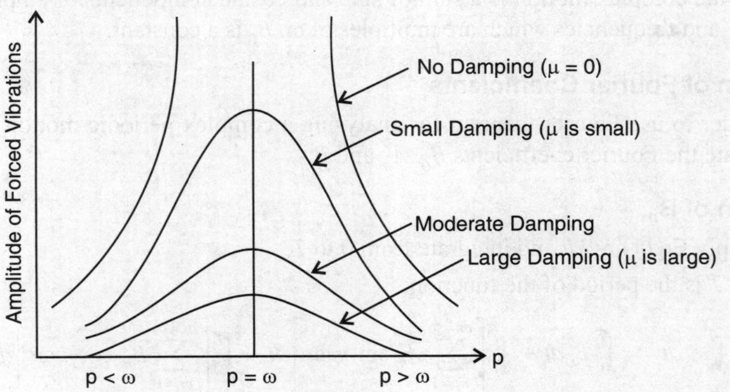

Fig. 11.5

Sharpness of Resonance : *Sharpness of resonance refers to the fall in amplitude with change in frequency on each side of the maximum amplitude.* The resonance is said to be sharp if the amplitude falls off rapidly with the deviation of ω from p. The degree of sharpness depends upon the magnitude of damping. For sharpness of resonance, μ (damping coefficient) must be small. In the sonometer experiment, we get sharp resonance due to the small damping coefficient of the wire.

In the experiment, for determining velocity of sound in air by the resonance column experiment, the resonance is flat due to the large damping coefficient of air.

Typical response curves showing *sharp resonance* and *flat resonance* are shown in Fig. 11.5.

(*i*) When the frictional forces are absent, *i.e.*, $\mu = 0$, the sharpness of resonance is maximum.

(*ii*) The sharpness of resonance decreases with increases in the value of μ.

11.5. FOURIER THEOREM

Statement : *Any single valued periodic function can be expressed as a sum of a number of simple harmonic terms which are multiples of the given function.*

Conditions : The theorem has the following two provisions:

(1) The displacement must be a single valued function and continuous. This condition is satisfied in the case of sound waves. A particle cannot actually have two different displacements simultaneously.

(2) The displacement must always have a finite value. This is true in the case of sound waves. A particle cannot have infinite displacement.

Explanation. The theorem deals with the synthesis of a complex periodic vibration from simple harmonic terms. It also gives a method to analyse a complex vibration into its component vibrations.

In acoustics, Fourier's theorem is applied for the analysis of musical notes. A complex musical *note* may be graphically represented as a periodic function. Fourier's theorem states that it is made up of a number of simple harmonic functions which represent pure tones.

11.6. FOURIER SERIES

Mathematically, Fourier's theorem can be expressed as :

$$y = f(\omega t) = B_0 + \sum_{m=1}^{m=\infty} A_m \sin m\omega t + \sum_{m=1}^{m=\infty} B_m \cos m\omega t \qquad ...(1)$$

Here, $y = f(\omega t)$ is the displacement of a complex periodic motion of angular frequency ω.

Thus the complex motion is a sum of sine and cosine components of amplitudes $A_1, A_2, \ldots$; $B_1, B_2, \ldots$ and frequencies which are multiples of ω. B_0 is a constant.

Evaluation of Fourier Coefficients

In order to use Fourier theorem for analysing a complex periodic motion or wave-form, we must evaluate the Fourier coefficients B_0, A_m and B_m.

Evaluation of B_0.

Multiply Eq.(1) by dt and integrate from 0 to T.

Here, T is the period of the function.

$$\int_0^T y\,dt = \int_0^T B_0 dt + \int_0^T \left[\sum_{m=1}^{m=\infty} A_m \sin m\omega t\right] dt + \int_0^T \left[\sum_{m=1}^{m=\infty} B_m \cos m\omega t\right] dt \qquad ...(2)$$

Here $$\int_0^T \left[\sum_{m=1}^{m=\infty} A_m \sin m\omega t \right] dt = 0$$

and $$\int_0^T \left[\sum_{m=1}^{m=\infty} B_m \cos m\omega t \right] dt = 0$$

$\therefore$ $$\int_0^T y\,dt = B_0 T$$

or $$B_0 = \frac{1}{T}\int_0^T y\,dt \qquad ...(3)$$

Evaluation of A_m

Multiply Eq.(1) by sin $m\,\omega t$ and integrate the terms for a complete cycle 0 to T. Then,

$$A_m = \frac{2}{T}\int_0^T y \sin(m\omega t)\,dt \qquad ...(4)$$

Evaluation of B_m

Multiply Eq. (1) by cos $m\omega t$ and integrate from 0 to T. Then we have

$$B_m = \frac{2}{T}\int_0^T y \cos(m\omega t)\,dt \qquad ...(5)$$

Application in Sound : The quality of a musical note depends upon the harmonics present in it. Hence Fourier's theorem can determine the quality of a note by analysing its oscillation into its harmonic components.

The determination of the Fourier series corresponding to a given complex vibration is called *harmonic analysis* of the vibration. The terms of this series, whose cyclic frequencies are equal to ω, 2ω, 3ω, ..., etc., are called the first (fundamental), second, third, ... etc., harmonics, respectively, of the complex vibration.

11.7. SAW-TOOTH WAVE

Let us apply Fourier's theorem to analyse a saw-tooth wave into its harmonic components. The displacement curve of a periodic saw-tooth wave is shown in Fig. 11.6.

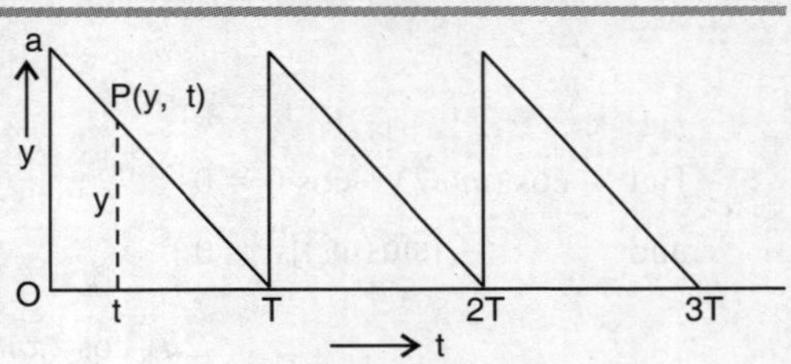

Fig. 11.6

The displacement falls off linearly from $y = a$ to $y = 0$ when t increases from 0 to T. Saw-tooth wave can be represented by the equation

$$y = f(\omega t) = a\left(1 - \frac{t}{T}\right) \text{ for } 0 < t < T \qquad ...(1)$$

According to Fourier Theorem,

$$y = f(\omega t) = B_0 + \sum_{m=1}^{m=\infty} A_m \sin m\,\omega t + \sum_{m=1}^{m=\infty} B_m \cos m\,\omega t \qquad ...(2)$$

Here, $$B_0 = \frac{1}{T}\int_0^T y\,dt \qquad ...(3)$$

$$A_m = \frac{2}{T}\int_0^T y \sin(m\,\omega t)\,dt \qquad ...(4)$$

and $$B_m = \frac{2}{T}\int_0^T y\cos(m\omega t)\,dt \qquad ...(5)$$

Let us obtain the values of the coefficients B_0, A_m and B_m.

(*i*) $$B_0 = \frac{1}{T}\int_0^T y\,dt$$

$$= \frac{1}{T}\int_0^T a\left(1-\frac{t}{T}\right)dt$$

$$= \frac{a}{T}\left[t-\frac{t^2}{2T}\right]_0^T$$

$$= \frac{a}{T}\left[T-\frac{T^2}{2T}\right] = \frac{a}{2}$$

$\therefore$ $$B_0 = \frac{a}{2} \qquad ...(6)$$

(*ii*) $$A_m = \frac{2}{T}\int_0^T y\sin(m\omega t)dt$$

$$= \frac{2}{T}\int_0^T a\left(1-\frac{t}{T}\right)\sin(m\omega t)dt$$

$$= \frac{2a}{T}\int_0^T \sin(m\omega t)dt - \frac{2a}{T^2}\int_0^T t\sin(m\omega t)dt$$

$$A_m = \frac{2a}{T}\left[-\frac{\cos m\omega t}{m\omega}\right]_0^T - \frac{2a}{T^2}\left[\left(-\frac{t\cos m\omega t}{m\omega}\right)_0^T + \int_0^T \frac{\cos m\omega t}{m\omega}dt\right]$$

$$A_m = \frac{2a}{Tm\omega}\left[-\cos m\omega T + \frac{\cos(m\omega\times 0)}{m\omega}\right]$$

$$+\frac{2a}{T^2}\left[\left(\frac{T\cos m\omega t}{m\omega}\right)\right] - \frac{2a}{T^2}\left[\frac{\sin m\omega t}{m^2\omega^2}\right]_0^T$$

But $-\cos(m\omega T) + \cos 0 = 0$

and $$[\sin(\omega t)]_0^T = 0 \qquad \left[\because T = \frac{2\pi}{\omega}\right]$$

$$A_m = \frac{2a}{T}\left[\frac{\cos 2\pi m}{m\omega}\right] = \frac{2a}{Tm\omega}$$

$\therefore$ $$A_m = \frac{a}{m\pi} \qquad ...(7)$$

(*iii*) $$B_m = \frac{2}{T}\int_0^T y\cos(m\omega t)dt$$

$$= \frac{2}{T}\int_0^T a\left(1-\frac{t}{T}\right)\cos(m\omega t)dt$$

$\therefore$ $$B_m = 0 \qquad ...(8)$$

Hence all the cosine terms of the Fourier series are absent.

Substituting the values of B_0, A_m and B_m in Eq. (2), we have

$$y = f(\omega t) = \frac{a}{2} + \sum_{m=1}^{m=\infty} \frac{a}{m\pi} \sin m\omega t \qquad ...(9)$$

$$\therefore \qquad y = \frac{a}{2} + \frac{a}{\pi}\left(\sin \omega t + \frac{1}{2} \sin 2\omega t + \frac{1}{3} \sin 2\omega t + ... + \frac{1}{m} \sin m\omega t .... \right) \qquad ...(10)$$

Fig. 11.7 shows how with the addition of successive terms, the resultant curve gradually approaches the shape of the saw-tooth wave.

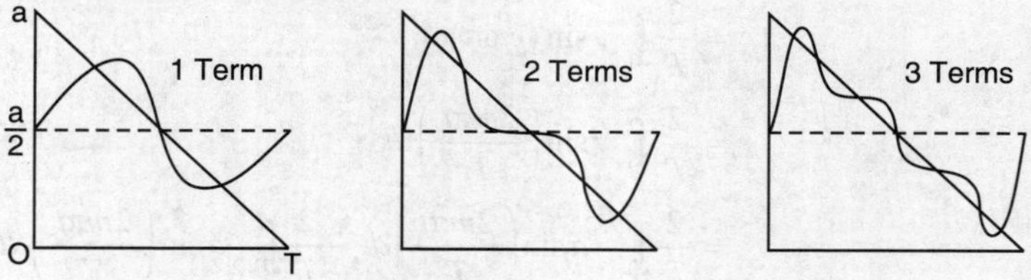

Fig. 11.7

11.8. SQUARE WAVE

Consider a square wave of period T shown in Fig. 11.8.

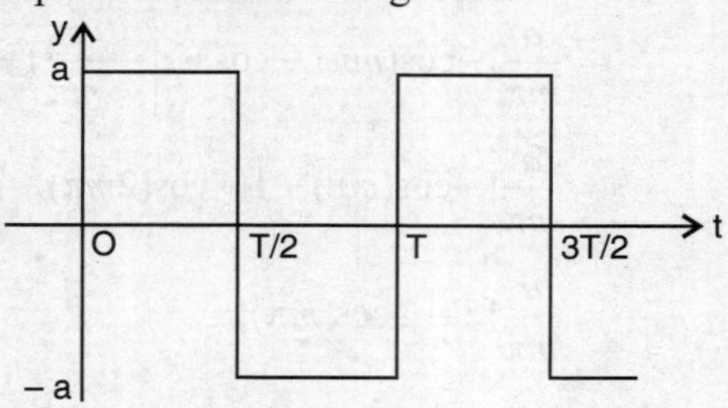

Fig. 11.8

Let $\qquad y = f(\omega t) = 0$ at $t = 0$ and $\frac{T}{2}$.

$$y = f(\omega t) = +a \text{ for } 0 < t < \frac{T}{2} \qquad ...(1)$$

$$y = f(\omega t) = -a \text{ for } \frac{T}{2} < t < T.$$

According to Fourier theorem,

$$y = f(\omega t) = B_0 + \sum_{m=1}^{m=\infty} A_m \sin m\omega t + \sum_{m=1}^{m=\infty} B_m \cos m\omega t \qquad ...(2)$$

Here, $$B_0 = \frac{1}{T} \int_0^T y\, dt \qquad ...(3)$$

$$A_m = \frac{2}{T} \int_0^T y \sin (m\omega t) dt \qquad ...(4)$$

$$B_m = \frac{2}{T} \int_0^T y \cos (m\omega t) dt \qquad ...(5)$$

Let us obtain the values of the coefficients B_0, A_m and B_m.

(i) $$B_0 = \frac{1}{T}\int_0^T y\,dt$$

Substituting the value of y from Eq. (1), we get

$$B_0 = \frac{1}{T}\int_0^{T/2} a\,dt - \frac{1}{T}\int_{T/2}^{T} a\,dt = 0$$

$\therefore$ $$B_0 = 0 \qquad ...(6)$$

(ii) $$A_m = \frac{2}{T}\int_0^T y\sin(m\omega t)dt$$

$$= \frac{2}{T}\int_0^T y\sin\left(\frac{2m\pi t}{T}\right)dt$$

$$= \frac{2}{T}\int_0^{T/2} a\sin\left(\frac{2m\pi t}{T}\right)dt - \frac{2}{T}\int_{T/2}^{T} a\sin\left(\frac{2m\pi t}{T}\right)dt$$

$$= \left(\frac{2a}{T}\right)\frac{T}{2m\pi}\left[-\cos\left(\frac{2m\pi t}{T}\right)\right]_0^{T/2} - \left(\frac{2a}{T}\right)\frac{T}{2m\pi}\left[-\cos\left(\frac{2m\pi t}{T}\right)\right]_{T/2}^{T}$$

$$= \frac{a}{m\pi}[-\cos(m\pi) + \cos 0)\left[-\frac{a}{m\pi}[-\cos(2m\pi) + \cos(m\pi)\right]$$

$$= \frac{a}{m\pi}[-\cos(m\pi) + 1 + \cos(2m\pi) - \cos(m\pi)]$$

$$= \frac{a}{m\pi}[2 - 2\cos(m\pi)] \qquad [\because \cos(2m\pi) = 1]$$

$$= \frac{2a}{m\pi}[1 - \cos(m\pi)]$$

$\therefore$ $A_m = 0$ for $m = 2, 4, 6, ...$ (even integers)

and $A_m = \dfrac{4a}{m\pi}$ for $m = 1, 3, 5$ (odd integers) ...(7)

(iii) $$B_m = \frac{2}{T}\int_0^T y\cos(m\omega t)dt$$

$$= \frac{2}{T}\int_0^T y\cos\left(\frac{2m\pi t}{T}\right)dt \qquad \left[\because \omega = \frac{2\pi}{T}\right]$$

Substituting the value of y from Eq. (1), we get

$$B_m = \frac{2}{T}\int_0^{T/2} a\cos\left(\frac{2m\pi t}{T}\right)dt - \frac{2}{T}\int_{T/2}^{T} a\cos\left(\frac{2m\pi t}{T}\right)dt$$

$$= \left(\frac{2a}{T}\right)\left(\frac{T}{2m\pi}\right)\left[\sin\left(\frac{2m\pi t}{T}\right)\right]_0^{T/2} - \left(\frac{2a}{T}\right)\left(\frac{T}{2m\pi}\right)\left[\sin\left(\frac{2m\pi t}{T}\right)\right]_{T/2}^{T}$$

$$= \left(\frac{a}{m\pi}\right)[\sin(m\pi) - \sin 0] - \frac{a}{m\pi}[\sin(2m\pi) - \sin(m\pi)]$$

$$= \left(\frac{a}{m\pi}\right)[2\sin(m\pi) - \sin(2m\pi)] = 0$$

$\therefore$ $$B_m = 0 \qquad ...(8)$$

Hence all the cosine terms in Eq. (2) are zero.

Substituting the values of B_0, A_m and B_m in Eq.(2), we have

$$y = f(\omega t) = \frac{4a}{\pi}\left[\sin\omega t + \frac{1}{3}\sin(3\omega t) + \frac{1}{5}\sin(5\omega t) + \ldots\right] \quad \ldots(9)$$

We take the first three terms of the above series representing the square wave and add them together. The result is shown in Fig. 11.9.

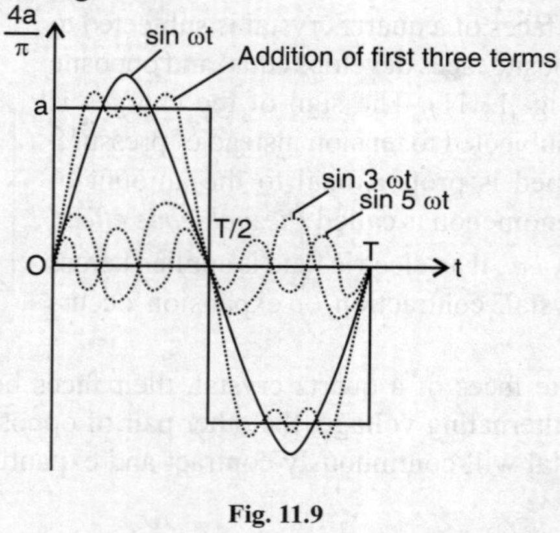

Fig. 11.9

The first harmonic has the frequency of the square wave. The higher frequencies build up the squareness of the wave.

The highest frequencies are responsible for the sharpness of the vertical sides of the wave.

The resultant of the first fifteen terms of Eq. (9) is shown in Fig. 11.10.

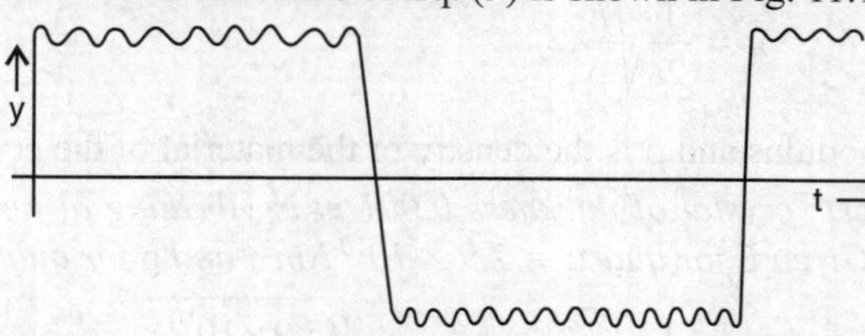

Fig. 11.10

This figure is a near approach of a square wave form but not an exact square form. However, addition of more and more terms will give a resultant curve nearly approaching a square wave form.

ULTRASONICS

11.9. INTRODUCTION

The human ear is sensitive to sound waves in the frequency range from 20 to 20,000 Hz. This range is called audible range. Sound waves of frequency more than 20,000 Hz are called *ultrasonics*. These frequencies are beyond the audible limit.

These waves also travel with the speed of sound (330 ms^{-1}).

These waves exhibit the properties of audible sound waves and also show some new phenomena.

Their wavelengths are small.

Example 1. *What is the wavelength of ultrasonic wave of frequency 330 kHz at 0°C ?* [**Given :** Velocity of sound at 0°C = 330 ms^{-1}.]

Solution. The relation between velocity (v), frequency (n) and wavelength (λ) is

$$v = n\lambda$$

$$\therefore \quad \lambda = \frac{v}{n} = \frac{330}{(330 \times 10^3)} = 10^{-3}\,\text{m} = \mathbf{1\,mm}$$

11.10. PIEZOELECTRIC EFFECT

If one pair of opposite faces of a quartz crystal is subjected to pressure, the other pair of opposite faces develops equal and opposite electric charges on them (Fig. 11.11). The sign of the charges is reversed when the faces are subjected to tension instead of pressure. The electric charge developed is proportional to the amount of pressure or tension. This phenomenon is called *Piezoelectric effect.*

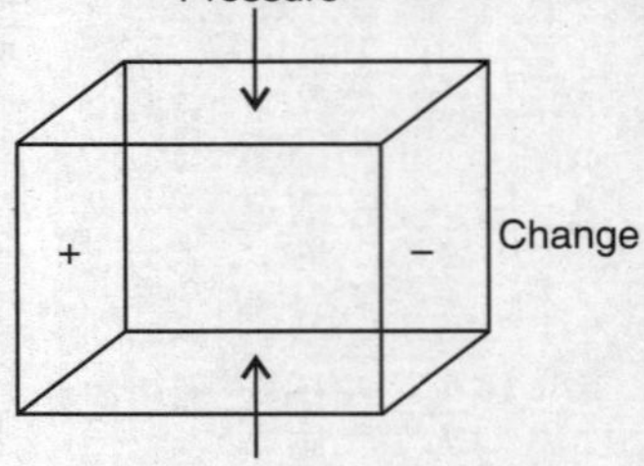

Fig. 11.11

The effect is *reversible, i.e.*, if an electric field is applied across one pair of faces of the crystal, contraction or expansion occurs across the other pair.

When the two opposite faces of a quartz crystal, their faces being cut perpendicular to the optic axis, are subjected to alternating voltage, the other pair of opposite faces experiences stresses and strains. The quartz crystal will continuously contract and expand. Elastic vibrations are set up in the crystal.

When the frequency of the alternating voltage is equal to the natural frequency of vibration of the crystal or its simple higher multiples, the crystal is thrown into resonant vibrations. The amplitude is large. These vibrations are longitudinal in nature.

Consider a X-cut crystal plate of thickness t. The fundamental frequency of vibration is given by

$$n = \frac{1}{2t}\sqrt{\frac{E}{\rho}}.$$

E is the Young's modulus and ρ is the density of the material of the crystal plate.

Example 1. *A quartz crystal of thickness 0.001 m is vibrating at resonance. Calculate the fundamental frequency. Given E for quartz = 7.9×10^{10} Nm^{-2} and ρ for quartz = 2650 kg m^{-3}.*

Solution. $$n = \frac{1}{2t}\sqrt{\frac{E}{\rho}} = \frac{1}{2 \times 0.001}\sqrt{\frac{(7.9 \times 10^{10})}{2650}} = \mathbf{2.73 \times 10^6\,Hz.}$$

Example 2. *A piezoelectric X-cut quartz plate has a thickness of 1.5 mm. If the velocity of propagation of longitudinal sound waves along the X direction is 5760 m/s, calculate the fundamental frequency of the crystal.*

Solution. For the fundamental mode of vibration,

$$\text{thickness} = \frac{\lambda}{2}$$

$$\therefore \quad \lambda = 2 \times \text{thickness} = 2 \times (1.5 \times 10^{-3})\,\text{m} = 3 \times 10^{-3}\text{m}$$

Frequency, $$n = \frac{v}{\lambda} = \frac{5760}{(3 \times 10^{-3})} = \mathbf{1.92 \times 10^6\,Hz}$$

11.11. PRODUCTION OF ULTRASONIC WAVES — PIEZOELECTRIC CRYSTAL METHOD

Principle. This is based on the inverse piezoelectric effect. When a quartz crystal is subjected to an alternating potential difference along the electric axis, the crystal is set into elastic vibrations along its mechanical axis. If the frequency of the electric oscillations coincides with the natural frequency of the crystal, the vibrations will be of large amplitude. If the frequency of the electric field is in the ultrasonic frequency range, the crystal produces ultrasonic waves.

Construction. The circuit diagram is shown in Fig. 11.12.

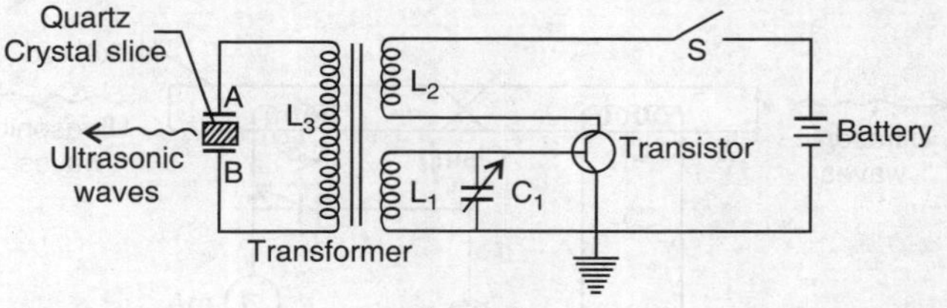

Fig. 11.12

It is a base tuned oscillator circuit. A slice of quartz crystal is placed between the metal plates A and B so as to form a parallel plate capacitor with the crystal as the dielectric. This is coupled to the electronic oscillator through primary coil L_3 of the transformer.

Coils L_2 and L_1 of oscillator circuit are taken from the secondary of the transformer. The collector coil L_2 is inductively coupled to base coil L_1. The coil L_1 and variable capacitor C_1 form the tank circuit of the oscillator.

Working. When the battery is switched on, the oscillator produces high frequency oscillations. An oscillatory e.m.f. is induced in the coil L_3 due to transformer action. So the crystal is now under high frequency alternating voltage.

The capacitance of C_1 is varied so that the frequency of oscillations produced is in resonance with the natural frequency of the crystal. Now the crystal vibrates with large amplitude due to resonance. Thus high power ultrasonic waves are produced.

Advantages

1. Ultrasonic frequencies as high as 500 MHz can be generated.
2. The output power is very high. It is not affected by temperature and humidity.
3. It is more efficient than magnetostriction oscillator.
4. The breadth of the resonance curve is very small. So we can get a stable and constant frequency of ultrasonic waves.

Disadvantages

1. The cost of the quartz crystal is very high.
2. Cutting and shaping the crystal is very complex.

11.12. PRODUCTION OF ULTRASONIC WAVES — MAGNETOSTRICTION METHOD

Principle. When a rod of ferromagnetic material like nickel is magnetised longitudinally, it undergoes a very small change in length. This is called *magnetostriction effect*.

A nickel rod placed in a rapidly varying magnetic field alternately expands and contracts with twice the frequency of the applied magnetic field. By adjusting the frequency of the alternating magnetic field to be equal to the natural frequency of longitudinal vibration of the rod, resonance is produced. Due to resonance, vibrations of large amplitude are produced in the rod. Ultrasonic waves are emitted from the ends of the rod if the frequency of the alternating magnetic field is more than 20 kHz.

The frequency of vibrations of the rod is

$$f = \frac{1}{2l}\sqrt{\frac{E}{\rho}}$$

Here, l = length of the rod,

E = Young's modulus of the material of the rod,

ρ = density of the material of the rod.

Construction. The circuit diagram of magnetostriction ultrasonic generator is shown in Fig. 11.13.

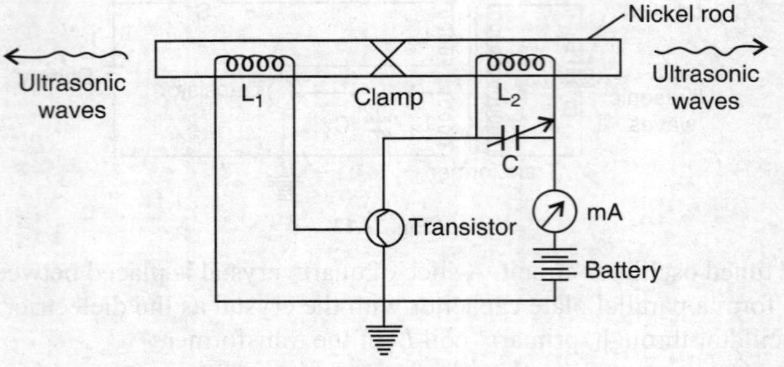

Fig. 11.13

A short permanently magnetised nickel rod is clamped in the middle between two knife edges. A coil L_2 is wound on the right hand portion of the rod. C is a variable capactior. L_2 and C form the resonant circuit of the collector tuned oscillator. Coil L_1 wound on the left hand portion of the rod is connected in the base circuit. The coil L_1 is used as a feedback loop.

Working. When the battery is switched on, the resonant circuit L_2C sets up an alternating current of frequency

$$f = \frac{1}{2\pi\sqrt{L_2C}}.$$

This current flowing round the coil L_2 produces an alternating magnetic field of frequency f along the length of the nickel rod. The rod starts vibrating due to magnetostrictive effect. The vibrations of the rod create ultrasonic waves.

The longitudinal expansion and contraction of the rod produces an e.m.f. in the coil L_1. This e.m.f. is applied to the base of the transistor. Hence the amplitude of high frequency oscillations in coil L_2 is increased due to positive feedback.

The developed alternating current frequency can be tuned with the natural frequency of the rod by adjusting the capacitor. The resonance condition is indicated by the rise in the collector current shown in the milliammeter.

Advantages

1. Magnetostriction oscillators are mechanically rugged.
2. The construction cost is low.
3. They are capable of producing large acoustical power with fairly good efficiency (*e.g.*, 60%).

Limitations

1. It can produce frequencies up to 3MHz only.
2. The frequency of oscillations depends greatly on temperature.
3. Breadth of the resonance curve is large. It is due to variations of elastic constants of ferromagnetic material with the degree of magnetisation. So we cannot get a constant single frequency.

11.13. DETECTION OF ULTRASONIC WAVES

1. **Quartz crystal method:** This method is based on the principle of Piezoelectric effect. When one pair of opposite faces of a quartz crystal is exposed to the ultrasonic waves, the other pair of opposite faces develops opposite charges. These charges are amplified and detected using an electronic circuit.
2. **Thermal detectors :** A platinum wire is placed in the region to be tested for ultrasonic waves. At nodes, due to alternate compressions and rarefactions, alternate heating and cooling is produced. Change in temperature at the node brings about changes in the electrical resistance of the platinum wire. This is detected by means of Wheatstone's bridge. No change in temperature occurs at the antinode.
3. **Kundt's tube method :** If the wavelength of the ultrasonic waves is greater than a few millimetres, a Kundt's tube can be used to form a stationary wave pattern with well defined nodes and antinodes. In the air or gas medium, lycopodium powder collects in the form of heaps at the nodes. In liquid medium, powdered coke is used to detect the position of nodes.

11.14. PROPERTIES OF ULTRASONIC WAVES

1. They have a high energy content.
2. Just like ordinary sound waves, ultrasonic waves get reflected, refracted and absorbed.
3. They show negligible diffraction due to their small wavelength. Hence they can be transmitted over long distances without any appreciable loss of energy.
4. They produce intense heating effect when passed through a substance.
5. If an arrangement is made to form stationary waves of ultrasounds in a liquid, it serves as a diffraction grating. It is called an acoustic grating.

11.15. DETERMINATION OF VELOCITY OF ULTRASONIC WAVES IN A LIQUID

Apparatus. A quartz crystal produces ultrasonic waves in a liquid contained in a glass vessel (Fig. 11.14).

These waves are reflected by the reflector. Due to superposition of direct and reflected waves, stationary waves are formed between the crystal and reflector. So nodal and antinodal planes are formed in the liquid. The density of the liquid is maximum at the nodal planes. The density is minimum at the antinodal planes. This arrangement behaves like a diffraction grating. It is called *acoustic grating*. The distance between two adjacent nodal planes is the grating element d.

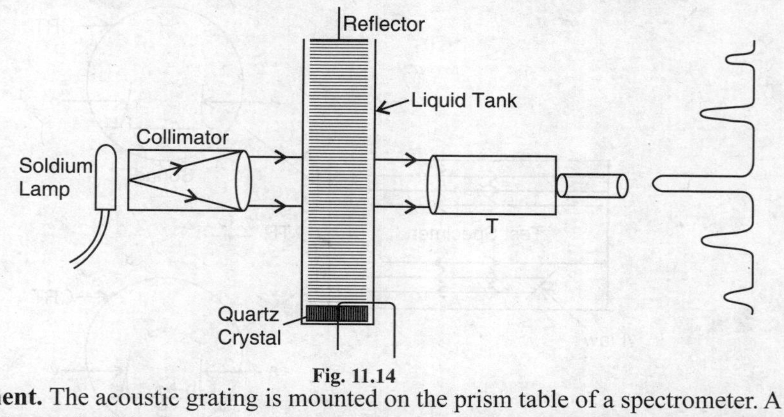

Fig. 11.14

Experiment. The acoustic grating is mounted on the prism table of a spectrometer. A parallel beam of monochromatic light from the collimator is normally incident on the acoustic grating. The diffracted beam of light is viewed through the telescope T. The diffraction pattern consists of a central maximum and principal maxima on either side.

Let θ be the angle of diffraction for the nth order principal maximum. Then,

$$d \sin \theta = n\lambda.$$

Here λ = wavelength of the light used.

Knowing θ, n and λ, the value of d can be calculated.

Let λ_u be the wavelength of ultrasonic waves in the liquid. Then,

$$d = \lambda_u/2 \quad \text{or} \quad \lambda_u = 2d.$$

The frequency N of ultrasonic waves produced by the quartz oscillator is found by means of a wavemeter.

The velocity of ultrasonic waves is calculated using the formula,

$$v = N\lambda_u$$

11.16. APPLICATIONS OF ULTRASONIC WAVES

(*i*) Science.

1. Investigation of structure of matter

We can determine the velocity of ultrasonics in liquids and gases and its variation with frequency and temperature. This study gives information about a number of properties of the medium such as its compressibility, absorption, concentration, specific heat capacity, chemical structure, arrangement of atoms in them etc.

2. Study of molecular energies

The frequencies of molecular vibrations are of the same order as the ultrasonic vibrations. So ultrasonic waves are used in the study of molecular energies. They are used in molecular acoustics for investigating structure and properties of substances.

3. Elastic symmetries of crystals

When ultrasonic waves are applied to certain crystals, they give rise to diffraction images. The diffraction images reveal the elastic symmetries of crystals.

(*ii*) Industrial Applications

1. Non-destructive testing (NDT)

Principle. *Whenever there is a change in medium, the ultrasonic waves will be reflected. Since the flaws can be detected without destroying the materials, it is called non-destructive testing.*

Working: The pulse echo system used to determine the various flaws like cracks, holes, air bubbles, laminations, etc., in the specimen is shown in Fig. 11.15.

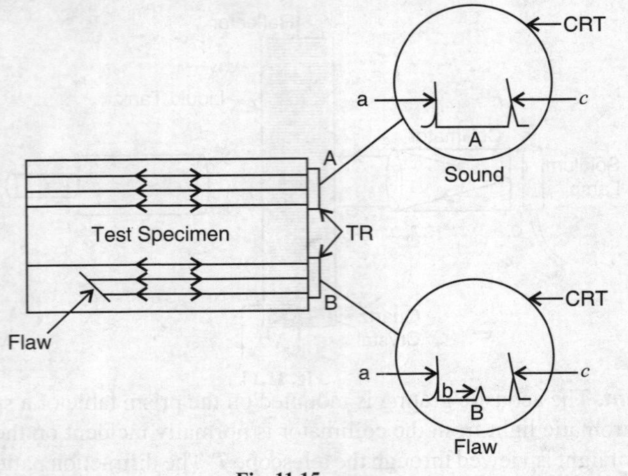

Fig. 11.15

a - Incident pulse

b - Pulse from flaw

c - Reflected pulse from the boundary of the specimen

TR - Transducer

CRT - Cathode Ray Tube.

Here short pulses of ultrasonic waves are transmitted into the material being tested. These pulses get reflected from discontinuities on their path or from any boundary of the material on which they strike. The received echoes are then displayed on a cathode ray tube screen. The CRT screen furnishes specific data as to the relative size of a discontinuity in terms of signal amplitude. The location of the discontinuity with respect to the scanning surface can be obtained by proper calibration of the CRT time base scale.

2. Ultrasonic soldering

Ultrasonic solders are used for soldering aluminium coil capacitors, aluminium wires and plates without using any fluxes.

An ultrasonic soldering iron consists of an ultrasonic generator having a tip fixed at its end. The tip is heated by an electrical heating element. The tip of the soldering iron melts solder on the aluminium. The ultrasonic vibrator removes the aluminium oxide layer. The solder thus gets fastened to the clear metal without any difficulty.

3. Ultrasonic welding

The properties of some metals change on heating. Therefore, they cannot be welded by electric or gas welding. In such cases, the metal sheets are welded together at room temperature using ultrasonic waves.

4. Ultrasonic drilling and cutting

Ultrasonics are used for making holes in very hard materials such as glass, diamond, gems and ceramics.

5. Ultrasonics in metallurgy

To irradiate molten metals which are in the process of cooling, so as to refine the grain size and to prevent the formation of cores and to release trapped gases, the ultrasonic waves are used.

6. Formation of alloys

The constituents of alloys, having widely different densities, can be mixed uniformly by a beam of ultrasonics. Thus it is easy to get alloys of uniform composition.

7. Acoustic holograms

Surface structures of various engineering materials used for space applications can be studied by using acoustic holograms.

8. Sound navigation and ranging (SONAR)

Ultrasonic waves sent from a point A travel through sea water and get reflected back from the bottom of the sea (Fig. 11.16). The reflected waves are received at the point B. Using a CRO, the time taken t, for the ultrasonic wave to travel to the bottom of the sea and reflected back to the surface is calculated.

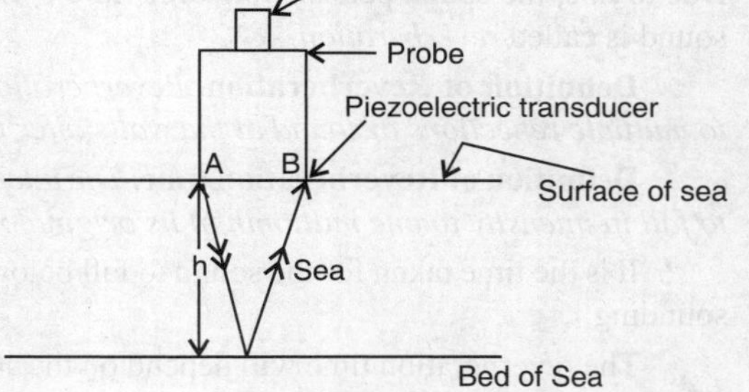

Fig. 11.16

Let v = velocity of ultrasonic wave in the sea water.

$$\text{Depth of the sea} = \frac{v \times t}{2}$$

The same method is used for finding the distance and direction of a submarine. The change in frequency of the echo signal due to Doppler effect helps to determine the velocity of the submarine and its direction. The whole system is called SONAR.

9. Chemical applications

1. Ultrasonic waves are used to form stable emulsions of even immiscible liquids like *water and oil* or *water and mercury*. This finds an application in the preparation of photographic films, face creams etc.
2. They are used to liquefy gels like aluminium hydroxide in the same way as they are liquefied by shaking.
3. They are used to coagulate fine solid or liquid particles in a gas ; for example, dust, smoke, mist etc. Ultrasonics thus find use in collecting factory dust and purifying the air.
4. Ultrasonics act like a catalytic agent and accelerate chemical reactions. Ultrasonic waves accelerate crystallisation.

(*iii*) Medical Applications

1. Disease treatment. Ultrasonic therapy can be used to treat diseases like bursites, abscesses, neuralgic and rheumatic pains etc.

2. Surgical use. Kidney stones and brain tumours can be removed without shedding any blood using ultrasonic waves. Also, any tissue in our body can be cut selectively during an operation using ultrasonics.

3. Diagnostic use. Ultrasonics are used for detecting tumours and other defects in human body. State of breast cancer can be identified using ultrasonics in a non-destructive manner. Also, the twins or any defect in the growth of foetus can be identified using ultrasonics before delivery.

4. Extraction of broken teeth. Dentists use ultrasonic waves to properly extract broken teeth.

5. Sterilization. Ultrasonic waves can kill bacteria. Therefore, they are used for sterilising milk.

6. Blood flow meters. Ultrasonic Doppler blood flow meters are used to study the blood flow velocities in blood vessels of our body.

ACOUSTICS OF BUILDINGS

11.17. REVERBERATION

When a source produces sound waves inside a closed building, the waves are generally reflected repeatedly by walls, ceiling and other materials present in the room. The intensity of the sound wave decreases at every reflection and finally the sound becomes inaudible. So the listener receives (*i*) direct waves and (*ii*) reflected waves due to multiple reflections, at the various surfaces. There is also a time gap between the direct wave received by the listener and the waves received by successive reflection. Due to this, the sound persists for some time even after the source has stopped. This persistence of sound is called *reverberation*.

Definition of Reverberation. *Reverberation is the persistence of sound in an enclosure due to multiple reflections of sound at the walls after the source has ceased to emit sound.*

Definition of Reverberation time. *The interval of time taken by sustained or continuous sound to fall in intensity to one millionth of its original value is called reverberation time* (T).

It is the time taken for the sound to fall below the minimum audibility after the source stopped sounding.

The reverberation time will depend on the size of the room or the auditorium, the nature of the reflecting material on the wall and the ceiling and the area of the reflecting surfaces.

Sabine's method of determining reverberation time

An organ pipe of frequency 512 Hz is sounded. At the same time, a stop watch is started. The time for the sound to become inaudible is measured. This gives the reverberation time.

11.18. DERIVATION OF SABINE'S FORMULA FOR REVERBERATION TIME

The main assumptions are :

(1) The average energy per unit volume is uniform. It is represented as σ.

(2) The energy is not lost in the auditorium. The energy lost is only due to the absorption of the material of the walls and ceiling and also due to the escape through the windows and ventilators. Both these factors are included in the term '**absorption**' of energy.

Suppose a source is producing sound continuously. This sound energy is propagated in all directions. Let σ be the energy contained in a unit volume, *i.e.*, energy density.

Energy contained in a solid angle $d\phi = \dfrac{\sigma.d\phi}{4\pi}$

Let this energy be incident on a unit surface area of the wall at an angle θ (Fig. 11.17).

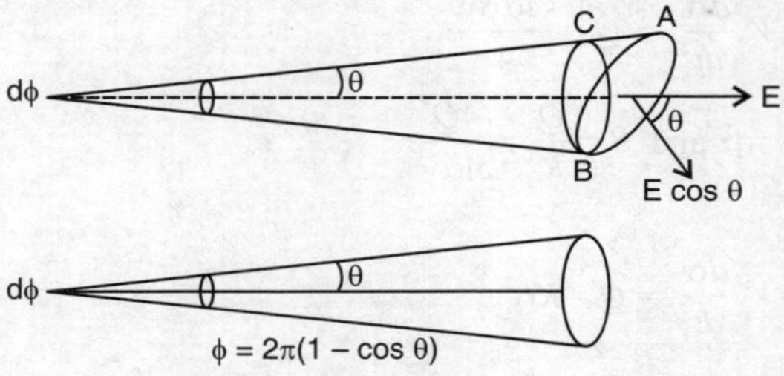

Fig. 11.17

Let $\quad v$ = Velocity of sound

Total energy falling per second on a unit surface area of the wall

$$= \left(\frac{\sigma.d\phi}{4\pi}\right)(\cos\theta).v$$

The total energy falling per second within a hemisphere

$$= \frac{\sigma v}{4\pi}.\int \cos\theta.d\phi \qquad ...(1)$$

But $\quad \phi = 2\pi(1 - \cos\theta)$

or $\quad d\phi = 2\pi \sin\theta.d\theta$

Substitute this value of $d\phi$ in Eq. (1).

The total energy falling per second within a hemisphere

$$= \frac{\sigma v}{4\pi}.\int_0^{\pi/2} 2\pi \sin\theta.\cos\theta.d\theta$$

$$= \frac{\sigma v}{2\pi}\left[\frac{-\cos^2\theta}{2}\right]_0^{\pi/2}$$

$$= \frac{\sigma v}{4}$$

Let $\quad \alpha$ = absorption coefficient of the walls.

The amount of energy absorbed per second per unit area $= \dfrac{\alpha\sigma v}{4}$.

Let A = area of the walls and the other absorbing materials including ceiling, windows and ventilators etc.

Total amount of energy absorbed per second $= \dfrac{A\alpha\sigma v}{4}$.

Let V be the volume of the auditorium.

Total sound energy present at any instant $= V\sigma$

$$\text{The rate of increase of energy} = \frac{d}{dt}(V\sigma) = v\frac{d\sigma}{dt} \quad ...(2)$$

Suppose, the source supplies energy at the rate of Q units per second.

$$\text{Rate of increase of energy} = Q - \frac{A\alpha\sigma v}{4} \quad ...(3)$$

Equating (2) and (3), we get

$$V.\frac{d\sigma}{dt} = Q - \frac{A\alpha\sigma v}{4} \quad ...(4)$$

Put $\dfrac{A\alpha v}{4} = K, \dfrac{K}{V} = \beta$ and $B = \dfrac{Q}{K} = \dfrac{4Q}{A\alpha v}$.

Eq. (4) becomes

$$V.\frac{d\sigma}{dt} = Q - K\sigma$$

$$\frac{d\sigma}{dt} = \frac{Q}{V} - \frac{K}{V}.\sigma \quad ...(5)$$

The general solution of this equation is

$$\sigma = B + be^{-\beta t}$$

When $t = 0, \sigma = 0.$

$$0 = B + b$$

or $b = -B$

$\therefore$ $\sigma = B - Be^{-\beta t}$

$$\sigma = B\,[1 - e^{-\beta t}]$$

Substituting the values of B and β,

$$\sigma = \frac{4Q}{A\alpha v}\left[1 - e^{-\left(\frac{A\alpha v}{4V}\right)t}\right] \quad ...(6)$$

Eq. (6) represents the rise of average sound energy per unit time from the time the source commences to produce sound.

The maximum value of average energy per unit volume

$$\sigma_{max} = \frac{4Q}{A\alpha v} \quad ...(7)$$

Similarly, after the source ceases to emit sound, the decay of the average energy per unit volume is given by

$$\sigma = \frac{4Q}{A\alpha v}e^{-\left(\frac{A\alpha v}{4V}\right)t} \quad ...(8)$$

$$\sigma = \sigma_{max}\, e^{-\left(\frac{A\alpha v}{4V}\right)t} \qquad ...(9)$$

Fig. 11.18 shows the decay of the energy density with time.

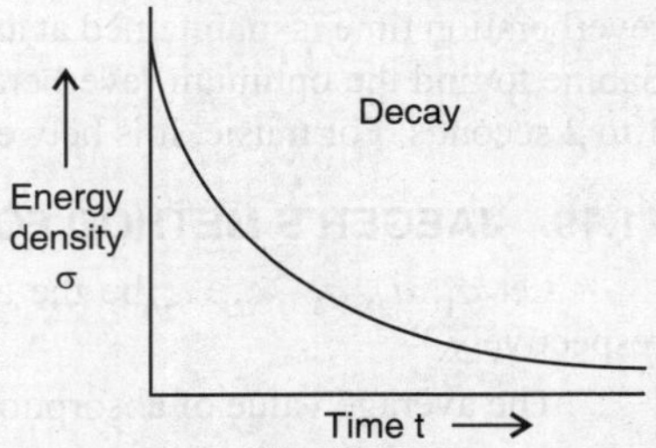

Fig. 11.18

Reverberation Time (*T*)

Reverberation time is defined as the time taken for the energy density to fall to one-millionth of its value before the cut-off.

Hence, substituting $\sigma = \dfrac{\sigma_{max}}{10^6}$ and $t = T$ in Eq. (9),

$$\frac{\sigma_{max}}{10^6} = \sigma_{max}\, e^{-\left(\frac{A\alpha v}{4V}\right)T}$$

$$10^{-6} = e^{-\left(\frac{A\alpha v}{4V}\right)T}$$

$$e^{\left(\frac{A\alpha v}{4V}\right)T} = 10^6$$

Taking logarithms,

$$\left(\frac{A\alpha v}{4V}\right)T = \log_e 10^6 = 6 \log_e 10 = 6 \times 2.3026$$

Take velocity of sound (v) approximately at room temperature as 350 ms^{-1}.

$$T = \frac{6 \times 2.3026 \times 4V}{A\alpha v} = \frac{6 \times 2.3026 \times 4}{350} \times \left(\frac{V}{A\alpha}\right)$$

$$T = \frac{0.158V}{A\alpha}$$

In general,

$$T = \frac{0.158V}{\Sigma A\alpha} \qquad ...(10)$$

Here, V and A are in metres.

Eq. (10) represents the Sabine's reverberation time formula.

Explanation of Sabine's formula

Sabine's formula for reverberation time is

$$T = \frac{0.158V}{\Sigma A\alpha} \qquad ...(1)$$

According to Eq.(1), the reverberation time is

(*i*) directly proportional to the volume of the auditorium,

(*ii*) inversely proportional to the area of the ceilings, area of the walls etc.; and

(*iii*) inversely proportional to the total absorption plus transmission through open surfaces.

Importance of Sabine's Formula

If the reverberation time is too small, the sound dies away almost instantaneously and gives the hall a '*dead*' effect. If the reverberation time is too long, each syllable continues to be heard even after the next syllable has been uttered. This makes the sound unintelligible. Therefore, the value of reverberation time is maintained at an *optimum value*. Rooms of different dimensions were used by Sabine to find the optimum reverberation time. For speech, the reverberation time should be about 1 to 2 seconds. For music, it is between 2 and 2.5 seconds.

11.19. JAEGER'S METHOD FOR DERIVATION OF SABINE'S FORMULA

Let a_1, a_2, a_3 .... a_n be the absorption coefficients of the surfaces of areas S_1, S_2, S_3,...S_n respectively.

The average value of absorption coefficient,

$$a = \frac{a_1S_1 + a_2S_2 + ... + a_nS_n}{S_1 + S_2 + S_3 + ... + S_n} = \frac{\Sigma a_i S_i}{S}$$

Here, S is the total area of the surfaces.

By statistical method, Jaeger showed that the sound travels an average distance $\frac{4V}{S}$ between two successive reflections.

Here, V is the volume of the hall.

Time between two successive reflections $= \frac{4V}{Sv}$.

Here, v is the velocity of sound.

Number of reflections in one second $= \frac{Sv}{4V}$.

The average number of reflections in time $t = \frac{Svt}{4V}$.

Let a be the fraction of sound absorbed at a single reflection.

Fraction of sound reflected $= (1 - a)$.

After two reflections, the fraction of sound reflected $= (1 - a)^2$.

After $\frac{Svt}{4V}$ reflections, the fraction of sound reflected $= (1 - a)^{Svt/4V}$

Let I_0 be the initial intensity of sound and I_t, the intensity after time t.

$$I_t = I_0 (1 - a)^{Svt/4V} \qquad ...(1)$$

Reverberation time is the time taken by sound to fall to one millionth of the intensity just before the source is cut off.

When $t = T, \quad I_T/I_0 = 10^{-6}$

Substituting these values in Eq.(1), we get

$$\frac{I_T}{I_0} = 10^{-6} = (1 - a)^{Svt/4V} \qquad ...(2)$$

Taking natural logarithms of both sides of Eq.(2),

$$\log_e 10^{-6} = \frac{SvT}{4V} \log_e (1 - a)$$

$$T = \log_e 10^{-6} \times \frac{4V}{Sv \log_e (1 - a)}$$

$$T = 2.3026 \times \log_{10} 10^{-6} \times \frac{4V}{Sv \log_e (1 - a)}$$

Put $v = 350\text{ms}^{-1}$

$$T = \frac{2.3026 \times (-6) \times 4V}{S \times 350 \times \log_e(1-a)}$$

$$= 0.158\frac{V}{-S\log_e(1-a)}$$

$$\log_e(1-a) = -a$$

$$T = \frac{0.158V}{Sa}$$

$$\therefore \quad T = \frac{0.158V}{\Sigma a_i S_i}$$

This is Sabine's formula for the reverberation time. It shows that the reverberation time depends on volume and total absorbing power of the hall.

Example 1. *A hall of volume 5500 metre^3 is found to have a reverberation time of 2.3s. The sound absorbing surface of the hall has an area of 750m^2. Calculate the average absorption coefficient.*

Solution. Here, $V = 5500\text{ m}^3$, $T = 2.3\text{s}$, $A = 750\text{ m}^2$.

Reverberation time, $T = \dfrac{0.158V}{A\alpha}$ (*Sabine's formula* for reverberation time)

$$\alpha = \frac{0.158V}{AT} = \frac{0.158 \times 5500}{750 \times 2.3} = \mathbf{0.504}$$

Example 2: *The volume of a room is 600 m^3. The wall area of the room is 220 m^2, the floor area is 120 m^2 and the ceiling area is 120 m^2. The average sound absorption coefficient, (i) for the walls is 0.03 ; (ii) for the floor is 0.06 and (iii) for the ceiling is 0.80. Calculate the average sound absorption coefficient and the reverberation time.*

Solution. Average absorption coefficient,

$$\alpha = \frac{\Sigma\alpha A}{\Sigma A} = \frac{\alpha_1 A_1 + \alpha_2 A_2 + \alpha_3 A_3}{A_1 + A_2 + A_3} = \frac{0.03 \times 220 + 0.06 \times 120 + 0.8 \times 120}{220 + 120 + 120} = 0.2387$$

Total sound absorption of the room = $\alpha\Sigma A = 0.2387 \times 460 = 109.8$ metric sabine

Reverberation time, $T = \dfrac{0.158V}{\alpha\Sigma A} = \dfrac{0.158 \times 600}{109.8} = \mathbf{0.8634s}$

11.20. ABSORPTION COEFFICIENT

First Definition. *The sound absorption coefficient (α) of a material is defined as the ratio of sound energy absorbed by it to the total energy incident on it.*

$$\textit{Absorption coefficient } (\alpha) = \frac{\text{Sound energy absorbed by the surface}}{\text{Total sound energy incident on the surface}}$$

Sabine chose 1 square metre of an open window as a standard unit of absorption since all sound waves falling on it pass through and can be said to be completely absorbed. This unit is called *sabine*. It is also called open window unit (O.W.U.).

Second Definition. *The absorption coefficient of a material is defined as the ratio of sound energy absorbed by it to that absorbed by an equal area of an open window.*

Determination of Absorption Coefficient

An audio frequency source of frequency 512 Hz is taken.

(*i*) The reverberation time T_1 in a hall is determined using a chronograph, without the absorbing material in the hall.

According to Sabine's formula,

$$T_1 = \frac{0.158V}{\Sigma \alpha A} \quad ...(1)$$

Here $(\Sigma \alpha A)$ is the absorption in the hall without the material.

V = Volume of the hall.

(*ii*) The absorbing material is put inside the room.

Let α_1 be the absorption coefficient of the material and s its surface area.

Now the reverberation time T_2 is measured.

$$T_2 = \frac{0.158V}{\Sigma \alpha A + \alpha_1 s} \quad ...(2)$$

$$\frac{1}{T_2} - \frac{1}{T_1} = \frac{\alpha_1 s}{0.158V}$$

$$\therefore \quad \alpha_1 = \frac{0.158V}{s}\left(\frac{1}{T_2} - \frac{1}{T_1}\right) \quad ...(3)$$

Knowing the values of V, s, T_1 and T_2, the absorption coefficient of the absorbing material α_1 is calculated.

11.21. ACOUSTICS OF BUILDINGS

The branch of Physics which deals with the design and construction of buildings with good acoustic properties is called "Acoustics of buildings." It deals basically with (*i*) reverberation control, (*ii*) noise insulation and reduction and (*iii*) sound distribution and absorption.

The following are the requirements of a *good auditorium*:

1. The sound should be sufficiently loud and intelligible in every part of the hall.
2. Sound of each syllable should soon decay so that the succeeding syllable may be heard distinctly. There must be no confusion due to overlapping of syllables. This means that the auditorium must be free from excessive reverberation.
3. No echoes should be present.
4. There should not be *undesirable focussing* of sound in any part of the hall. There should not be any *zones of silence* or regions of poor audibility anywhere in the hall.
5. Interference, reflection and resonance effects should be avoided.
6. All extraneous noises must be shut out as far as possible. The boundaries should be sufficiently soundproof to exclude extraneous noise.
7. There should be no Echelon effect.
8. The quality of sound must be unaltered, *i.e.*, the relative intensity of the components of the complex sound must be preserved.

Fig. 11.19 shows the usual design of the hall.

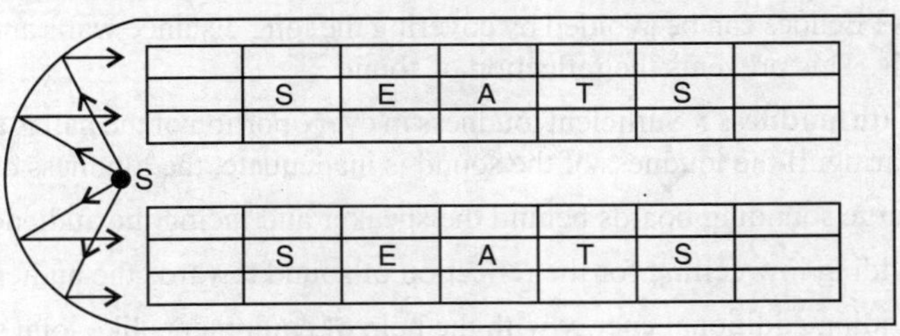

Fig. 11.19

It has a parabolic shape at the speaker's end *S*. As the speaker speaks, sound is reflected from the back almost as a parallel beam. So there is uniform distribution of sound intensity throughout the hall.

11.22. FACTORS AFFECTING THE ACOUSTICS OF BUILDINGS

The factors affecting the acoustics of a hall are :

(*i*) Reverberation time

(*ii*) Presence of echoes

(*iii*) Adequate loudness

(*iv*) Focussing and interference effects

(*v*) Resonances within the building

(*vi*) Echelon effect

(*vii*) Extraneous noises

(*viii*) Inside noise.

(*i*) **Reverberation.** Reverberation is one of the important single factors that affect the acoustics of a room or a hall.

The reverberation of sound in an auditorium is due to multiple reflections taking place at various surfaces present within the auditorium.

If the reverberation time is too small, the sound dies away almost instantaneously and gives the hall a '*dead*' effect. If the reverberation time is too long, each syllable continues to be heard even after the next syllable has been uttered. This makes the sound unintelligible. Therefore, the value of reverberation time is maintained at an *optimum value*. Rooms of different dimensions were used by Sabine to find the optimum reverberation time. For speech, the reverberation time should be about 1 to 2 seconds. For music, it is between 2 and 2.5 seconds.

To adjust the reverberation time to the best optimum value, the absorption of sound in the hall is increased in the following ways :

(*a*) providing windows and openings,

(*b*) covering ceiling and walls with sound absorbing materials like felt, fibre board, celotex, asbestos etc.,

(*c*) using heavy curtains with folds,

(*d*) covering the floor with carpets,

(*e*) by upholstering the seats so that the absorption is approximately the same with or without the audience,

(*f*) by having a good audience because one listener is equivalent to about 0.5 m^2 area of an open window, and

(*g*) by decorating the walls with pictures, maps etc.

(ii) **Echoes :** Echoes can be avoided by covering the long distance walls and high ceiling with absorbent materials. This prevents the reflection of sound.

(iii) **Adequate loudness :** Sufficient loudness in every portion of the hall is an important factor for satisfactory hearing. If the loudness of the sound is inadequate, the loudness can be increased by

(*a*) using large sounding boards behind the speaker and facing the audience,

(*b*) by providing low ceiling for the reflection of sound towards the audience, and

(*c*) by providing additional energy with the help of equipments like loud speakers.

(iv) **Focussing :** If there is any concave surface in the hall, sound is concentrated at its focus region. There may be dead space at some other regions. Hence such surfaces must be avoided. If curved surfaces are present, they should be covered with absorbent materials.

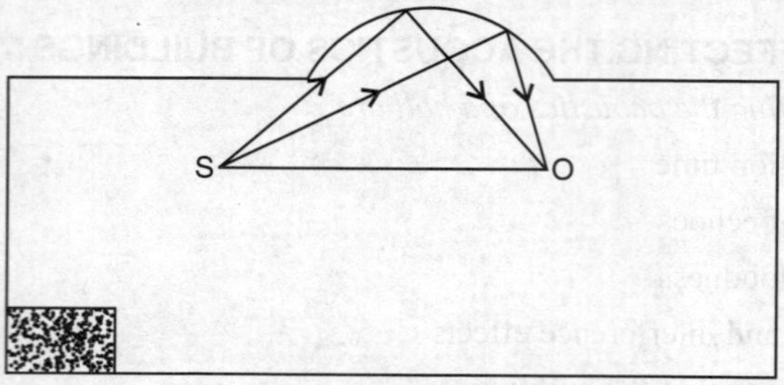

Fig. 11.20

In Fig. 11.20, the observer at *O* receives sound from the speaker along the direct path *SO*. The observer also receives the sound waves after reflection from the ceiling. Thus the intensity of sound at *O* is comparatively higher than at other positions in the auditorium. There should be no interference of direct and reflected waves. Good absorbing materials avoid this defect.

Suppose the direct and the reflected waves are in opposite phase at *O*. This results in minimum intensity of sound at *O*. Further, the direct and the reflected waves may form a stationary wave pattern. This causes uneven distribution of sound intensity.

(v) **Resonance Effects :** Window panels, sections of wooden portions which are loosely fitted start vibrating, creating extra sounds. If there is resonance for any audio frequency note, the intensity of the note will be entirely different from the intensity desired. Such resonant vibrations should be damped.

(vi) **Echelon effect :** Suppose there is regular structure similar to a flight of stairs in the hall. The sound produced in front of such a structure may produce a *musical note* due to regular successive echoes of sound reaching the observer. Such an effect is called *echelon effect*. If the frequency of this note is within the audible range, the listener will hear only this note prominently.

The remedy is to cover such surfaces with absorbing materials like carpets.

(vii) **Extraneous noises :** They are classified as *air-borne* and *structure- borne* noises.

(a) **Air-borne noise :** The noise which reaches the hall from outside through open windows, doors and ventilators is called *Air borne noise*. It is reduced by the following methods:

1. Using double doors and windows with separate frames and having insulating materials between them.

2. Making the hall as an air-conditioned hall where the hall is completely closed.

(b) **Structure-borne noise :** The noise which is conveyed through the structure of the building is called *Structure borne noise*. This noise can be minimised by using double walls with air space between them.

(viii) **Inside noise :** The noise which is produced inside the hall is called *inside noise*. It is produced by machinery like air-conditioners, typewriters, etc., in the hall. This noise is minimised by the following methods:

1. Covering the floor with carpet.
2. Walls, floor and ceiling should be provided with sound absorbing materials.
3. Placing the working machineries like air-cooler or air-conditioning machine on sound absorbent pads.
4. The rotating or impacting machines must be fixed using a fixed bed.

11.23. SOUND DISTRIBUTION IN AN AUDITORIUM

Sound distribution describes how the sound pressure level varies with position in an enclosure. The design of an auditorium requires smooth decay and growth of sound. In an auditorium, the sound must be distributed or diffused over the whole area. To ensure these factors, acoustic treatment is given, *viz.*, scattering effect of objects, irregularities on the wall surfaces, fixing absorptive material on the walls etc. Model analysis with light rays, ultrasonic waves or ordinary audio frequency sound is used to study sound distribution. Graphical construction of first reflection of the sound waves at various cross-sections can also be used as in Fig. 11.21. It is clear from the figure that the reflected sound is distributed evenly in the whole auditorium, viz., the main floor and the balcony. This design enables an even distribution of sound intensity.

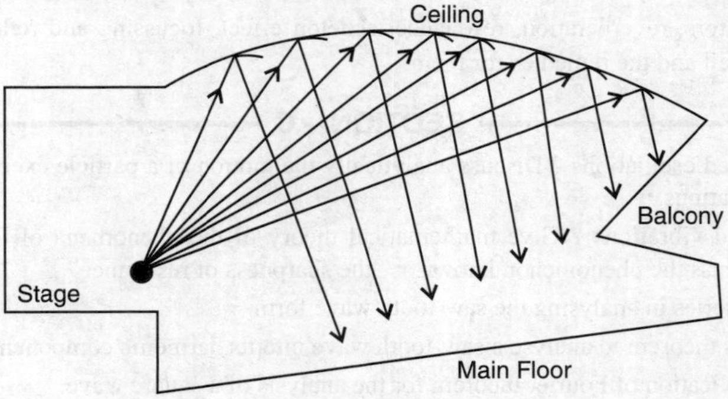

Fig. 11.21

EXERCISE XI

SECTION – A

1. Explain : Free vibrations, forced vibrations and resonance. Illustrate your answer by giving an example of each.
2. What are free, damped and forced vibrations ?
3. State the Fourier theorem and obtain expressions for Fourier coefficients.
4. What are ultrasonic waves ?
5. What is piezoelectric effect ? How is inverse piezoelectric effect used to produce ultrasonic wave ?
6. Explain the principle of magnetostriction method of producing ultrasonics.
7. What are the usual methods employed to detect ultrasonics ?
8. Mention the properties of ultrasonics.
9. Give important applications of ultrasonics.
10. What is meant by 'reverberation' and 'reverberation time'? Explain clearly what causes reverberation in a hall and how it can be minimised.
11. State and explain Sabine's formula for Reverberation Time of a hall.

12. Define 'absorption coefficient of a material' and 'time of reverberation of a hall'.

13. What are the factors affecting the acoustic quality of a building ?

14. Enumerate the features that an auditorium should have for good acoustics.

15. Write a brief account of the acoustical analysis and its correction.

Ans. Acoustic analysis is to check the prevailing condition of the hall for good acoustics. For this, it is necessary to determine the absorption and distribution of sound in the hall. If the absorption is inadequate, then by adding or removing absorbing materials, the absorption in the hall is corrected. By adjusting the shape and interior surfaces of the hall, uniform sound distribution is ensured.

SECTION – B

16. Explain what is meant by free vibrations of a body. Give a mathematical treatment of free oscillations of a particle executing simple harmonic motion.

17. What are damped vibrations ? Derive the differential equation of damped oscillatory motion and give its general solution.

18. Discuss the phenomenon of sharpness of resonance and show how it depends upon the damping factor.

19. Describe the piezoelectric method of producing ultrasonic waves.

20. Explain how ultrasonics are produced in a magnetostriction oscillator.

21. Write any four applications of ultrasonics.

22. Describe with necessary theory a method of measuring the absorption coefficient of a material.

23. Discuss the factors, reverberation, resonance, echelon effect, focussing and reflection that affect the acoustics in a hall and the remedies for them.

SECTION – C

24. What are damped oscillations ? Discuss analytically the motion of a particle executing damped simple harmonic oscillations.

25. What are forced vibrations ? Give mathematical theory of the phenomena of forced vibrations and resonance. Discuss the phenomenon known as "the sharpness of resonance".

26. Apply Fourier series in analysing the saw-tooth wave form.

27. Apply Fourier's theorem to analyse a saw-tooth wave into its harmonic components.

28. Discuss the application of Fourier theorem for the analysis of a square wave.

29. What are ultrasonics ? Describe in detail one method of their production and detection. Give some of their important applications.

30. What are ultrasonic waves ? How are they produced and detected ? Give an account of their properties and uses.

31. Define the reverberation time. Derive Sabine's formula for reverberation time. Explain its significance.

32. Discuss the salient points associated with acoustics of auditorium.

PROBLEMS

33. A room has dimensions 6 × 4 × 5 metres. Calculate (*i*) the mean free path of the sound waves in the room, (*ii*) the number of reflections made per second by the sound wave with the walls of the room. Velocity of sound in air = 350 ms^{-1}. [**Ans.** (*i*) 3.243 m, (*ii*) 107.9]

34. A hall has a volume of 2250 m^3. Its total absorption is equivalent to 100 m^2 of open window. What will be the effect on the reverberation time if audience fills the hall and thereby increases the absorption by another 100 m^2 ? [**Ans.** Reverberation time is reduced from 3.622 seconds to 1.811 seconds]

12 CHAPTER
Composition of Simple Harmonic Motions

12.1. COMPOSITION OF TWO SIMPLE HARMONIC MOTIONS IN A STRAIGHT LINE

The two simple harmonic vibrations are represented by the equations

$$y_1 = a_1 \sin (\omega t + \alpha_1) \quad ...(1)$$

and

$$y_2 = a_2 \sin (\omega t + \alpha_2) \quad ...(2)$$

y_1 and y_2 are the displacements of a particle due to the two vibrations.

a_1 and a_2 are the amplitudes of the two vibrations.

α_1 and α_2 are the epoch angles.

The angular frequency ω is the same for both vibrations. The resultant displacement y of the particle is given by,

$$\begin{aligned} y &= y_1 + y_2 \\ &= a_1 \sin (\omega t + \alpha_1) + a_2 \sin (\omega t + \alpha_2) \\ &= a_1 (\sin \omega t \cos \alpha_1 + \cos \omega t \sin \alpha_1) + \\ &\quad a_2 (\sin \omega t \cos \alpha_2 + \cos \omega t \sin \alpha_2) \\ y &= (a_1 \cos \alpha_1 + a_2 \cos \alpha_2) \sin \omega t + (a_1 \sin \alpha_1 + a_2 \sin \alpha_2) \cos \omega t \quad ...(3) \end{aligned}$$

We can write the displacement of the resultant motion as

$$\begin{aligned} y &= A \sin (\omega t + \phi) \\ y &= A \sin \omega t \cos \phi + A \cos \omega t \sin \phi \quad ...(4) \end{aligned}$$

A is the amplitude of the resultant motion and ϕ is the initial phase.

Equating the coefficients of $\sin \omega t$ and $\cos \omega t$ in Eqs. (3) and (4),

$$A \cos \phi = \alpha_1 \cos \alpha_1 + a_2 \cos \alpha_2 \quad ...(5)$$

$$A \sin \phi = a_1 \sin \alpha_1 + a_2 \sin \alpha_2 \quad ...(6)$$

Squaring Eqs. (5) and (6) and adding,

$$A^2 = a_1^2 + a_2^2 + 2a_1 a_2 \cos (\alpha_1 - \alpha_2) \quad ...(7)$$

Dividing equation (6) by (5),

$$\tan \phi = \frac{a_1 \sin \alpha_1 + a_2 \sin \alpha_2}{a_1 \cos \alpha_1 + a_2 \cos \alpha_2} \quad ...(8)$$

Equations (7) and (8) give the values of A and ϕ.

12.2. COMPOSITION OF TWO SIMPLE HARMONIC MOTIONS OF EQUAL TIME PERIODS AT RIGHT ANGLES

Suppose the same particle is simultaneously subjected to two S.H. motions along two mutually perpendicular directions (X and Y axes). Let the displacements of the S.H. motions be

$$x = a \sin(\omega t + \alpha) \quad ...(1)$$

and $$y = b \sin \omega t \quad ...(2)$$

α = The phase difference betwean the two S.H. motions.

The resultant motion is obtained by eliminating t from (1) and (2).

From equation (2),

$$\sin \omega t = \frac{y}{b}.$$

$$\therefore \quad \cos \omega t = \sqrt{1 - \frac{y^2}{b^2}}$$

From equation (1), $\frac{x}{a} = [\sin \omega t \cos \alpha + \cos \omega t \sin\alpha]$ ...(3)

Substituting the values of sin ωt and cos ωt in equation (3),

$$\frac{x}{a} = \left[\frac{y}{b}\cos\alpha + \sqrt{1 - \frac{y^2}{b^2}}.\sin\alpha\right]$$

or $$\frac{x}{a} - \frac{y}{b}\cos\alpha = \left(\sqrt{1 - \frac{y^2}{b^2}}\right)\sin\alpha$$

$$\frac{x^2}{a^2} + \frac{y^2}{b^2}\cos^2\alpha - \frac{2xy}{ab}\cos\alpha = \left(1 - \frac{y^2}{b^2}\right)\sin^2\alpha$$

or $$\frac{x^2}{a^2} + \frac{y^2}{b^2}[\sin^2\alpha + \cos^2\alpha] - \frac{2xy}{ab}\cos\alpha = \sin^2\alpha$$

$$\therefore \quad \frac{x^2}{a^2} + \frac{y^2}{b^2} - \frac{2xy}{ab}\cos\alpha = \sin^2\alpha \quad ...(4)$$

This represents the general equation of an ellipse.

Hence the resultant motion is in general elliptical.

Special cases:

(*i*) If $\alpha = 0$, $\cos\alpha = 1$; $\sin\alpha = 0$

$$\frac{x}{a} - \frac{y}{b} = 0$$

or $$y = \frac{b}{a}x.$$

This represents a straight line *AB* (Fig. 12.1) of slope *b*/*a*.

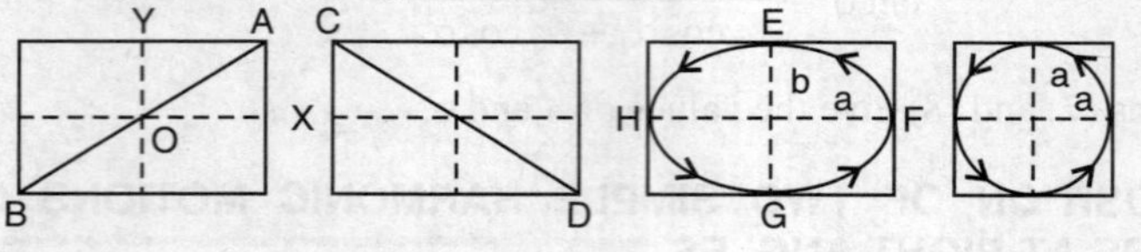

Fig. 12.1

(*ii*) If $\alpha = \pi$, $\sin\alpha = 0$; $\cos\alpha = -1$; Hence $y = -\frac{b}{a}x$.

The resultant motion is represented by the straight line *CD* passing through the origin and having a negative slope.

(*iii*) If $\alpha = \frac{\pi}{2}, \frac{x^2}{a^2} + \frac{y^2}{b^2} = 1.$

This represents an ellipse *EFGH* whose major and minor axes coincide with the coordinate axes.

(*iv*) If $\alpha = \pi/2$ and $a = b$, then $x^2 + y^2 = a^2$.

This represents a circle of radius *a*. Hence, the resultant motion is circular.

Lissajous Figures

When a particle is acted upon simultaneously by two S.H. motions at right angles to each other, the resultant path traced by the particle is called *Lissajous figures*. The shape of the Lissajous' Figure depends upon (*i*) the amplitudes, (*ii*) the periods of the two component motions and (*iii*) the phase difference between them.

12.3. EXPERIMENTAL METHODS FOR OBTAINING LISSAJOUS' FIGURES

1. Optical method.

A and *B* are two tuning forks with frequencies in the ratio of 2 : 1 (Fig. 12.2).

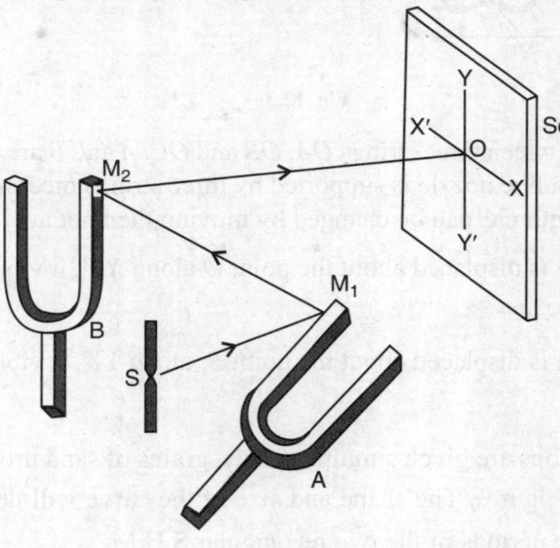

Fig. 12.2

The prongs of A vibrate in a horizontal plane. The prongs of *B* vibrate in a vertical plane. M_1 and M_2 are two small plane mirrors attached to the prongs of *A* and *B* respectively. It is adjusted that the spot of light after reflection form M_1 and M_2 is obtained at *O*, the centre of the screen.

• When only the tuning fork *A* vibrates, the spot of light moves along *XX′*.

• When only the tuning fork *B* vibrates, the spot of light moves along *YY′*.

• When *A* and *B* vibrate simultaneously and are in phase, the spot of light traces the figure of eigtht.

2. Blackburn's Pendulum

Fig. 12.3 shows a simple form of this pendulum.

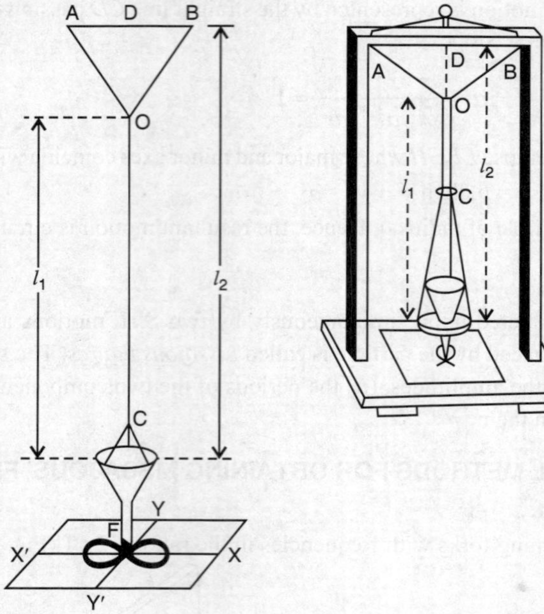

Fig. 12.3

A knot is made at O between three strings OA, OB and OC. A and B are fixed to hooks in a rigid frame. A funnel (F) having a fine nozzle is supported by three strings meeting at C. Dry fine sand is taken in the funnel. The length OC can be changed by moving the knot at O.

- When the pendulum is displaced about the point O along XX', it vibrates with a time period $t_1 = 2\pi\sqrt{l_1 / g}$.
- When the pendulum is displaced about the point B, along YY', it vibrates with a time period $t_2 = 2\pi\sqrt{l_2 / g}$.
- When both the motions are given simultaneously, grains of sand dropping out of the funnel trace out Lissajous' figures. The shape and size of the curve will depend upon the relative amplitudes and time periods of the two rectangular S.H.M.'s. If $l_1 = l_2/4$, then $t_1/t_2 = \sqrt{1/4}$ $= 1/2$, giving us a figure of '8'. If the ratio between time periods is not exactly 1:2, the resultant motion will gradually pass through various phases shown in Fig. 12.1.

12.4. USES OF LISSAJOUS FIGURES

(1) The form of a Lissajous figure enables us to determine the ratio of the time periods of component vibrations. From the Lissajous' figure, the number of times the curve touches the horizontal and the vertical sides of a rectangle bounding the Lissajous figure, is found. If the curve touches the horizontal side p times and the vertical side q times, the ratio of the time periods is $p : q$.

(2) The Lissajous figures are used for the accurate determination of the frequency of a tuning fork.

(3) Lissajous figures are used in obtaining beautiful designs for calico printing in textile industry.

EXERCISE XII

1. Two simple harmonic motions having the same period but differing in phase are acting in the same direction on a partice. Show that the resultant motion is simple harmonic and deduce expression for the resulting amplitude and phase.
2. Find the resultant of two S.H.M.'s of equal period when they act at right angles to one another. Discuss the different important cases.
3. What are Lissajous' figures? Show how they may be produced and demonstrated. Mention their uses.

13 CHAPTER
Beats

13.1 SUPERPOSITION OF HARMONIC WAVES OF NEARLY THE SAME FREQUENCY

When two notes of nearly the same frequency and amplitude are sounded together, they combine to form a single note whose intensity alternately rises and falls at regular intervals. This phenomenon of the waxing and waning of sound is called *beats.* The number of beats heard per second is equal to the difference in frequency between the two sounding bodies.

- The human ear cannot perceive more than 16 beats per second.
- When a tuning fork is loaded, its frequency decreases.
- When a tuning fork is filed, its frequency increases.

13.2. DEMONSTRATION OF BEATS

Two tuning forks of the same frequency are taken. One prong of one of the forks is loaded with wax so as to decrease its frequency slightly. The two forks are excited and their shanks pressed on two sounding boxes. (Fig. 13.1). It is found that the intensity of sound increases and decreases periodically.

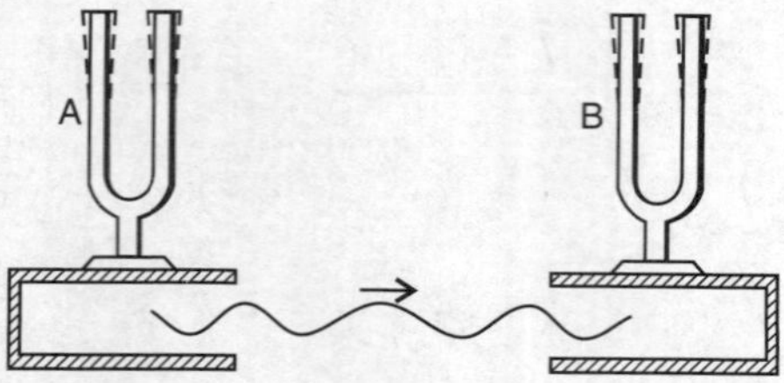

Fig. 13.1

13.3. ANALYTICAL TREATMENT OF BEATS

Consider two wave trains of frequencies n_1 and n_2 such that $(n_1 - n_2)$ is small. Let a and b be the amplitudes of the waves respectively. It is assumed that the two waves are in phase at any point in the medium at $t = 0$.

The displacements y_1 and y_2 due to each wave are given by

$$y_1 = a \sin 2\pi\, n_1 t \quad ...(1)$$

and

$$y_2 = b \sin 2\pi\, n_2 t \quad ...(2)$$

The resultant displacement is given by

$$y = y_1 + y_2$$

$$y = a \sin 2\pi\, n_1 t + b \sin 2\pi n_2 t$$

$$y = a \sin 2\pi\, n_1 t + b \sin 2\pi\, [n_1 - (n_1 - n_2)\, t]$$

$$y = a \sin 2\pi n_1 t + b [\sin 2\pi n_1 t \cos 2\pi (n_1 - n_2) t - \cos 2\pi n_1 t \sin 2\pi [(n_1 - n_2) t] \quad ...(3)$$

$$y = \sin 2\pi n_1 t [a + b \cos 2\pi (n_1 - n_2) t] - \cos (2\pi n_1 t) [b \sin 2\pi (n_1 - n_2) t]$$

Take $\quad a + b \cos 2\pi (n_1 - n_2) t = A \cos \theta$

and $\quad b \sin 2\pi (n_1 - n_2) t = A \sin \theta$

$\therefore \quad y = A \sin (2\pi n_1 t - \theta)$

Here
$$\tan \theta = \frac{b \sin 2\pi (n_1 - n_2) t}{a + b \cos 2\pi (n_1 - n_2) t} \quad ...(4)$$

and
$$A = \sqrt{a^2 + b^2 + 2ab \cos 2\pi (n_1 - n_2) t} \quad ...(5)$$

- From Eq. (4), it is evident that the phase angle θ changes with respect to time.
- From Eq. (5), it is evident that the amplitude of the resultant vibration also changes with time.

(*i*) Let us calculate the interval between two successive maxima.

When $2\pi (n_1 - n_2) t = 2K\pi$ where $K = 0, 1, 2, 3...$ etc., the resultant amplitude,

$$A = (a + b) \quad ...(6)$$

The resultant amplitude is maximum, when

$$t = \frac{K}{(n_1 - n_2)}$$

i.e., at the time instants, $0, \frac{1}{(n_1 - n_2)}, \frac{2}{(n_1 - n_2)}, \frac{3}{(n_1 - n_2)}$ and so on.

$\therefore$ Time interval between two maxima $= \frac{1}{(n_1 - n_2)}$ sec.

The frequency of maxima $= (n_1 - n_2)$ per sec.

The intensity of sound is directly proportional to the square of the amplitude.

So the maximum intensity of sound will be heard at the insants,

$$0, \frac{1}{(n_1 - n_2)}, \frac{2}{(n_1 - n_2)}, \frac{3}{(n_1 - n_2)} \text{... etc.}$$

(*ii*) Let us calculate the interval between two successive minima.

When $2\pi (n_1 - n_2) t = (2K + 1) \pi$

where $\quad K = 0, 1, 2, 3,$ etc.

the resultant amplitude,

$$A = (a - b).$$

The resultant amplitude is minimum when,

$$t = \frac{(2K + 1)}{2(n_1 - n_2)}$$

i.e., at the instants,

$$\frac{1}{2(n_1 - n_2)}, \frac{3}{2(n_1 - n_2)}, \frac{5}{2(n_1 - n_2)} \text{... etc.}$$

The interval between two minima $= 1/(n_1 - n_2)$ sec.

The frequency of minima $= (n_1 - n_2)$ per sec.

Thus, the maxima and minima occur alternately after equal intervals of time $\frac{1}{2(n_1 - n_2)}$.

The time interval between two successive maxima or between two successive minima is $\frac{1}{(n_1 - n_2)}$.

The number of beats produced per second $= (n_1 - n_2)$.

13.4. APPLICATIONS OF THE PHENOMENON OF BEATS

(1) *To determine the frequency of a tuning fork:*

Take a standard tuning fork A of frequency N. The frequency of the tuning fork B *is* unknown. The two tuning forks are sounded together. Find the number of beats produced per second. Let it be n. Then the frequency of B is $N \pm n$. The fork B is now slightly loaded with wax, thus reducing its frequency. The two forks are now sounded together. If the number of beats per second increases, the frequency of B is $N - n$. If the number of beats per second decreases, the frequency of B is $N + n$.

(2) Beats are made use of to tune two notes to be in unison. For example, the note produced by a sonometer wire can be tuned to be in unison with that of a tuning fork by observing beats. If the frequencies of the sounds emitted by them are not equal, then beats will be heard. The length of the sonometer wire is adjusted until the beats disappear. The frequencies of the notes emitted by the sonometer wire and the fork are now exactly equal.

EXERCISE XIII

1. What are beats? How are beats produced? Give a mathematical treatment of beats and point out their important uses.

CHAPTER 14 Intensity of Sound

14.1. ACOUSTIC INTENSITY (I)

Definition : *Acoustic intensity of a sound wave is defined as the average power transmitted per unit area in the direction of propagation of the wave.*

14.2. TO SHOW THAT THE ACOUSTIC INTENSITY AT A POINT IS GIVEN BY $I = (P_{MAX})^2/2\rho C$

Let ρ = density of the medium

C = velocity of sound in the medium and

P_{max} = pressure maximum.

Instantaneous power per unit area = Instantaneous pressure (P) × Instantaneous particle velocity (v).

Averae power per unit area measures the acoustic intensity.

∴ Acoustic intensity,

$$I = \frac{1}{T}\int_0^T Pv\, dt \qquad ...(1)$$

For harmonic plane acoustic wave propagation in the positive *x* direction, particle displacement is expressed as

$$u = A\cos(\omega t - kx) \qquad ...(2)$$

Particle velocity,

$$v = \frac{du}{dt} = -\omega A\sin(\omega t - kx) \qquad ...(3)$$

$$\frac{du}{dx} = Ak\sin(\omega t - kx) \qquad ...(4)$$

Acoustic pressure,

$$P = -\rho C^2\left(\frac{du}{dx}\right) = -\rho C^2\,[Ak\sin(\omega t - kx)]$$

But $kC = \omega$

∴ $$P = -\rho CA\,\omega\sin(\omega t - kx) \qquad ...(5)$$

Substituting the values of P and v in Eq. (1),

$$I = \frac{1}{T}\int_0^T [-\rho CA\,\omega\sin(\omega t - kx)][-\omega A\sin(\omega t - kx)]dt$$

$$= \frac{\rho C\omega^2 A^2}{T}\int_0^T [\sin(\omega t - kx)]^2\, dt$$

$$= \frac{\rho C \omega^2 A^2}{T} \int_0^T (\cos^2 kx \sin^2 \omega t + \sin^2 kx \cos^2 \omega t - \frac{1}{2} \sin 2\omega t \sin 2kx) dt$$

$$= \frac{\rho C \omega^2 A^2}{T} \left(\frac{T}{2} \right)$$

$$I = \frac{1}{2} \rho C \omega^2 A^2 \quad ...(6)$$

Since $P = -\rho CA\omega \sin(\omega t - kx)$

$$P_{max} = -\rho CA\omega$$

$$\therefore \quad I = \frac{(P_{max})^2}{2\rho C} \quad ...(7)$$

Root mean square value of pressure,

$$P_{RMS} = \frac{P_{max}}{\sqrt{2}} \quad ...(8)$$

$$I = \frac{\left(\frac{P_{max}}{\sqrt{2}} \right)^2}{\rho C}$$

or

$$I = \frac{P^2_{RMS}}{\rho C} \quad ...(9)$$

14.3. MEASUREMENT OF INTENSITY OF SOUND-DECIBEL AND BEL

The absolute intensity of a sound wave is measured in Wm^{-2}. The lowest intensity of sound (at 1000 Hz) to which a normal human ear can respond is 10^{-12} Wm^{-2}. This standard intensity is called *threshold of audibility.* This is the 'zero' or 'standard' of intensity.

The ratio of intensity of a sound to the standard intensity is called *intensity level* or *relative intensity* of sound.

The relative intensity or intensity level of a sound wave having intensity I is given by

$$\text{Intensity Level} = \log_{10} \left[\frac{I}{I_0} \right]$$

The unit for intensity level is 'bel'.

Definition of Bel. *Bel* is the intensity level of a sound whose intensity is ten times the standard intensity.

Whenever the intensity of sound increases by a factor of 10, the increase in intensity is said to be 1 bel.

Decibel. In practice, bel is a large unit. Hence, another unit called decibel (dB) is used. When the intensity increases by a factor $10^{0.1}$, the increase in intensity is 0.1 bel or 1 decibel.

1 bel = 10 dB.

$$\text{Intensity level} = \log_{10} \left[\frac{I}{I_0} \right] \text{bel} = 10 \log_{10} \left[\frac{I}{I_0} \right] dB$$

Here, I is the intensity of sound and I_0 the standard intensity.

Example 1. *Show that a 26% change in intensity alters the sound level by I dB.*

Sol. Intensity level= $10\log_{10}\left[\frac{I}{I_0}\right] dB$

If intensity level is 1 dB, $1 = 10\log_{10}\left[\frac{I}{I_0}\right]$

$$\log_{10}\left[\frac{I}{I_0}\right] = \frac{1}{10} = 0.1$$

$$\frac{I}{I_0} = 10^{0.1} = 1.26$$

$$\therefore \quad \mathbf{I = 1.26\, I_0}$$

Hence 26% change in intensity corresponds to 1 dB of difference in intensity levels.

14.4. LOUDNESS OF SOUND

The **loudness** *of a sound relates to the degree of sensation depending on the intensity of sound and the sensitiveness of the ear.*

It does not increase proportionally with intensity but more nearly as its logarithm.

According to Weber and Fechner's Law,

$$L \propto \log I.$$

Here, L = sensation of loudness and I = intensity of sound.

14.5. NOISE POLLUTION

Unwanted sound is called noise. Pollution produced in the air due to noise is called *noise pollution.*

Noise Sources. The principal sources of noise may be classified as surface transportation noise, aircraft noise, industrial noise, noise due to construction sites and noise at house.

There are three types of noises.

(*a*) Air-borne noise, (*b*) Structure-borne noise and (*c*) Inside noise.

The prevention of the transmission of noise inside or outside the hall is called *sound insulation.*

(*a*) Air-borne noise : The noise carried by air that enters into the hall from outside through open windows, doors and ventilators is called *air-borne noise*.

Sound insulation is achieved for air-borne noise by the following methods :

1. Very heavy glasses are used in doors, windows and ventilators.

2. The windows and doors are arranged such that outside noises are prevented from entering the room.

3. Double doors with absorbing materials in between are used.

4. Openings for pipes and ventilators are avoided.

(*b*) Structure-borne noise : Structure-borne noises are produced inside the hall itself and conveyed through the structure. Any loose fitting inside the building vibrates and produces noises. Other activities such as drilling, hammering, operating machinery inside the building, moving of furniture etc., also produce structure-borne noise.

Sound insulation for the reduction of structure-borne noise is done in the following ways:

1. By breaking the continuity of pipelines etc., by interposing sound insulation materials.

2. By arresting vibrations of loose fittings.

3. It can be done by insulating the machinery and having carpets, rubber sheets, etc., on the floor.

(*c*) **Inside noise :** The noises which are produced inside the hall are called *inside noises.* They are produced due to machinery like television sets, automatic washers, air-conditioners, typewriters etc., in the hall.

Following methods are used for sound insulation of inside noise:

1. The walls, floor and ceiling of the room are padded with sound absorbing materials.
2. Working machineries are placed on absorbent pads.
3. Any engine inside the hall should be fitted on the floor with a layer of wood or felt between them.

Effects of Noise Pollution

1. Noise affects work efficiency.
2. It produces mental fatigue and irritation.
3. A person's hearing becomes damaged when he is prolongedly exposed for a long time to sounds of high intensity (140-150 dB).
4. Noise produces a change in bodily activity. The results are indigestion, sleeplessness, nervousness and nausea.

EXERCISE XIV

1. Deduce an expression for the intensity of sound in terms of pressure amplitude.
2. Define intensity and loudness of sound. Also define bel and decibed.
3. Write a note on noise pollution.

CHAPTER 15 Progressive Waves

15.1. EQUATION OF A PLANE PROGRESSIVE WAVE

Suppose the wave is travelling from left to right. The particle situated at the origin acts as the source of disturbance *i.e.*, the wave starts from O (Fig. 15.1).

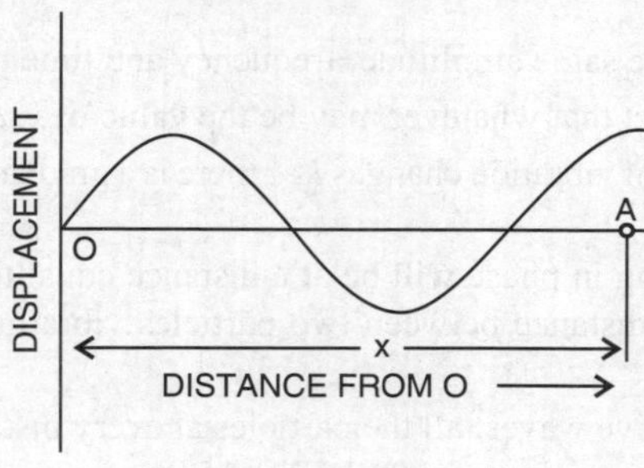

Fig. 15.1

The particle at O is executing S.H.M. Then each particle of the medium will execute similar S.H.M.

We measure the time from the instant when the particle O passes through its mean position in the positive direction. The equation of motion of the particle O is given by,

$$y = a \sin \omega t \qquad ...(1)$$

Here, a is the amplitude of the particle, y is the displacement after time t and ω is the angular velocity.

Consider another particle A at a distance x from the particle O to its right *i.e.*, $OA = x$.

Let the wave be travelling with velocity v along the positive direction of x-axis.

The displacement at A is given by

$$y = a \sin (\omega t - \alpha) \qquad ...(2)$$

Here, α is the phase difference between the particles O and A.

For a phase difference of 2π, the path difference is λ.

Suppose for a phase difference of α, the path difference is x.

$$\frac{\alpha}{x} = \frac{2\pi}{\lambda}$$

$$\therefore \qquad \alpha = \frac{2\pi x}{\lambda}$$

Also $$\omega = \frac{2\pi}{T} = \frac{2\pi v}{\lambda}$$

Substituting the values of α and ω in equation (2),

$$y = a\sin\left(\frac{2\pi vt}{\lambda} - \frac{2\pi x}{\lambda}\right)$$

$$y = a\sin\frac{2\pi}{\lambda}(vt - x) \qquad ...(3)$$

Equation (3) represents the equation for a simple harmonic wave.

Similarly for a particle at a distance x in the negative direction (*i.e.,* to the left of O), the equation for displacement is,

$$y = a\sin\frac{2\pi}{\lambda}(vt + x) \qquad ...(4)$$

15.2. PROPERTIES OF PROGRESSIVE WAVES

1. The particles of the medium vibrate simple harmonically along the direction of propagation of the wave.
2. All the particles have the same amplitude, frequency and time period.
   This is seen from the fact that whatever may be the value of x, a and ω are the same.
3. As x changes, the phase of vibration changes *i.e.*, there is a gradual phase difference between the successive particles.
4. All the particles vibrating in phase will be at a distance equal to $n\lambda$. Here $n = 1, 2, 3$ etc. It means the minimum distance between two particles vibrating in phase is equal to the wavelength.
5. In longitudinal progressive waves, all the particles at every instant are in a straight line in the form of condensations and rarefactions (Fig. 15.2).

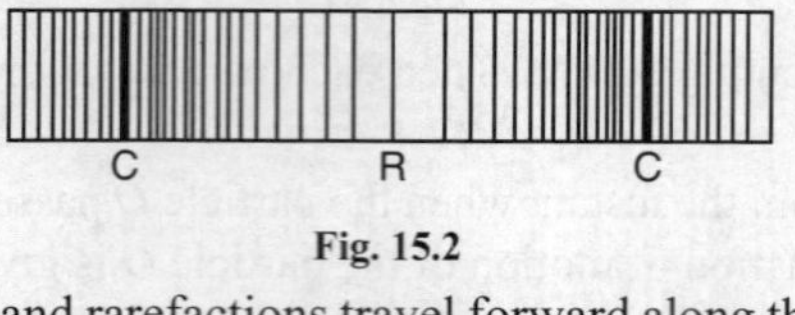

Fig. 15.2

   These compressions and rarefactions travel forward along the wave and transfer energy in the direction of propagation of the wave.
6. The velocity of the particle is maximum at their mean position and it is zero at their extreme positions.

EXERCISE XV

1. Obtain an expression for a progressive plane wave of the simple harmonic type.
2. Describe briefly the properties of longitudinal progressive waves.

16 CHAPTER Stationary Waves

16.1. STATIONARY WAVES

When two simple harmonic waves of the same amplitude, frequency and time period travel in *opposite directions* in a straight line, the resultant wave obtained is called a stationary or a standing wave. The formation of stationary waves is due to the superposition of the two waves on the particles of the medium.

Demonstration of Stationary Vibrations in Strings :

The Melde's experiment illustrates the formation and maintenance of *stationary vibrations* in strings. T is an electrically maintained tuning fork. One end of string PQ is attached to one prong of the tuning fork T. The string passes over a pulley and has a small weight W attached to its other end (Fig.16.1). When the fork is excited, the vibration of its prongs acts as a source of transverse vibrations.

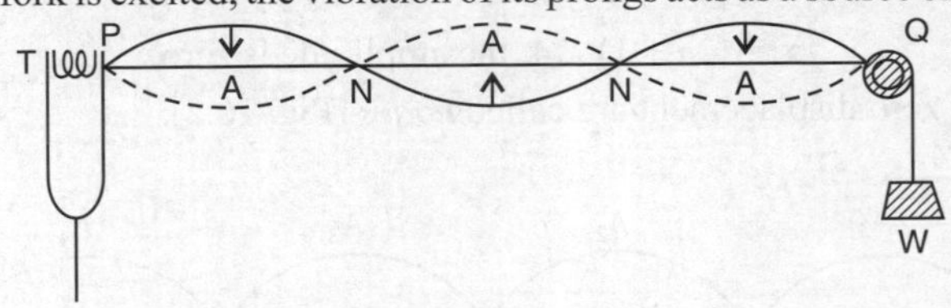

Fig. 16.1

The transverse waves thus set up in the string will travel along it with a velocity depending on its tension and linear density. The waves will then be reflected from the pulley and travel backwards.

Thus, in the string there will be formed two sets of identical waves travelling in opposite directions with the same velocity. These will give rise to stationary waves in the string having fixed positions for nodes and antinodes. In the figure, the string PQ is vibrating in three equal segments with two nodal points marked N, N, lying along the length between the end points P and Q, which are also *nodes*. At the centre of the vibrating loops, we have the antinodes marked A, A, A. The motions of alternate segments are in opposite directions as indicated by the vertical arrows.

16.2. EQUATION FOR A STATIONARY WAVE

Consider two plane simple harmonic waves of the same amplitude and wavelength travelling along the positive x-axis and negative x-axis respectively with the same velocity. Let the displacements due to the two waves be y_1 and y_2 respectively. The waves may be represented by

$$y_1 = a \sin \frac{2\pi}{\lambda}(vt - x) \quad \textbf{Incident wave} \qquad ...(1)$$

and

$$y_2 = a \sin \frac{2\pi}{\lambda}(vt + x) \quad \textbf{Reflected wave} \qquad ...(2)$$

Here a is the amplitude, v the velocity and λ the wavelength of the waves. We assume that *reflection occurs at a free boundary.*

The resultant displacement

$$y = y_1 + y_2$$

or
$$y = a \sin \frac{2\pi}{\lambda}(vt - x) + a \sin \frac{2\pi}{\lambda}(vt + x) \quad ...(3)$$

$$y = 2a \cos \frac{2\pi x}{\lambda} \sin \frac{2\pi vt}{\lambda} \quad ...(4)$$

The resultant vibration of a particle represents a simple harmonic wave of the same time period. The amplitude of the resultant wave is given by

$$A = 2a \cos \frac{2\pi x}{\lambda} \quad ...(5)$$

The resultant amplitude is a function of x.

16.3. PROPERTIES OF STATIONARY LONGITUDINAL WAVES

(1) Equation (5) shows that the amplitude is a function of x and so it varies in different positions.

Nodes. The positions where the amplitude of vibration is zero are called *Nodes*.

At points where x is such that

$$\cos \frac{2\pi x}{\lambda} = 0, \text{ i.e., } \frac{2\pi x}{\lambda} = (2n+1)\frac{\pi}{2}$$

or $x = (2n + 1)\, \lambda/4$, the amplitude is zero.

These points of zero displacement are called *nodes* (Fig. 16.2).

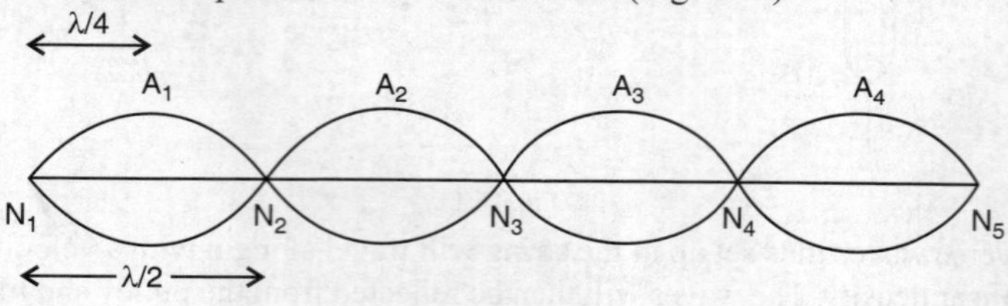

Fig. 16.2

- The distance between successive nodes is $\lambda/2$.
- At the nodes, the amplitude of vibration of the particles is zero; the energy is zero; the pressure is maximum (in the case of longitudinal waves).

Antinodes. At places where x is such that

$$\cos \frac{2\pi x}{\lambda} = \pm 1,$$

the amplitude of the resultant wave is maximum. These points of maximum amplitudes are called *Antinodes*. At such points $2\pi x/\lambda = n\pi$, where n is zero or an integer. Hence $x = n\lambda/2$.

$$\left.\begin{array}{l}\text{The distance between}\\ \text{two successive antinodes}\end{array}\right\} = \left[(n+1)\frac{\lambda}{2} - n\frac{\lambda}{2}\right] = \frac{\lambda}{2}.$$

The distance between a node and the adjacent antinode is $\lambda/4$.

Between two nodes there is an antinode and *vice versa.*

(2) Every particle (*except at the nodes*) executes S.H.M. with the same period as the component waves. But the amplitude varies continuously along the line assuming + ve, zero and – ve values.

(3) The medium splits up into segments. All points between two nodes vibrate in the same phase. Particles on either side of a node or antinode vibrate in opposite phase.

(4) Condensations (crests) or rarefactions (troughs) do not advance as they do in a progressive wave. Consequently there is no advance of wave energy in either direction, *i.e., the transmission of energy across any plane is zero.*

(5) The arrangement of the particles at any instant is a sine curve but twice in a period all particles have zero displacement and the curve shrinks to a straight line.

(6) The instantaneous form repeats at intervals of λ. Particles separated by an integral multiple of a wavelength have the same displacement, velocity and acceleration at every instant.

Distinction between Progressive and Stationary Waves.

Progressive Waves	Stationary Waves
1. A progressive wave advances continuously through the medium,	A stationary wave neither advances nor recedes through the medium. It remains confined to the limited space in which it is produced.
2. All particles pass through their mean positions with the same maximum velocity but one after the other.	All particles pass through their mean positions simultaneously but with different maximum velocities.
3. All particles of the medium undergo the same changes of pressure and density but one after the other	The changes of pressure and density are the maximum at the nodal points and zero at the antinodal points but occur simultaneously at all points.
4. There is a regular transference of energy across every section of the medium.	There is no transference of energy across any section of the medium.

EXERCISE XVI

1. Give an analytical treatment of the formation of stationary waves and indicate their main characteristics.

17 Determination of Frequency

CHAPTER

17.1. MELDE'S EXPERIMENT

Experimental arrangement: One end of a long string is connected to the prong of an electrically maintained tuning fork. The other end of the string passes over a frictionless pulley and is connected to a scalepan. The tension in the string can be varied by placing different weights in the pan. The distance between the tuning fork and the pulley can be adjusted (Fig. 17.1).

(1) Transverse mode of vibration

The tuning fork vibrates perpendicular to the length of the string (Fig. 17.1).

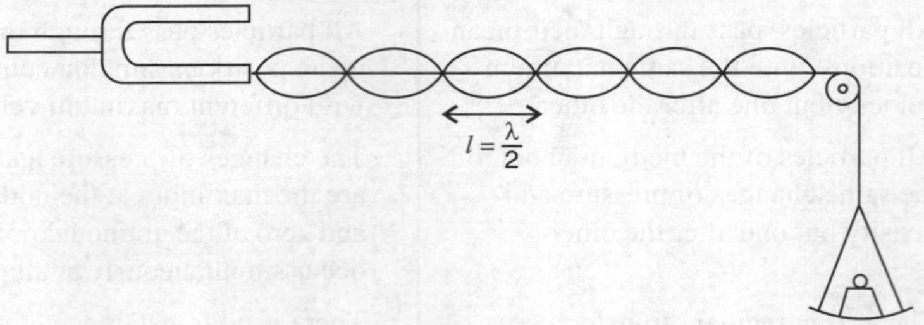

Fig. 17.1

The length of the string is adjusted until an even number of loops is clearly formed. The length of a known number of loops is measured and the length of a loop l is found. The tension (T) is the weight in the pan together with the weight of the pan. The linear density (m) is found by weighing 10 metres of the string.

In this case, the frequency of vibration of the string = the frequency of the tuning fork.

$\therefore$ The frequency of the tuning fork $= \frac{1}{2l}\sqrt{\frac{T}{m}}$.

(2) Longitudinal mode of vibration

The tuning fork vibrates along the direction of the length of the string (Fig. 17.2).

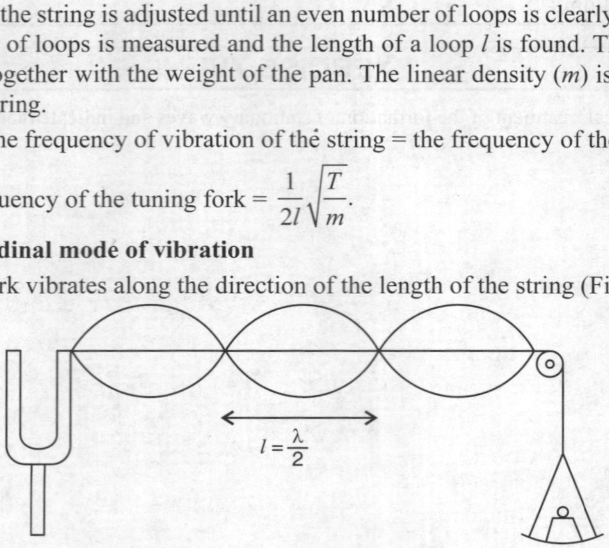

Fig. 17.2

Relation between the frequency of the string and frequency of the fork

In this case, for one complete vibration of the tuning fork, the string completes half vibration. The frequency of the fork is twice the frequency of vibration of the string.

The vibratory stages of the prong and the string are depicted in the diagrams of Fig. 17.3.

Stage (*i*) : When the prongs are farthest apart, the amount of sag in the string is maximum.

Stage (*ii*) : The prongs move inwards, the string begins to tighten and becomes horizontal when the prongs are in their nearest position. But the string has a tendency to move upwards.

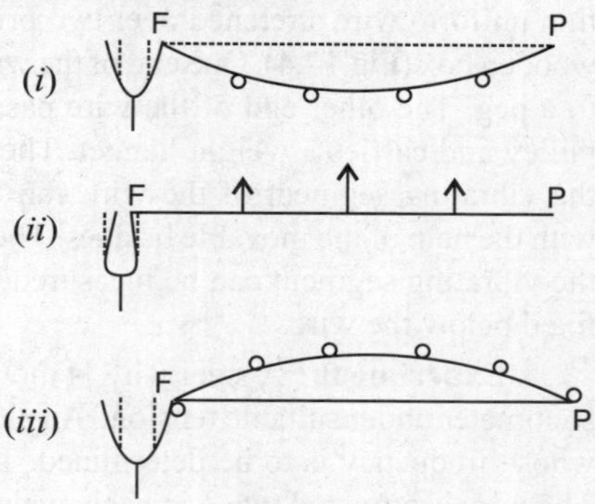

Fig. 17.3

Stage (*iii*) : The prongs again separate out and the string moves upward because of upward momentum. In the extreme position, the string is in its position of maximum upward displacement.

This shows that for every complete oscillation of the fork, the string makes one half of a complete oscillation.

Experiment : The length of the string is adjusted until an even number of loops is clearly formed. The length of a known number of loops is measured and the length of a loop l is found. The tension (T) is the weight in the pan together with the weight of the pan. The linear density (m) is found by weighing 10 metres of the string.

$$\text{Frequency of the string} = \frac{1}{2l}\sqrt{\frac{T}{m}}.$$

Frequency of the fork is twice this value.

$$\therefore \text{ The frequency of the tuning fork} = \frac{1}{l}\sqrt{\frac{T}{m}}$$

Example 1: *Find the tension needed to produce stationary waves with four loops in a string one metre long and 0.5 gm in weight fixed to a tuning fork of frequency 200 Hz when the prongs of the fork are vibrating perpendicular to the string.*

Solution: l = length of a loop = 0.25 m, m = 0.0005 kg/m; n = 200 Hz, T = ?

$$n = \frac{1}{2l}\sqrt{\frac{T}{m}}$$

$$T = 4l^2n^2\, m = 4(0.25)^2(200)^2\,(0.0005) = 5N = 0.5102\ kgf.$$

Example 2: *Calculate the frequency of the fundamental mode of a string 1 m long weighing 2 g when loaded with 40 g in Melde's string experiment (longitudinal mode).*

Solution : $l = 1$ m, $m = 2 \times 10^{-3}\ kg\ m^{-1}$, $T = (40 \times 10^{-3} \times 9.8)\ N = 0.392\ N.$

$$\text{Frequency of the string} = \frac{1}{2l}\sqrt{\frac{T}{m}} = \frac{1}{2\times 1}\sqrt{\frac{0.392}{2\times 10^{-3}}} = 7\text{Hz}.$$

Frequency of the tuning fork is twice this value.

∴ The frequency of the tuning fork = 14 Hz.

17.2. A.C. FREQUENCY MEASUREMENT USING SONOMETER

The frequency of the alternating current mains in the laboratory can be determined using a sonometer.

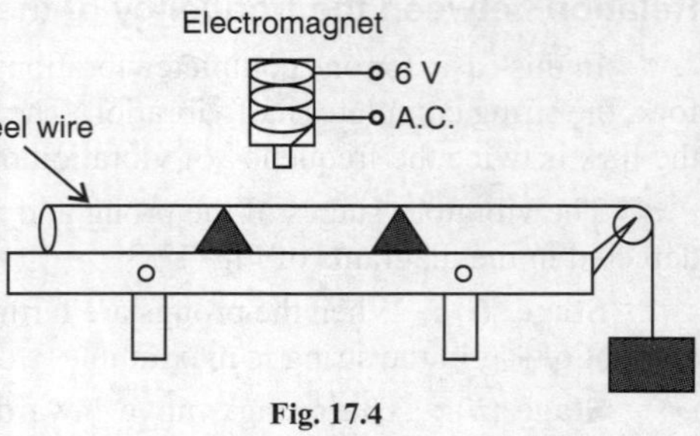

Fig. 17.4

Description : A sonometer consists of a thin uniform wire stretched over two bridges on a wooden box (Fig.17.4). One end of the wire is fixed to a peg. The other end of the wire passes over a pulley and carries a weight hanger. The length of the vibrating segment of the wire can be altered with the help of the movable bridges. The length of the vibrating segment can be measured by a scale fixed below the wire.

Experiment : A steel wire is mounted on a sonometer under suitable tension. An electromagnet is excited by the low voltage alternating current whose frequency is to be determined. The electromagnet is placed just above the sonometer wire. The wire is attracted twice in each cycle.

A small paper rider is placed on the wire. The length of the wire is adjusted until the paper rider placed at the centre of the vibrating segment is thrown off. The length of the vibrating segment (l) is measured. The experiment is repeated for different tensions. The readings are tabulated as shown below:

No.	*Tension* (T)	*Length of the vibrating segment l*	$\frac{\sqrt{T}}{l}$

The mean value of $\frac{\sqrt{T}}{l}$ is found.

The mass per unit length of the wire is determined by finding the mass of a given length of the wire.

Calculation : The frequency of sonometer wire is

$$n = \frac{1}{2l}\sqrt{\frac{T}{m}}.$$

During both the positive peak and the negative peak of the A.C., the wire is pulled by the electromagnet. So the wire vibrates twice, for each cycle of the A.C.

The frequency of the A.C. supply is given by

$$f = \frac{n}{2}.$$

Hence, the frequency of A.C. mains is calculated.

For Brass or Copper wire

The ends of the secondary of a transformer are connected to the two ends A and B of the wire (Fig. 17.5). The P.D. at the ends of the transformer should be about 6 volts. The wire is set between the poles of a powerful horse shoe magnet or the opposite poles of two equal bar magnets, so that the magnetic field is in a horizontal plane and at right angles to the length of the wire.

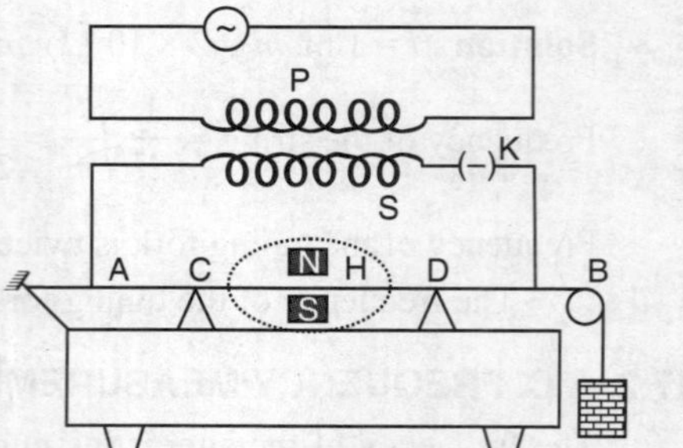

Fig. 17.5

The sonometer wire is subjected to a load of 150 gm. The wire is thrown into clear vibrations. The length of the wire is adjusted by moving the bridges *C* and *D*, till a single loop is seen between them with maximum amplitude. The distance *CD* is measured with a metre scale. The experiment is repeated for loads of 200, 250, 300, 350, 400,450 and 500 gm.

The length *CD* is measured in each case.

The frequency of the wire $n = \frac{1}{2l}\sqrt{\frac{Mg}{m}}$.

In this arrangement, the frequency of the wire is equal to the frequency of the A.C. mains.

EXERCISE XVII

1. Describe Melde's string method to determine the frequency of an electrically maintained tuning fork by transverse and longitudinal modes.
2. Explain how the A.C. frequency is measured using sonometer.

18 Motion in a Vertical Circle

CHAPTER

18.1. MOTION IN A VERTICAL CIRCLE

Suppose a body of mass m is rotated in a vertical circle of radius R (Fig. 18.1). The motion is circular but *not* uniform, since the body speeds up while coming down and slows down while going up.

The forces acting on the body at any instant are its weight $\vec{mg}$ directed vertically downward, and the tension $\vec{T}$ in the string directed radially inward.

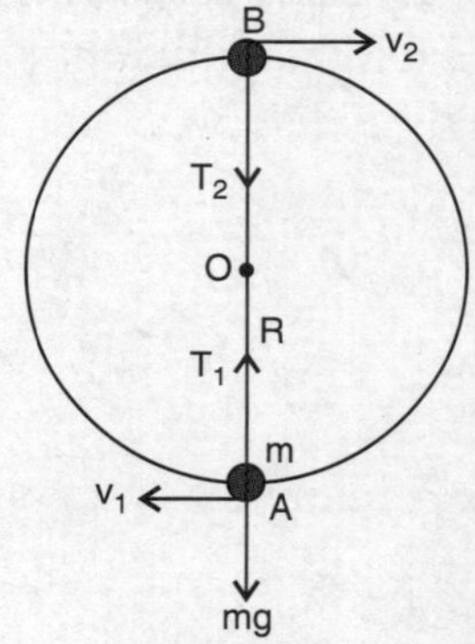

Fig. 18.1

(*i*) When the body is in the lowermost position *A*

Let v_1 be the velocity of the body while it passes the lowest point A.

The tension in the string in the upward direction = T_1.

The weight mg acts in the downward direction.

$$\therefore \qquad T_1 - mg = \frac{mv_1^2}{R}$$

$$\text{or} \qquad T_1 = \frac{mv_1^2}{R} + mg \qquad \text{...(1)}$$

(*ii*) When the body is in the topmost position *B*

At the highest point B, the velocity is v_2 and the tension in the string is T_2.

$$\therefore \qquad T_2 + mg = \frac{mv_2^2}{R}$$

$$\text{or} \qquad T_2 = \frac{mv_2^2}{R} - mg \qquad \text{...(2)}$$

Critical velocity. From Eq. (2) it is clear that if v_2 decreases, the tension T_2 in the string also decreases and becomes zero at a certain minimum velocity. This particular velocity is called *critical velocity*. Thus putting $v_2 = v_c$ and $T_2 = 0$ in Eq. (2), we get

$$0 = \frac{mv_c^2}{R} - mg$$

$$\text{or} \qquad v_c = \sqrt{gR} \qquad \text{...(3)}$$

If the velocity of the body at highest point becomes less than the critical velocity, the body can not continue motion on the whole vertical circle, but will leave the vertical circle at certain point of path.

To find the minimum velocity at the bottom of the circle so that the body can loop a loop.

In order that the velocity v_2 at the top is not below the critical velocity $\sqrt{gR}$, the minimum velocity v_1 at the bottom should be such that v_2 becomes equal to $\sqrt{gR}$ at the top.

$$AB = \text{diameter of circle} = 2R.$$

Assuming zero potential energy at lowest position, we have

(Kinetic energy + potential energy) at A = (Kinetic energy + potential energy) at B

$$\therefore \quad \frac{1}{2}mv_1^2 = mg(2R) + \frac{1}{2}mv_2^2$$

$$v_1^2 = 4gR + v_2^2$$

But $$v_2^2 = gR$$

$$\therefore \quad v_1^2 = 4gR + gR = 5gR$$

or $$v_1 = \sqrt{5gR}$$

Therefore while "looping the loop" the minimum velocity at the bottom should be $\sqrt{5gR}$.

EXERCISE 18

1. Discuss the motion of a particle in a vertical circle and show that the minimum velocity at the bottom of the circle should be more than $(5gR)^{1/2}$, so that the body can loop a loop.
2. A particle is moving in a vertical circle. Find an expression for its critical speed. Hence explain why the water does not spill, out of a bucket full of water, rotating in a vertical circle, even when the bucket is completely inverted.

19 CHAPTER Systems of Particles

19.1. CENTRE OF MASS

Definition : Consider the motion of a system consisting of a large number of particles. There is one point in it which behaves as though the entire mass of the system were concentrated there and all the external forces were acting at this point. This point is called the *centre of mass* of the system.

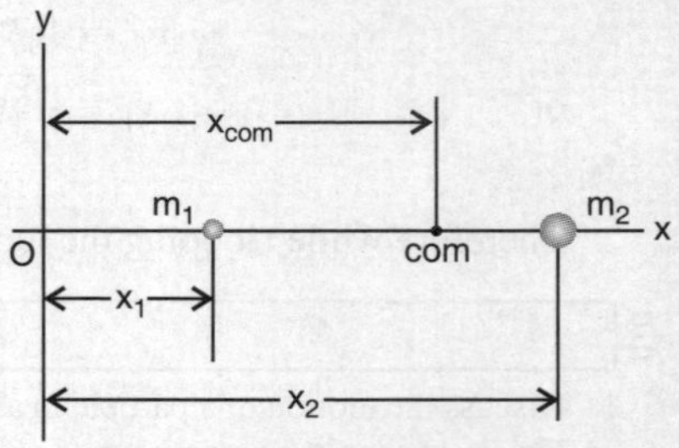

Fig. 19.1

Centre of Mass of two-particle system

Consider two particles of masses m_1 and m_2 (Fig. 19.1) Their x coordinates are x_1 and x_2.

We define the position of the center of mass (com) as

$$x_{com} = \frac{m_1x_1 + m_2x_2}{m_1 + m_2} = \frac{m_1x_1 + m_2x_2}{M} \quad ...(1)$$

Here, $M = m_1 + m_2$ = total mass of the system.

We can extend this equation to a more general situation in which n particles are strung out along the x axis.

Then, the location of the center of mass is

$$x_{com} = \frac{m_1x_1 + m_2x_2 + m_3x_3 + ... + m_nx_n}{M} = \frac{1}{M}\sum_{i=1}^{n} m_ix_i. \quad ...(2)$$

Here, $M = m_1 + m_2 + ... + m_n$ = total mass of the system.

If the particles are distributed in three dimensions, the center of mass must be identified by three coordinates.

$$x_{com} = \frac{1}{M}\sum_{i=1}^{n} m_ix_i,\ y_{com} = \frac{1}{M}\sum_{i=1}^{n} m_iy_i,\ z_{com} = \frac{1}{M}\sum_{i=1}^{n} m_iz_i. \quad ...(3)$$

Position Vector of the Centre of Mass

Let us consider a system of n particles of masses $m_1, m_2, ... m_n$ with position vectors $\boldsymbol{r}_1, \boldsymbol{r}_2, ..., \boldsymbol{r}_n$ relative to a fixed origin (Fig. 19.2).

The position of a particle at coordinates $x_i\ y_i$, and z_i is given by a position vector:

$$\vec{r}_i = x_i\hat{\text{i}} + y_i\hat{\text{j}} + z_i\hat{\text{k}}, \quad ...(4)$$

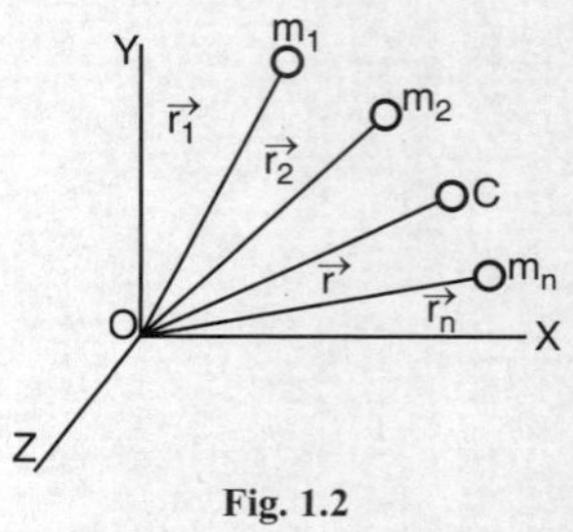

Fig. 1.2

The position of the center of mass of a system of particles is given by a position vector:

$$\vec{r}_{com} = x_{com}\hat{i} + y_{com}\hat{j} + z_{com}\hat{k} \quad ...(5)$$

The three scalar equations of Eq. (3) can now be replaced by a single vector equation,

$$\vec{r}_{com} = \frac{1}{M}\sum_{i=1}^{n} m_i \vec{r}_i, \quad ...(6)$$

Here, $M = \sum_{i=1}^{n} m_i$ = total mass of the system.

For a continuous body, we suppose that the body is formed of a large number of infinitesimal mass elements. Let dm be the mass of such an element at position (x, y, z). Then, the co-ordinates of the centre of mass are defined as

$$x_{com} = \frac{1}{M}\int x\, dm,\ y_{com} = \frac{1}{M}\int y\, dm,\ z_{com} = \frac{1}{M}\int z\, dm, \quad ...(7)$$

Example 1. *Three particles of masses m_1 = 1.2 kg, m_2 = 2.5 kg, and m_3 = 3.4 kg form an equilateral triangle of edge length a = 140 cm. Where is the center of mass of this three-particle system?*

Sol. We choose the x and y axes so that one of the particles (m_1) is located at the origin and the x axis coincides with one of the triangle's sides (Fig. 19.3).

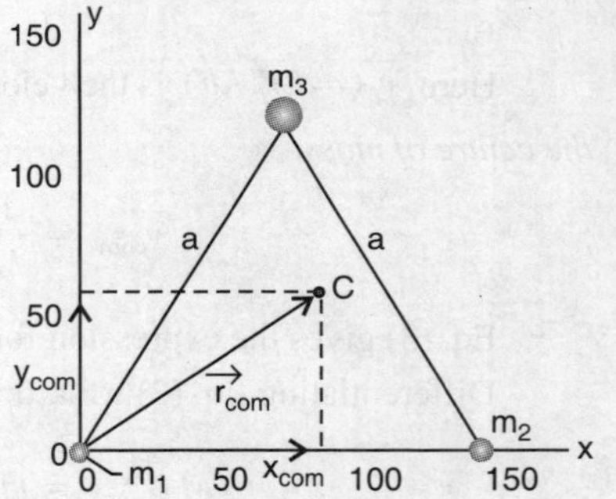

Fig. 19.3

The three particles then have the following coordinates:

Particle	Mass (kg)	*x (cm)*	*y (cm)*
1	1.2	0	0
2	2.5	140	0
3	3.4	70	121

The total mass M of the system is 7.1 kg.

The center of mass is located by the position vector $\vec{r}_{com}$

The coordinates of the center of mass C are

$$x_{com} = \frac{1}{M}\sum_{i=1}^{3} m_i x_i = \frac{m_1x_1 + m_2x_2 + m_3x_3}{M}$$

$$= \frac{(1.2\text{ kg})(0) + (2.5\text{ kg})(140\text{ cm}) + (3.4\text{ kg})(70\text{ cm})}{7.1\text{ kg}}$$

$$= 83\text{ cm}$$

and

$$y_{com} = \frac{1}{M}\sum_{i=1}^{3} m_i y_i = \frac{m_1y_1 + m_2y_2 + m_3y_3}{M}$$

$$= \frac{(1.2\text{ kg})(0) + (2.5\text{ kg})(0) + (3.4\text{ kg})(121\text{ cm})}{7.1\text{ kg}}$$

$$= 58\text{ cm.}$$

In Fig. 19.3, the center of mass is located by the position vector $\vec{r}_{com}$, which has components x_{com} and y_{com}.

19.2. MOTION OF THE CENTRE OF MASS OF A SYSTEM OF PARTICLES EXPRESSIONS FOR THE VELOCITY AND ACCELERATION OF CENTRE OF MASS

Consider a system of n particles of masses $m_1, m_2, ... m_n$ having position vectors $\mathbf{r}_1, \mathbf{r}_2, ..., \mathbf{r}_n$ relative to some arbitrary origin (Fig. 19.4).

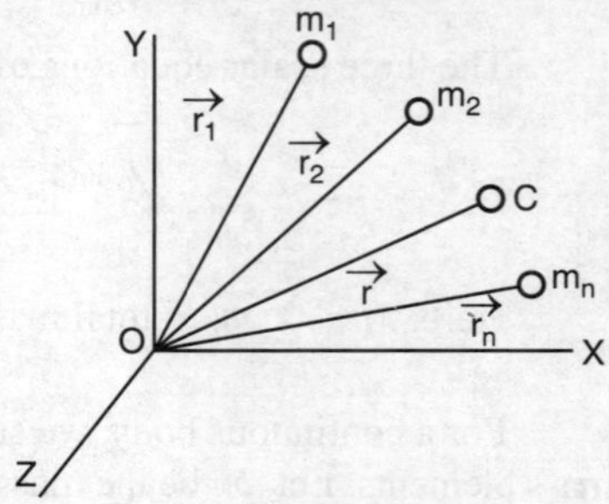

Fig. 19.4

For a system of n particles, we have,

$$M\vec{r}_{\text{com}} = m_1\vec{r}_1 + m_2\vec{r}_2 + m_3\vec{r}_3 + ... + m_n\vec{r}_n, \quad ...(1)$$

Here, M is the system's total mass and $\vec{r}_{\text{com}}$ is the vector locating the position of the system's center of mass.

Differentiating Eq. (1), w.r.t. time t, we get

$$M\vec{v}_{\text{com}} = m_1\vec{v}_1 + m_2\vec{v}_2 + m_3\vec{v}_3 + ... + m_n\vec{v}_n. \quad ...(2)$$

Here $\vec{v}_i\,(= d\vec{r}_i/dt)$ is the velocity of the ith particle, and $\vec{v}_{\text{com}}\,(= d\vec{r}_{\text{com}}/dt)$ is the *velocity of the centre of mass*.

$$\vec{v}_{\text{com}} = \frac{1}{M}[m_1\vec{v}_1 + m_2\vec{v}_2 + m_3\vec{v}_3 + ... + m_n\vec{v}_n] = \frac{1}{M}\sum_{i=1}^{n} m_i\vec{v}_i \quad ...(3)$$

Eq. (3) gives the expression for the velocity of the centre of mass.

Differentiating Eq. (2) w.r.t. time t, we get

$$M\vec{a}_{\text{com}} = m_1\vec{a}_1 + m_2\vec{a}_2 + m_3\vec{a}_3 + ... + m_n\vec{a}_n \quad ...(4)$$

Here $\vec{a}_i\,(= d\vec{v}_i/dt)$ is the acceleration of the ith particle, and

$\vec{a}_{\text{com}}\,(= d\vec{v}_{\text{com}}/dt)$ is the acceleration of the center of mass.

$$\therefore \quad \vec{a}_{\text{com}} = \frac{1}{M}[m_1\vec{a}_1 + m_2\vec{a}_2 + m_3\vec{a}_3 + ... + m_n\vec{a}_n] \quad ...(5)$$

Eq. (5) gives the expression for the acceleration of the centre of mass of the system.

Newton's Second Law for a System of Particles

From Newton's second law, $m_i\vec{a}_i$ is equal to the resultant force $\vec{F}_i$ that acts on the ith particle. Thus, we can rewrite Eq. (4) as

$$M\vec{a}_{\text{com}} = \vec{F}_1 + \vec{F}_2 + \vec{F}_3 + ... + \vec{F}_n \quad ...(6)$$

Among the forces that contribute to the right side of Eq. (6) will be forces that the particles of the system exert on each other (internal forces) and forces exerted on the particles from outside the system (external forces). By Newton's third law, the internal forces form third-law force pairs and cancel out in the sum. What remains is the vector sum of all the *external* forces that act on the system.

$$\therefore \quad \vec{F}_{\text{net}} = \mathbf{M}\vec{\boldsymbol{a}}_{\textbf{com}} \quad ...(7)$$

This equation is Newton's second law for the motion of the center of mass of a system of particles. $\vec{F}_{\text{net}}$ is the net force of *all external forces* that act on the system.

Thus, *the centre of mass of a system of particles moves as if all the mass of the system were concentrated at it and all the 'external' forces were applied at it.*

Eq. (7) is equivalent to three equations involving the omponents of $\vec{F}_{net}$ and $\vec{a}_{com}$ along the three coordinate axes. These equations are

$$F_{net.x} = Ma_{com.x} \quad F_{net.y} = Ma_{com.y} \quad F_{net.z} = Ma_{com.z} \qquad ...(8)$$

19.3. SYSTEMS WITH VARYING MASS: A ROCKET

Principle. In a rocket, the fuel is burnt in a combustion chamber (Fig. 19.5). The large quantity of the heat of combustion produced greatly raises the pressure inside the chamber. As a result, the burnt up gases issue out of an orifice at the tail end of the rocket in the form of a high velocity stream called the *JET.* In consequence, the rocket is propelled forwards (*i.e*, opposite to the direction of the jet), since the momentum lost by the jet of fuel gases must be equal to the momentum gained by the rocket.

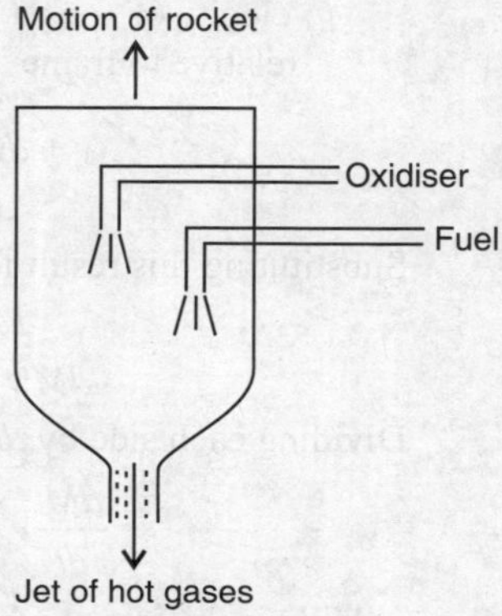

Fig. 19.5

Most of the mass of a rocket on its launching pad is fuel. Finally all fuel is burnt up. We handle the variation of the mass of the rocket as the rocket accelerates by applying Newton's second law, not to the rocket alone but to the rocket and its ejected combustion products taken together. The mass of *this* system does *not* change as the rocket accelerates.

Acceleration of the Rocket at any Instant

Let M be the mass of the rocket and v its velocity at time t, as seen from an inertial reference frame (Fig. 19.6 a).

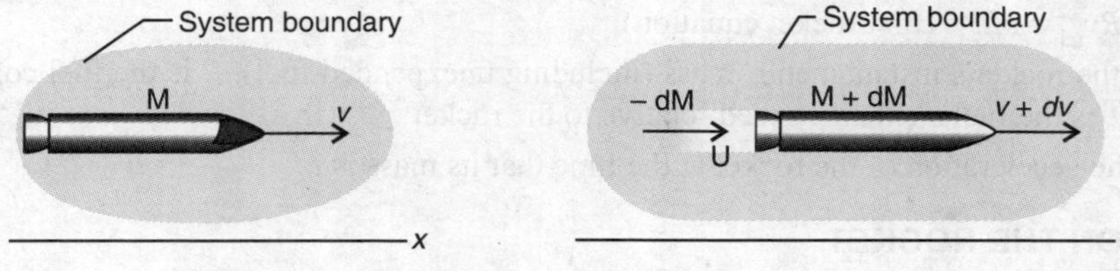

Fig. 19.6

We assume that no gravitational or atmospheric drag forces (air resistance) are acting on the rocket.

The linear momentum of the system at time t is

$$P_i = Mv \qquad ...(1)$$

Fig. 19.6*b* shows our system at time $t + dt$.

The system includes both the rocket and its exhaust gases released during interval dt.

The rocket now has velocity $v + dv$ and mass $M + dM$.

The change in mass dM is a *negative quantity*.

($-dM$) is the mass of the exhaust products released by the rocket during interval dt.

U is the exhaust velocity of the gases, relative to our inertial reference frame.

The linear momentum at time $t + dt = P_f =$

Linear momentum of the exhaust products released during interval dt +

Linear momentum of the rocket at the end of interval dt.

$$P_f = -dM\,U + (M + dM)(v + dv), \qquad ...(2)$$

The system is *closed* and isolated.

So the linear momentum of the system must be conserved during *dt;* that is,

$$P_i = P_f, \qquad ...(3)$$

$$Mv = -dM\,U + (M + dM)(v + dv) \qquad ...(4)$$

We can simplify Eq. (4) by using the relative velocity v_{rel} between the rocket and the exhaust products.

$$\begin{pmatrix}\text{velocity of rocket}\\ \text{relative to frame}\end{pmatrix} = \begin{pmatrix}\text{velocity of rocket}\\ \text{relative to products}\end{pmatrix} + \begin{pmatrix}\text{velocity of products}\\ \text{relative to frame}\end{pmatrix}$$

$$(v + dv) = v_{rel} + U,$$

$$U = v + dv - v_{rel}, \qquad ...(5)$$

Substituting this result for U into Eq. (4), we get

$$Mv = (-dM)(v + dv - v_{rel}) + (M + dM)(v + dv),$$

$$-dM\,v_{rel} = M\,dv \qquad ...(6)$$

Dividing each side by *dt* gives us

$$-\frac{dM}{dt}v_{rel} = M\frac{dv}{dt}. \qquad ...(7)$$

(dM/dt) is the rate at which the rocket loses mass.

Put $$\frac{dM}{dt} = -R.$$

Here, R is the (positive) mass rate of fuel consumption. dv/dt is the acceleration of the rocket $= a$

Eq. (7) becomes

$$Rv_{rel} = Ma = \text{(first rocket equation)}, \qquad ...(8)$$

M is the rocket's instantaneous mass (including unexpended fuel), R is the fuel consumption rate, and v_{rel} is the fuel's exhaust speed relative to the rocket.

a is the acceleration of the rocket at the time that its mass is M.

THRUST ON THE ROCKET

The term Rv_{rel} is the *thrust* of the rocket engine.

It depends only on design characteristics of the rocket engine, namely, the rate R at which it consumes fuel mass and the speed v_{rel} with which that mass is ejected relative to the rocket.

$\therefore$ Thrust on the rocket $= T = Rv_{rel}$ ...(9)

For a rocket, the rocket designer's aim is to make the thrust as large as possible. Inspection of equation (9) shows that this requires that the rocket eject as much mass per unit time as possible and that the exhaust velocity of the gases relative to the rocket be as high as possible.

Velocity of the Rocket at any Instant

The velocity of a rocket change as it consumes its fuel.

From Eq. (6) we have

$$dv = -v_{rel}\frac{dM}{M}.$$

Integrating leads to

$$\int_{v_i}^{v_f} dv = -v_{\text{rel}} \int_{M_i}^{M_f} \frac{dM}{M}.$$

Here, M_i is the initial mass of the rocket and M_f its final mass.

$$v_f - v_i = v_{\text{rel}} \ln \frac{M_i}{M_f} \quad \text{(second rocket equation)} \qquad ...(10)$$

$(v_f - v_i)$ is the increase in the speed of the rocket during the change in mass from M_i to M_f.

EXERCISE XIX

1. Define centre of mass. Deduce expressions for the velocity and acceleration of the centre of mass of a system of particles.
2. Explain the motion of a rocket. Derive expressions for thrust and velocity in rocket motion.
3. Find an expression for velocity of rocket in terms of exhaust speed of gases, initial and final masses.

20 CHAPTER Centre of Gravity

20.1. INTRODUCTION

Definition : *The centre of gravity of a body is the point at which the resultant of the weights of all the panicles of the body acts, whatever may be the orientation of the body. The total weight of the body may be supposed to act at its centre of gravity.*

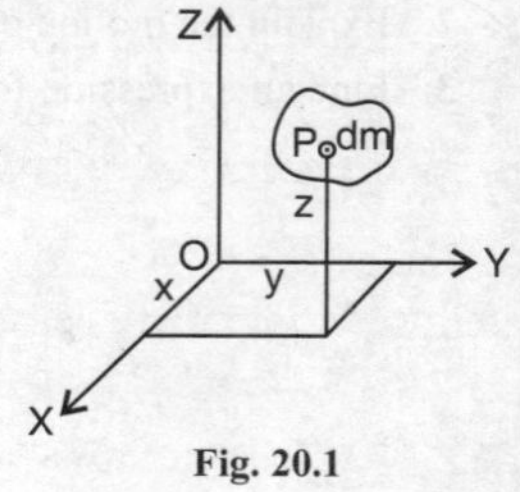

Fig. 20.1

Suppose the particles $A, B, C,$ .... of a body have masses $m_1, m_2, m_3,$ .... Let their coordinates in a rectangular cartesian coordinate system be $(x_1, y_1, z_1), (x_2, y_2, z_2)$.... (x_n, y_n, z_n).

Then, the coordinates of the centre of gravity G of the body are

$$\bar{x} = \frac{\sum m_n x_n}{\sum m_n}; \bar{y} = \frac{\sum m_n y_n}{\sum m_n}; \bar{z} = \frac{\sum m_n z_n}{\sum m_n};$$

Suppose an element P of the body has a mass dm (Fig. 20.1) and its coordinates are x, y, z. Then,

$$\bar{x} = \frac{\int x\, dm}{\int dm} = \frac{1}{M}\int x\, dm; \bar{y} = \frac{1}{M}\int y\, dm; \bar{z} = \frac{1}{M}\int z\, dm$$

Here, the integrals extend over all elements of the body, and

$M = \int dm$ = Total mass of the body.

20.2. CENTRE OF GRAVITY OF A RIGHT SOLID CONE

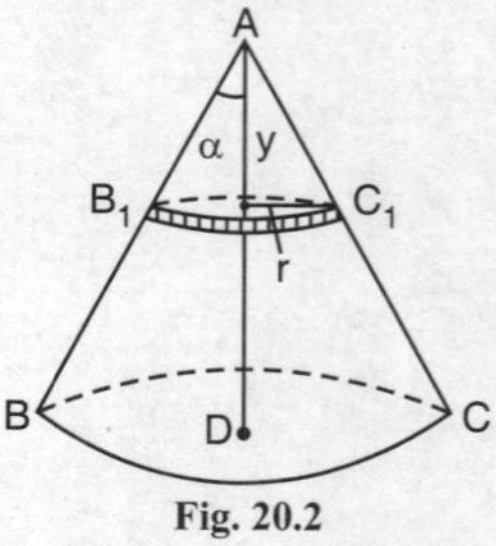

Fig. 20.2

Let ABC represent a solid cone of height h and semi-vertical angle α (Fig. 20.2). The cone may be considered to be made up of a large number of circular discs parallel to the base. The centre of gravity of each disc lies at its centre. Therefore, the C. G., of the cone should lie along the axis AD of the cone.

Consider a disc B_1C_1 of thickness dy at a distance y below the vertex A. If r is the radius of the disc, then

$$r = y \tan \alpha$$

Volume of the disc = Area × thickness = $\pi y^2 \tan^2 \alpha\, dy$

Mass of the disc = $dm = \pi y^2 \rho \tan^2 \alpha\, dy.$

where ρ = density of the cone.

The distance of the *C. G.* of the cone from the vertex is given by

$$\bar{y} = \frac{\int y\, dm}{\int dm} = \frac{\int_0^h \pi y^3 \rho \tan^2 \alpha\, dy}{\int_0^h \pi y^2 \rho \tan^2 \alpha\, dy} = \frac{\int_0^h y^3\, dy}{\int_0^h y^2\, dy} = \frac{3}{4}h.$$

Therefore, the *C. G.*, of the cone is along its axis at a distance of $\frac{3}{4}h$ from the vertex.

20.3. CENTRE OF GRAVITY OF A SOLID HEMISPHERE

Let *ABC* represent a solid hemisphere of radius *r*, centre *O* and density ρ (Fig. 20.3). Consider an elementary slice of the hemisphere with radius *y* and thickness *dx*, at a distance *x* from *O*.

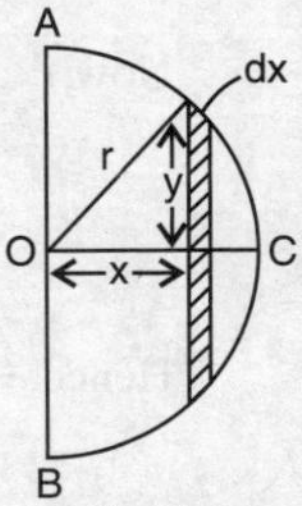

Fig. 20.3

$$\text{Volume of the slice} = \pi y^2\, dx = \pi(r^2 - x^2)\, dx.$$

$$\text{Mass of the slice} = dm = \rho\pi\,(r^2 - x^2)\, dx.$$

The distance of the *C. G.*, of the hemisphere from *O* is given by

$$\bar{x} = \frac{\int x\, dm}{\int dm} = \frac{\int_0^r x\,\rho\pi\,(r^2 - x^2)dx}{\int_0^r \rho\pi\,(r^2 - x^2)dx} = \frac{\int_0^r (r^2 x - x^3)dx}{\int_0^r (r^2 - x^2)dx}$$

$$\therefore \quad \bar{x} = \frac{3}{8}r.$$

Hence, the *C. G.*, of the solid hemisphere is on its axis at a distance $\frac{3}{8}r$ from the centre.

20.4. CENTRE OF GRAVITY OF A HOLLOW HEMISPHERE

Let *ACB* be a section of a hemisphere of radius *r*, centre *O* and surface density ρ [Fig, 20.4]. Imagine the surface of the hemisphere to be divided into slices like PQQ_1P_1 by planes parallel to *AB*. If $\angle POC = \theta$ and $\angle POQ = d\theta$, then

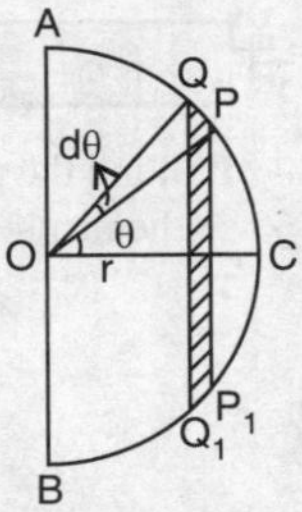

Fig. 20.4

$$\text{radius of the ring} = r \sin\theta$$

$$\text{width of the ring} = r\, d\theta$$

$$\text{Area of the ring} = 2\pi r \sin\theta.\, r d\theta$$

$$\therefore \quad \text{mass of the ring} = dm = 2\pi r^2\, \rho \sin\theta\, d\theta.$$

The *C. G.*, of this ring is at the centre of the ring at a distance $r\cos\theta$ from *O*.

The distance of the *C. G.*, of the hollow hemisphere from *O* is given by

$$\bar{x} = \frac{\int x\, dm}{\int dm} = \frac{\int_0^{\pi/2} (r\cos\theta)\, 2\pi r^2 \rho \sin\theta\, d\theta}{\int_0^{\pi/2} 2\pi r^2 \rho \sin\theta\, d\theta} = \frac{\int_0^{\pi/2} r\sin\theta\cos\theta\, d\theta}{\int_0^{\pi/2} \sin\theta\, d\theta}$$

$$\therefore \quad \bar{x} = r/2$$

The *C.* G., of a hollow hemisphere is on its axis at a distance *r*/2 from the centre, *i.e.*, the centre of gravity is at the mid point of the radius *OC*.

20.5. CENTRE OF GRAVITY OF A SOLID TETRAHEDRON

Let *ABCD* be the tetrahedron and G_1, the centre of gravity of the base *BCD* (Fig. 20.5). Let *h* be the altitude of the tetrahedron and ρ its density. Suppose the tetrahedron is divided into thin slices by planes parallel to the base *BCD*. Consider one such slice $B_1C_1D_1$ of thickness *dx* at a depth *x* below *A*. Let *S* be the area of the triangular base *BCD*. Then we have, $\frac{B_1C_1}{BC} = \frac{x}{h}$.

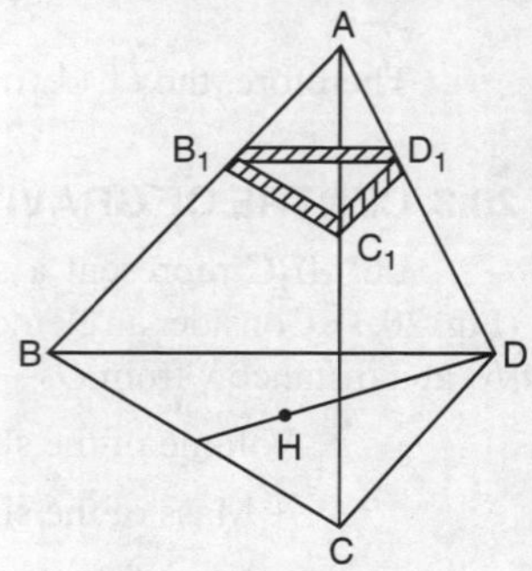

Fig. 20.5

If a_1 and a are the altitudes of triangles $B_1C_1D_1$ and BCD respectively,

$$\frac{a_1}{a} = \frac{x}{h}.$$

Now, area of $\Delta B_1C_1D_1 = \frac{1}{2} B_1C_1 \times a_1$

$$\text{Area of } \Delta BCD = \frac{1}{2} BC \times a = S.$$

Hence, $\frac{\text{Area of } \Delta B_1C_1D_1}{S} = \frac{B_1C_1}{BC} \times \frac{a_1}{a} = \frac{x^2}{h^2}$

$\therefore$ Area of $\Delta B_1C_1D_1 = Sx^2/h^2$

Volume of the slice $B_1C_1D_1 = Sx^2dx/h^2$

Mass of the slice $= dm = \rho Sx^2dx/h^2$

The distance of the centre of gravity of the tetrahedron from A is given by

$$\bar{x} = \frac{\int x\, dm}{\int dm} = \frac{\int_0^h x \rho S x^2 dx/h^2}{\int_0^h \rho S x^2 dx/h^2} = \frac{\int_0^h x^3 dx}{\int_0^h x^2 dx} = \frac{3}{4}h$$

Hence, the *C.G.*, of a uniform tetrahedron lies at a point G on the line AH such that $AG : GH = 3 : 1$.

EXERCISE XX

1. Find the position of C.G., in the following cases: (*i*) Right solid cone, (*ii*) solid hemisphere, (*iii*) hollow hemisphere and (*iv*) solid tetrahedron.

21 Centre of Pressure

CHAPTER

21.1. PRESSURE AND THRUST

Since a liquid possesses weight, it exerts force on all bodies in contact with it. The ratio between the small force δF and the area δA on which it acts gives the pressure. Thus, pressure = $\delta F/\delta A$.

$$\text{Pressure at a point} = \underset{\delta A \to 0}{\text{Limit}}\left(\frac{\text{Force}}{\text{Area}}\right) = \underset{\delta A \to 0}{\text{Limit}}\left(\frac{\delta F}{\delta A}\right) = \frac{dF}{dA}.$$

Unit of pressure is Nm^{-2}. It can be shown that the hydrostatic pressure due to a liquid column of density ρ at a depth h from the surface = $h\rho g$.

The total force exerted by a liquid column on the whole of the area in contact with it is called thrust. Thus, *thrust* = *pressure* × *area.* Unit : Newton (N). The thrust is always normal to the plane area.

21.2. THRUST ON A PLANE SURFACE IMMERSED IN A LIQUID AT REST

Consider a plane lamina of area S immersed in a liquid of density ρ (Fig. 21.1). Divide the area into a very large number of small elements. Let dS be an elementary area at a depth h below the free surface of the liquid.

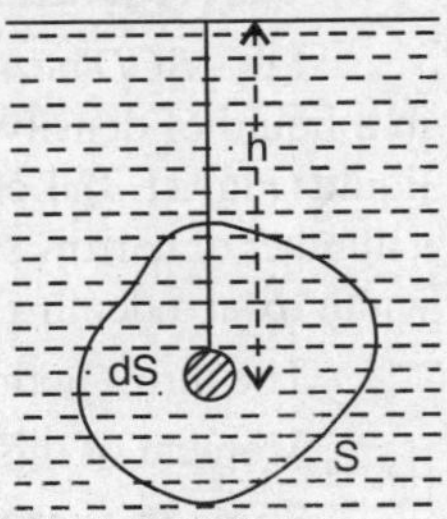

Fig. 21.1

Thrust on dS at right angles to it = $h\rho g\ dS$.

∴ the resultant thrust on the whole surface

$$= \int h\rho g\ dS = \rho g \int h\ dS.$$

Let $\bar{h}$ be the depth of the C.G., from the liquid surface.

From the theorem of moments,

or $h_1\ dS_1 + h_2\ dS_2 + h_3\ dS_3 + \ldots\ldots\ldots\ldots = \bar{h}\ S$

or $\int h\ dS = \bar{h} S$

∴ resultant thrust on the immersed plane surface = $\bar{h}\rho g \times S$

= pressure at C.G. of area × area of the plane.

21.3. CENTRE OF PRESSURE

We know that the liquid pressure acts *normally* at every point of the immersed area. The force acting on an elementary area like dS is $h\rho g\ dS$. The thrusts on different elements of the plane form a set of *like parallel forces.* All these parallel forces can be compounded into a resultant acting at some definite point on the plane area. This point is called the centre of pressure.

The centre of pressure of a plane surface in contact with a fluid is the point on the surface through which the line of action of the resultant of the thrusts on the various elements of the area passes.

Determination of Centre of pressure—General case

Consider a plane surface of area S immersed vertically in a liquid of density ρ. Let XY be the surface of the liquid (Fig. 21.2).

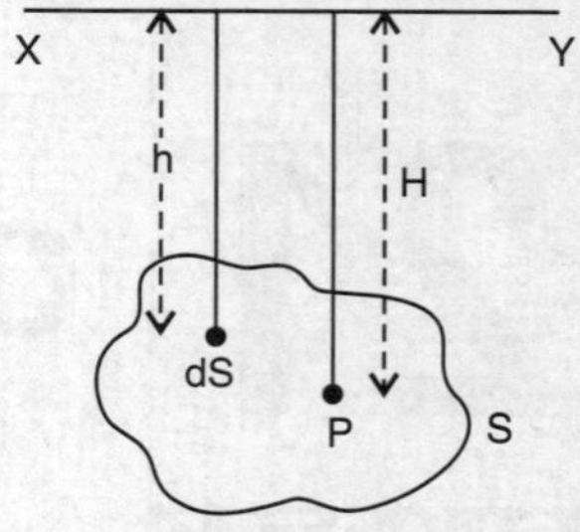

Fig.21.2

Thrust on an elementary area dS at a depth $h = h\rho g dS$

Moment of this thrust about XY

$$= (h\rho g dS) \times h = h^2 \rho g dS$$

Resultant moment of all thrusts $= \int h^2 \rho\, g dS$

where the integration is carried over all the elements of the plane area.

Resultant thrust on the plane area $= \int h\rho g dS$

Let the centre of pressure of the plane area be at the point P.

Let the distance of P from XY be H.

Moment of the resultant thrust about $XY = H\int h\rho g\, dS$

By definition of the resultant of several forces, we get

Moment of resultant force = resultant of the moments of the forces.

or $$H\int h\rho g\, dS = \int h^2 \rho g\, dS$$

or $$H = \frac{\int h^2\, dS}{\int h\, dS}$$

The result holds good for any inclined position of the plane also.

21.4. CENTRE OF PRESSURE OF A RECTANGULAR LAMINA IMMERSED VERTICALLY IN A LIQUID WITH ONE EDGE IN THE SURFACE OF THE LIQUID

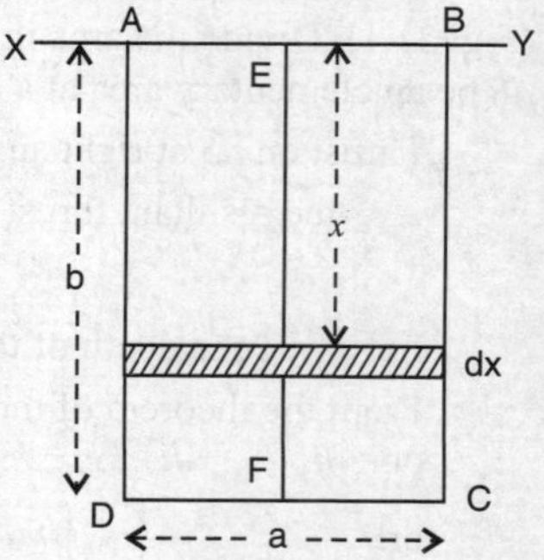

Fig. 21.3

Let $ABCD$ be a plane rectangular lamina immersed vertically in a liquid of density ρ with one edge AB in the surface XY of the liquid (Fig. 21.3). Let $AB = a$ and $AD = b$. Divide the rectangle into a number of narrow strips parallel to AB. Consider one such strip of width dx at a depth x below the surface of the liquid.

The thrust acting on the strip $= (x\rho g) \times (adx) = x\rho ga\, dx$

Moment of this thrust about $AB = (x\rho ga\, dx) \times x = x^2\rho ga\, dx$

Sum of the moments of the thrusts on all the strips $= \int_0^b x^2 \rho ga\, dx$

Resultant of the thrusts on all the strips $= \int_0^b x\rho ga\, dx$

Moment of the resultant thrust about $AB = H\int_0^b x\rho ga\, dx$

where H = depth of the centre of pressure below AB.

$$H\int_0^b x\rho ga\, dx = \int_0^b x^2 \rho ga\, dx$$

or $$H\rho ga\frac{b^2}{2} = \rho ga\frac{b^3}{3} \quad \text{or} \quad H = \frac{2}{3}b.$$

The thrust on every elementary strip acts through its midpoint. Hence the centre of pressure will lie on EF where E and F are the mid-points of AB and DC.

21.5. CENTRE OF PRESSURE OF A TRIANGULAR LAMINA IMMERSED VERTICALLY IN A LIQUID WITH ITS VERTEX IN THE SURFACE OF THE LIQUID AND ITS BASE HORIZONTAL

Let ABC be a triangular lamina immersed vertically in a liquid with its vertex A in the surface XY of the liquid and with its base BC horizontal (Fig. 21.4). $BC = a$. Let the depth of the base of the lamina be b from the free surface of the liquid. Divide the triangle into a number of elementary strips of width dx parallel to the base BC. Consider one such strip B_1C_1 of width dx at a depth x below the surface XY.

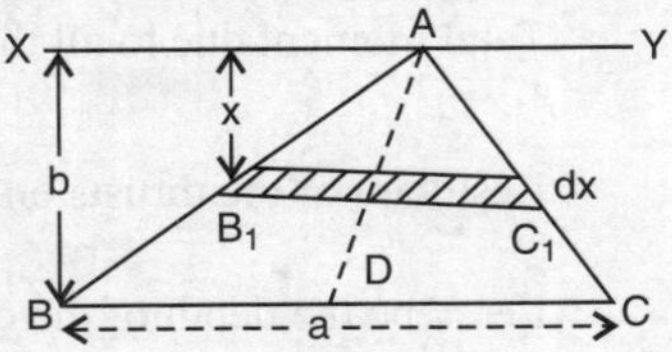

Fig. 21.4

Area of the strip $B_1C_1 = B_1C_1\, dx = (ax/b)\, dx$

Thrust on the strip $B_1C_1 = (x\rho g) \times (ax/b)\, dx$

Moment of this thrust about $XY = \left(\dfrac{ax^3\rho g}{b} dx\right)$

Total moment due to all the strips $= \displaystyle\int_0^b \frac{a\rho g}{b} x^3 dx.$

Resultant of the thrusts on all the strips $= \displaystyle\int_0^b \frac{a\rho g}{b} x^2 dx.$

Moment of the resultant thrust about $XY = H \displaystyle\int_0^b \frac{a\rho g}{b} x^2 dx.$

Here H = the depth of the centre of pressure below XY.

Since the two moments are equal,

$$\int_0^b \frac{a\rho g}{b} x^3 dx = H \int_0^b \frac{a\rho g}{b} x^2 dx.$$

or
$$\frac{a\rho g}{b}\left(\frac{b^4}{4}\right) = H \frac{a\rho g}{b}\left(\frac{b^3}{3}\right).$$

or
$$H = \frac{3}{4} b.$$

The centre of pressure lies on the line joining the mid-points of the strips. i.e., lies on the median AD at a depth $3b/4$ below the surface XY.

21.6. CENTRE OF PRESSURE OF A TRIANGULAR LAMINA IMMERSED IN A LIQUID WITH ONE SIDE IN THE SURFACE, WHEN THERE IS NO EXTERNAL PRESSURE

Let ABC be a triangular lamina immersed in a liquid with its base $BC = a$ in the surface XY of the liquid. Let AD be a median of the triangle. Let b be the depth of the vertex A below the surface XY. Divide the triangle into a number of elementary strips of width dx parallel to the base BC. Consider one such strip B_1C_1 at a depth x below BC.

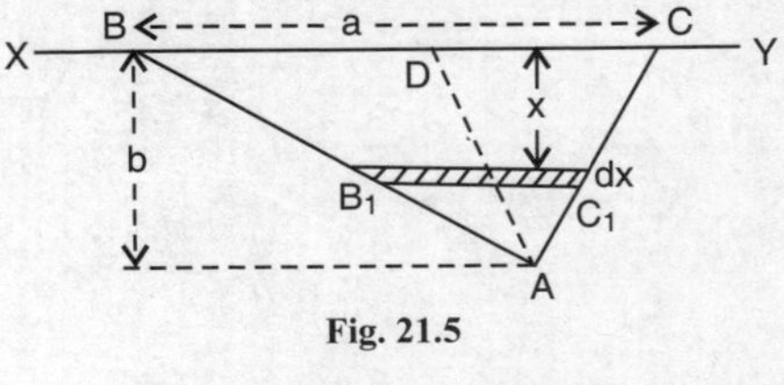

Fig. 21.5

Area of the strip $\quad B_1C_1 = B_1C_1\, dx = \dfrac{a(b-x)}{b} dx \qquad \left(\text{Since } \dfrac{B_1C_1}{a} = \dfrac{b-x}{b}\right)$

Thrust on the strip $\quad B_1C_1 = x\rho g \dfrac{a(b-x)}{b} dx$

Moment of this thrust about $XY = x^2\rho g \dfrac{a(b-x)}{b} dx.$

Total moment due to all the strips $= \int_0^b x^2\rho g \dfrac{a(b-x)}{b} dx.$

Resultant of the thrusts on all the strips $= \int_0^b x\rho g \dfrac{a(b-x)}{b} dx.$

Let H be the depth of the centre of pressure below XY.

Moment of the resultant thrust about $XY = H\int_0^b x\rho g \dfrac{a(b-x)}{b} dx.$

$$\therefore \quad \int_0^b x^2\rho g \frac{a(b-x)}{b} dx = H\int_0^b x\rho g \frac{a(b-x)}{b} dx.$$

$$\text{or} \quad H\int_0^b x(b-x)dx = \int_0^b x^2(b-x)dx$$

$$\text{or} \quad H\left[\frac{b^3}{2}-\frac{b^3}{3}\right] = \left[\frac{b^4}{3}-\frac{b^4}{4}\right]$$

$$\text{or} \quad H = \frac{b}{2}$$

The centre of pressure is on the median AD at a depth $b/2$ below XY.

EXERCISE XXI

1. Define 'centre of pressure'. Derive an expression for the resultant thrust, and the depth of centre of pressure, of a plane lamina immersed in a liquid.
2. Determine the position of centre of pressure for a triangular lamina of height h immersed vertically with (*a*) its apex, and (*b*) its base in the surface.
3. Determine the position of the centre of pressure of a rectangular lamina immersed vertically in a liquid with one edge in the surface of the liquid.

22 CHAPTER Friction

22.1. INTRODUCTION

Suppose we push a book along a table, giving it some velocity. After we release it, it slows down and finally stops. This indicates that a force is opposing the motion. Whenever one surface moves in contact with another, a force is generated that tends to retard the motion. This force is called friction.

Friction plays a vital role in our life. For example, man is able to walk on the road because of friction between his feet and the ground. In machines friction reduces efficiency. Lubrication and ball bearings reduce friction in machines.

22.2. STATIC, DYNAMIC, ROLLING AND LIMITING FRICTION

The frictional forces acting between surfaces at rest with respect to each other are called forces of *static friction.* The maximum value of the frictional force between two bodies in contact is called *limiting friction.* The forces acting between surfaces in relative motion are called *kinetic* (or *dynamic) friction.* The frictional force between two surfaces when one rolls over the other is called *rolling friction.* It is easy to roll a cylindrical body than to slide it. Thus, the rolling friction is less than the sliding friction. We require more force to start the motion of a body than to keep it in uniform motion. Thus, static friction is larger than dynamic friction.

22.3. LAWS OF STATIC FRICTION

1. The direction of the frictional force is always opposite to the direction in which one body tends to slide over another.

2. The magnitude of the force of friction when there is equilibrium between two bodies is just sufficient to prevent the motion of one body over the other. The frictional force attains a maximum value when one body is just on the point of sliding over the other. The maximum value of the force of friction is called limiting friction.

3. The magnitude of the force of limiting friction bears a constant ratio to the normal reaction between the two bodies. This ratio is called *coefficient of friction* and is denoted by μ. If F is the limiting friction and R the normal reaction between the two bodies, then $\mu = F/R \cdot \mu$ depends only on the nature of surfaces in contact.

4. The limiting friction is independent of the extent and shape of the surfaces in contact, provided the normal reaction is unaltered.

5. When a body is in motion, the direction of friction is still opposite to the direction of motion of the body and is independent of the velocity. But the ratio of the force of friction to the normal reaction is slightly less than that when the body is just on the point of motion.

Angle of friction: Let F be the force of limiting friction and R, the normal reaction. Let S be the resultant of these two forces (Fig. 22.1). Then the angle which this resultant force makes with the normal reaction is called the *angle of friction*. It is denoted by λ.

Then, $\tan\lambda = \frac{F}{R} = \frac{\mu R}{R} = \mu \left(\because \mu = \frac{F}{R} \right)$

Cone of Friction: Consider a cone with the point of contact of two bodies as the vertex, the normal reaction as axis and semi-vertical angle λ. Then the resultant reaction (S) may lie anywhere within or on the surface of the cone (Fig.22.1). This imaginary cone is called the *cone of friction.*

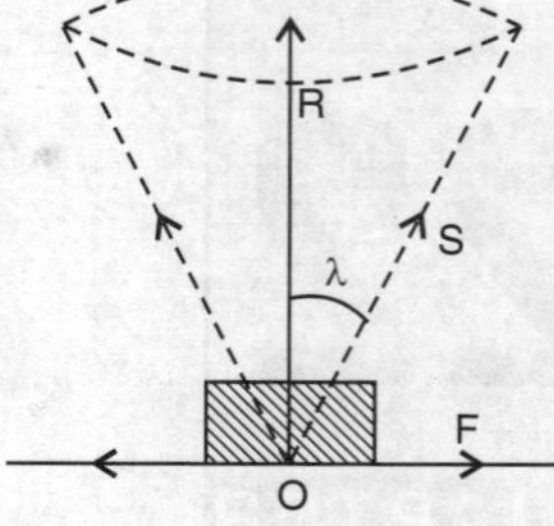

Fig. 22.1

22.4. EXPERIMENTAL METHOD FOR DETERMINING COEFFICIENT OF FRICTION BETWEEN TWO SURFACES

Suppose a body of weight *mg* is placed on a rough inclined plane (Fig. 22.2). The inclination is increased till a position is attained when the body just slides. At this position let the inclination of the plane to the horizontal be α. The forces acting on the body are: (*i*) the weight *mg* acting vertically downwards, (*ii*) the normal reaction *R* perpendicular to the plane and (*iii*) the force of limiting friction μ*R* acting up the plane. Resolve *mg* into a component *mg* sin α down the plane and a component *mg* cos α at right angles to the plane. When the equilibrium is limiting,

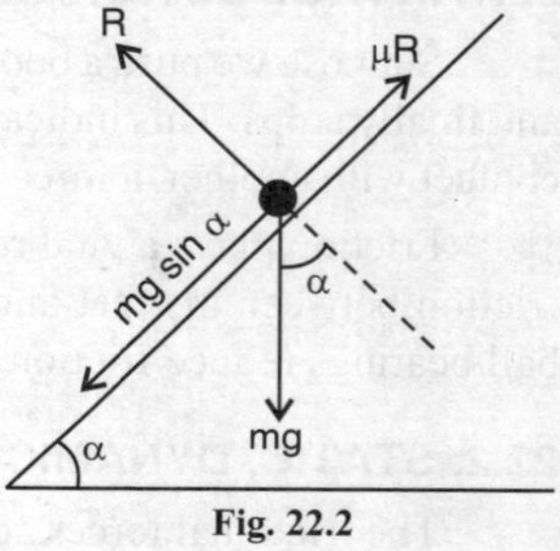

Fig. 22.2

$$mg \sin\alpha = \mu R \quad ...(i)$$

$$mg \cos\alpha = R \quad ...(ii)$$

Dividing (*i*) by (*ii*), $\tan\alpha = \mu$

22.5. EQUILIBRIUM OF A BODY ON A ROUGH INCLINED PLANE ACTED UPON BY AN EXTERNAL FORCE

Proposition: *A body of weight w is in equilibrium on a rough inclined plane of angle α > λ under the action of an external force inclined upwards at an angle θ with the plane. Find the value of P for limiting equilibrium.*

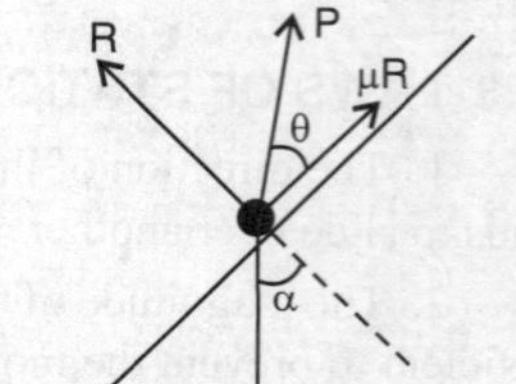

Fig. 22.3

Case 1: Let the body be just on the point of sliding down the plane. Let *P* be the magnitude of the external force, applied at an angle θ with the plane. The forces acting on the body are: (*i*) The weight of the body (*w*) acting vertically down, (*ii*) The normal reaction (*R*) acting perpendicular to the plane (*iii*) The force of limiting friction (μ*R*) acting up the plane (*iv*) The external force (effort) *P* making an angle θ with the line of gratest slope of the inclined plane (Fig. 22.3). Resolving the forces parallel and perpendicular to the plane.

$$P\cos\theta + \mu R = w\sin\alpha \quad ...(1)$$

$$P\sin\theta + R = w\cos\alpha \quad ...(2)$$

Multiplying (2) by μ and subracting from (1),

$$P(\cos\theta - \mu\sin\theta) = w(\sin\alpha - \mu\cos\alpha)$$

or
$$P = w\frac{(\sin\alpha - \mu\cos\alpha)}{(\cos\theta - \mu\sin\theta)}$$

But μ = tan λ, where λ is the angle of friction.

Hence $$P = w\frac{(\sin\alpha - \tan\lambda\cos\alpha)}{(\cos\theta - \tan\lambda\sin\theta)} = w\frac{(\sin\alpha\cos\lambda - \sin\lambda\cos\alpha)}{(\cos\theta\cos\lambda - \sin\lambda\sin\theta)}$$

$$\therefore \quad P = \frac{w\sin(\alpha-\lambda)}{\cos(\theta+\lambda)} \quad \text{...(3)}$$

Case 2: Let the body be just on the point of sliding up the plane. Let P_1 be the magnitude of the external force. In this case, the force of limiting friction (μR) acts down the plane.

Resolving the forces parallel and perpendicular to the plane,

$$P_1\cos\theta = w\sin\alpha + \mu R \quad \text{...(4)}$$

$$P_1\sin\theta + R = w\cos\alpha \quad \text{...(5)}$$

Simplifying, we get, $$P_1 = \frac{w\sin(\alpha+\lambda)}{\cos(\theta-\lambda)} \quad \text{...(6)}$$

Corollary 1: P_1 is a minimum when $\cos(\theta - \lambda)$ is maximum *i.e.*, when $\cos(\theta - \lambda) = 1$ i.e., when $\theta = \lambda$. Hence force required to move the body up the plane will be least when it is applied in a direction making with the inclined plane an angle equal to the angle of friction.

Corollary 2: Let a body be at rest on a rough inclined plane whose inclination to the horizontal $\alpha > \lambda$. Let it be acted upon by an external force applied parallel to the plane. Here $\theta = 0$. From (3) and (6),

$$P = \frac{w\sin(\alpha-\lambda)}{\cos\lambda} \quad \text{...(7)}$$

and $$P_1 = \frac{w\sin(\alpha+\lambda)}{\cos\lambda} \quad \text{...(8)}$$

EXERCISE XXII

1. Define friction. What is the cone of friction ?
2. Define limiting friction, coefficient of friction, angle of friction and cone of friction. Show that coefficient of friction is related to angle of friction by $\mu = \tan\lambda$. *(Madras 1995, 1992)*
3. State the laws of friction.
4. A block is placed on an inclined plane of inclination θ. Show that to keep the block at rest, the external force P must lie between $\frac{w\sin(\theta\pm\lambda)}{\cos\lambda}$ where λ is angle of friction.

23 CHAPTER Classical Mechanics

23.1. MECHANICS OF A SYSTEM OF PARTICLES

(a) Conservation theorem for linear momentum:

The net linear momentum of a system of *n*-particles is

$$P = \sum_{i=1}^{n} p_i = \sum_{i=1}^{n} m_i v_i$$

From Newton's second law, $F^{ext} = \frac{dP}{dt}$

i.e., the rate of change of linear momentum of a system of particles is equal to the net external force acting on the system.

If $F^{ext} = 0$, $\frac{dP}{dt} = 0$. Integrating, P = constant.

This gives the *theorem* for conservation of linear momentum of the system according to which *"If the sum of external forces acting on the system of particles is zero, the total linear momentum of the system is constant or conserved."*

(b) Conservation theorem for angular momentum.

The angular momentum of *i*th particle of the system about any point *O*, from definition is given by

$$\mathbf{J}_i = \mathbf{r}_i \times \mathbf{p}_i, \qquad \text{...(1)}$$

where $\boldsymbol{r}_i$ is the radius vector of *i*th particle from the point *O* and $\boldsymbol{p}_i$ its linear momentum.

For a *system of n particles,* we have

$$\mathbf{J} = \sum_i \mathbf{J}_i = \sum_i \mathbf{r}_i \times \mathbf{p}_i \qquad \text{...(2)}$$

$$\frac{d\mathbf{J}}{dt} = \sum_i \mathbf{r}_i \times \frac{d\mathbf{p}_i}{dt} = \sum_i \mathbf{r}_i \times \mathrm{F}_i \qquad \left(\because \frac{d\mathbf{r}_i}{dt} \times \mathbf{p}_i = \mathbf{v}_i \times \mathbf{p}_i = 0\right)$$

Here, $\mathbf{F}_i = \frac{d}{dt}\mathbf{p}_i$ = net force acting on *i*th particle.

Internal forces occur in equal and opposite pairs. Hence the net internal force acting on the system of particles is zero. Thus,

$$\frac{d\mathbf{J}}{dt} = \sum_i \mathbf{r}_i \times \mathbf{F}_i^{ext} = \tau^{ext}$$

Here, $\tau^{ext} = \sum_i \mathbf{r}_i \times \mathbf{F}_i^{ext}$ is the torque arising due to external forces only.

If $$\tau^{ext} = 0, \frac{d\mathbf{J}}{dt} = 0 \text{ or } \mathbf{J} = \text{constant}.$$

Thus, *if external torque acting on a system of particles is zero, the angular momentum of the system remains constant.*

This is the conservation theorem for angular momentum of a system of particles.

(*c*) Conservation of Energy. If the work done by a force is independent of path, the force is said to be conservative.

If the forces acting on the system of particles are conservative, the total energy of the system of particles which is the sum of the total kinetic energy and the total potential energy of the system is conserved.

This is the energy conservation theorem.

On the other hand *if the forces are non-conservative, the total energy of universe* (mechanical energy + chemical energy + sound energy + light energy + heat energy etc.) *remains constant.*

23.2. BASIC CONCEPTS

Degrees of Freedom. *The number of mutually independent variables required to define the state or position of a system is the number of degrees of freedom possessed by it.*

For example, the position of a simple ideal mass-point can be defined completely by the three cartesian coordinates *x, y, z.* So it has three degrees of freedom. Extending this idea, for a system of *N* particles moving independently of each other, the number of degrees of freedom is 3 *N.*

Constraints. *Constraints are restrictions imposed on the position or motion of a system, because of geometrical conditions.*

Examples. (1) The beads of an abacus are constrained to one-dimensional motion by the supporting wires.

(2) Gas molecules within a container are constrained by the walls of the vessel to move only *inside* the container.

(3) The motion of rigid bodies is always such that the distance between any two particles remains unchanged.

(4) A particle placed on the surface of a solid sphere is restricted by the constraint so that it can only move on the surface or in the region exterior to the sphere.

For a particle constrained to move on a plane, only two variables *x, y* or *r,* θ *are sufficient to describe its motion and the particle is said to have two degrees of freedom. Thus, the constraint on the motion of the particle in a plane reduces the number of degrees of freedom by one.*

Very often, we can express constraints in terms of certain equations. For example, the equation of constraint in the case of a particle moving on or outside the surface of a sphere of radius *a* is $x^2 + y^2 + z^2 \geq a^2$ if the origin of the coordinate system coincides with the centre of the sphere.

(*i*) Holonomic and non-holonomic constraints.

If the constraints can be expressed as equations connecting the co-ordinates of the particles (and possibly time) in the form

$$f(\mathbf{r}_1, \mathbf{r}_2, \mathbf{r}_3 \ldots\ldots, \mathbf{r}_n, t) = 0 \qquad \ldots(1)$$

then the constraints are said to be *holonomic.*

Examples. (1) *The constraints involved in the motion of rigid bodies* in which the distance between any two particular points is always fixed, are *holonomic* since the conditions of constraints are expressed as

$$(\mathbf{r}_i - \mathbf{r}_j)^2 - c_{ij}^2 = 0.$$

(2) *The constraints involved* when a particle is restricted to move along a curve or surface are holonomic. Here the equation defining the curve or surface is the equation of constraint.

If the constraints cannot be expressed in the form of Eq. (1), they are called *non-holonomic constraints.*

Examples. (1) The constraints involved in the motion of the particle placed on the surface of a solid sphere are non-holonomic. The conditions of constraints in this case are expressed as

$$r^2 - a^2 \geq 0,$$

where a is the radius of sphere. This is an inequality and hence not in the form of Eq. (1).

(2) The walls of the gas vessel constitute a non-holonomic constraint.

(3) An object rolling on a rough surface without slipping is also an example of non-holonomic constraint.

***(ii)* Scleronomic and Rheonomic Constraints.** If the constraints are independent of time, they are called *scleronomic.*

If the constraints are explicitly dependent on time, they are called *rheonomic.*

The constraint in the case of rigid body motion is scleronomous. Ahead sliding on a moving wire is an example of rheonomic constraint.

In the solution of mechanical problems, the constraints introduce two types of difficulties:

(1) The co-ordinates $\boldsymbol{r}_i$ are connected by the equations of constraints. Therefore, they are not independent.

(2) The forces of constraint are not *a priori* known. In fact, they cannot be estimated till a complete solution of the problem is obtained.

The first problem can be solved by introducing *generalized co-ordinates,* whereas the second is practically an insurmountable problem. We therefore reformulate the problem such that the forces of constraint disappear.

23.3. GENERALISED CO-ORDINATES

A system consisting of N particles, free from constraints, has $3N$ independent co-ordinates or degrees of freedom. If the sum of the degrees of freedom of all the particles is k, then the system may be regarded as a collection of free particles subjected to $(3N-k)$ independent constraints. So only k co-ordinates are needed to describe the motion of the system. These new co-ordinates $q_1, q_2, q_3, ...$ q_k are called the *Generalised Co-ordinates of Lagrange.* Generalised co-ordinates may be lengths or angles or any other set of independent quantities which define the position of the system.

Definition. *The generalised co-ordinates of a material system are the independent parameters* $q_1, q_2, q_3, ...q_k$ *which completely specify the configuration of the system, i.e., the position of all its particles with respect to the frame of reference.*

Example. Consider the simple pendulum of mass m_1 with fixed length r_1 (Fig. 23.1). The single co-ordinate θ_1 will determine uniquely the position of m_1 since the simple pendulum is a system of one degree of freedom. Since the only variable involved is θ_1, it can be chosen as the generalised co-ordinate. Thus $q = \theta_1$. The two coordinates x_1 and y_1 could also be used to locate m_1 but would require the inclusion of the *equation of the constraint* $x_1^2 + y_1^2 = r_1^2$. Since x_1 and y_1 are not independent, they are not generalised co-ordinates.

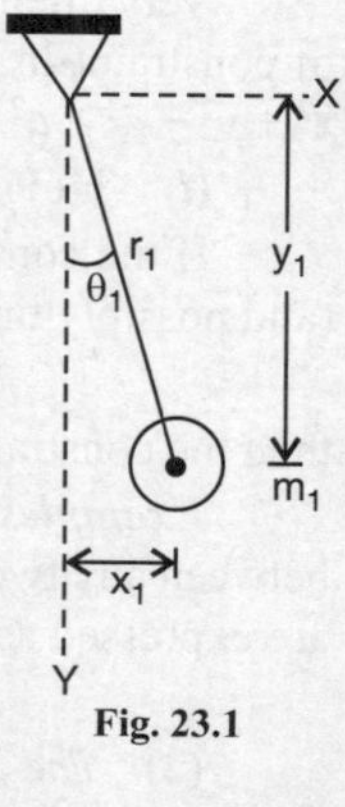

Fig. 23.1

Generalised velocities : The generalised velocities of a system are the total time derivatives of the generalised co-ordinates of the system.

Thus $$\dot{q}_i = \frac{dq_i}{dt} \; (i = 1, 2, 3, ..., k).$$

23.4. TRANSFORMATION EQUATIONS

The rectangular cartesian co-ordinates can be expressed as the functions of generalised co-ordinates. Let x_i, y_i and z_i be the cartesian co-ordinates of *i*th particle of the system. Let *t* denote the time. Then, these cartesian co-ordinates can be expressed as functions of generalised co-ordinates $q_1, q_2, q_3.., q_k$ *i.e.*,

$$\left.\begin{aligned} x_i &= x_i(q_1, q_2, ..., q_k, t) \\ y_i &= y_i(q_1, q_2, ..., q_k, t) \\ z_i &= z_i(q_1, q_2, ..., q_k, t) \end{aligned}\right\} \quad ...(1)$$

Let $\mathbf{r}_i$ be the position vector of *i*th particle, i.e., $\mathbf{r}_i = \mathbf{i}x_i + \mathbf{j}y_i + \mathbf{k}z_i$ ...(2)

Then $$\mathbf{r}_i = \mathbf{r}_i\,(q_1, q_2, q_3, ..., q_k, t),$$

Eq. (2) is the vector form of Eq. (1).

The equations like (1) and (2) are called *transformation equations.* The functions and their derivatives in the above two equations are supposed to be continuous. The equations also contain the constraints explicitly.

23.5. CONFIGURATION SPACE

In the case of motion of a single particle we can represent its trajectory in the three dimensional space by specifying its variables. For a system of *N* particles described by *3N* space coordinates with $(3N - k)$ equations of constraint in the real space, it is difficult to visualise the motion of the entire system. It is convenient to describe the motion of a system having *k* coordinates in a hypothetical *k* dimensional space. The instantaneous configuration of the system is described by the values of the *k* generalised coordinates $q_1, q_2, q_3......q_k$, and corresponds to a particular point in a cartesian hyperspace where the *q's* form the *k* coordinate axes. The point is called the system point and the *k* dimensional space is known as the *Configuration space.* At some later instant, the state of the system changes and it will be represented by some other point in the configuration space. Thus, the system point moves in the configuration space tracing out a curve. This curve represents "the path of motion of the system". "The motion of the system", as used above, then refers to the motion of the system point along this path in *configuration space.* Time can be considered formally as a parameter of the curve since each point in the configuration space has one or more values of time associated with it.

Configuration space has nothing in common with the three-dimensional space which we can visualise physically. It is a purely geometric structure by means of which the laws of the variation of the state of a system can be formulated in geometrical language.

23.6. PRINCIPLE OF VIRTUAL WORK

Consider a system described by *n* generalized coordinates q_j $(j = 1, 2, 3,..., n)$. Suppose the system undergoes a certain displacement in the configuration space in such a way that it does not take any time and that it is consistent with the constraints on the system. Such displacements are called *virtual* because they do not represent actual displacements of the system. Since there is no actual motion of the system, the work done by the forces of constraint in such a virtual displacement is zero.

Let the virtual displacement of the *i*th particle of the given system be $\delta\mathbf{r}_i$. If the given system is in equilibrium, the resultant force acting on the *i*'th particle of the system must be zero, *i.e.*, $\mathbf{F}_i = 0$. It is, then, obvious that *virtual work* $\mathbf{F}_i.\ \delta\mathbf{r}_i = 0$ for the *i*th particle and hence it is also zero for all the particles of the system.

Thus $$dW = \sum_i \mathbf{F}_i.\delta\mathbf{r}_i = 0 \quad ...(1)$$

The resultant force $\mathbf{F}_i$ acting on the *i*th particle is,

$$\mathbf{F}_i = \mathbf{F}_i^a + \boldsymbol{f}_i \quad ...(2)$$

Here, $\mathbf{F}_i^a$ is the applied force and f_i is the force of constraint.

Eq. (1) then becomes

$$\sum_i \mathbf{F}_i^a.\delta\mathbf{r}_i + \sum_i \mathbf{f}_i.\delta\mathbf{r}_i = 0 \quad ...(3)$$

We now consider systems for which the virtual work done by the forces of constraints is zero, *i.e.,*

$$\sum_i \mathbf{f}_i.\delta\mathbf{r}_i = 0 \quad ...(4)$$

Then Eq. (3) becomes

$$\sum_i \mathbf{F}_i^a.\delta\mathbf{r}_i = 0 \quad ...(5)$$

This equation is termed as *principle of virtual work.*

23.7. D' ALEMBERT'S PRINCIPLE

Most of the systems we come across in mechanics are not in static equilibrium. Hence the principle must be modified to include dynamic systems as well.

According to Newton's second law of motion,

$$\mathbf{F}_i = \dot{\mathbf{p}}_i \quad \text{or} \quad \mathbf{F}_i - \dot{\mathbf{p}}_i = 0 \quad ...(6)$$

According to the above equation, a moving system of particles can be considered to be in equilibrium under the force $(\mathbf{F}_i - \dot{\mathbf{p}}_i)$, *i.e.*, the actual applied force $\mathbf{F}_i$ plus an additional force $-\dot{\mathbf{p}}_i$ which is known as *reversed effective force* on *i*th particle. Let us again assume that *the forces of constraint do no work.* Then, we can generalize Eq. (5) by the use of Eq. (6) to the form

$$\sum_i (\mathbf{F}_i - \dot{\mathbf{p}}_i).\delta\mathbf{r}_i = 0 \quad ...(7)$$

Eq. (7) is the mathematical statement of D' Alembert's principle.

It is to be noted here that we have restricted ourselves to the systems where the virtual work done by the forces of constraints disappears. With this in mind we can drop the superscript *a* in equation (2) *i.e.,* D'Alembert's principle may be written as

$$\sum_i (\mathbf{F}_i - \dot{\mathbf{p}}_i).\delta\mathbf{r}_i = 0 \quad ...(8)$$

23.8. DERIVATION OF LAGRANGE'S EQUATION OF MOTION

Lagrange's Equations from D'Alembert's Principle

Consider a system of particles whose position vectors are expressed as functions of generalized coordinates $q_1, q_2, q_3, ... q_k ... q_f$ and the time *t*.

Consider any particle of the system (*i*th particle) of mass m_i and acted upon by an external force $\mathbf{F}_i$.

According to D'Alembert's principle,

$$\sum_i (\mathbf{F}_i - \dot{\mathbf{p}}_i).\delta\mathbf{r}_i = 0 \quad ...(1)$$

Here $\dot{\mathbf{p}}_i$ is the inertial force for *i*th particle and $\delta\mathbf{r}_i$ is the virtual displacement of *i*th particle due to action of force $\mathbf{F}_i$.

In general, $\mathbf{r}_i = \mathbf{r}_i\,(q_1, q_2, q_3, ... q_k, ... q_f, t)$ ...(2)

$$\dot{\mathbf{r}}_i = \frac{\delta\mathbf{r}_i}{\delta t} = \frac{\partial\mathbf{r}_i}{\partial q_1}\dot{q}_1 + \frac{\partial\mathbf{r}_i}{\partial q_2}\dot{q}_2 + ... + \frac{\partial\mathbf{r}_i}{\partial q_k}\dot{q}_k + ... + \frac{\partial\mathbf{r}_i}{\partial q_f}\dot{q}_f + \frac{\partial\mathbf{r}_i}{\partial t}$$

$$= \sum_k \frac{\partial\mathbf{r}_i}{\partial q_k}\dot{q}_k + \frac{\partial\mathbf{r}_i}{dt} \quad ...(3)$$

The virtual displacement $\delta \mathbf{r}_i$ in terms of generalised co-ordinates is given by

$$\delta \mathbf{r}_i = \sum_k \frac{\partial \mathbf{r}_i}{\partial q_k} \delta q_k. \qquad ...(4)$$

Now, $\dot{\mathbf{p}}_i = m_i \ddot{\mathbf{r}}_i$

Therefore Eq. (1) becomes $\sum_i (\mathbf{F}_i - m_i \ddot{\mathbf{r}}_i) . \delta \mathbf{r}_i = 0$ ...(5)

or $$\sum_i m_i \ddot{\mathbf{r}}_i \cdot \delta \mathbf{r}_i = \sum_i \mathbf{F}_i \cdot \delta \mathbf{r}_i \qquad ...(6)$$

Putting the value of $\delta \mathbf{r}_i$ from Eq. (4) in (6), we get

$$\sum_i \sum_k m_i \ddot{\mathbf{r}}_i . \frac{\partial \mathbf{r}_i}{\partial q_k} \delta q_k = \sum_i \sum_k \mathbf{F}_i . \frac{\partial \mathbf{r}_i}{\partial q_k} \delta q_k \qquad ...(7)$$

Now, $$\frac{d}{dt}\left(\dot{\mathbf{r}}_i . \frac{\partial \mathbf{r}_i}{\partial q_k} \right) = \ddot{\mathbf{r}}_i . \frac{\partial \mathbf{r}_i}{\partial q_k} + \dot{\mathbf{r}}_i . \frac{d}{dt}\left(\frac{\partial \mathbf{r}_i}{\partial q_k} \right),$$

$$\therefore \quad \ddot{\mathbf{r}}_i . \frac{\partial \mathbf{r}_i}{\partial q_k} = \frac{d}{dt}\left(\dot{\mathbf{r}}_i . \frac{\partial \mathbf{r}_i}{\partial q_k} \right) - \dot{\mathbf{r}}_i . \frac{d}{dt}\left(\frac{\partial \mathbf{r}_i}{\partial q_k} \right) \qquad ...(8)$$

Putting the value of $\ddot{\mathbf{r}}_i . \frac{\partial \mathbf{r}_i}{\partial q_k}$ from above equation in (7), we get

$$\sum_i \sum_k m_i \left[\frac{d}{dt}\left(\dot{\mathbf{r}}_i . \frac{\partial \mathbf{r}_i}{\partial q_k} \right) - \dot{\mathbf{r}}_i . \frac{d}{dt}\left(\frac{\partial \mathbf{r}_i}{\partial q_k} \right) \right] \delta q_k = \sum_i \sum_k \mathbf{F}_i . \left(\frac{\partial \mathbf{r}_i}{\partial q_k} \right) \delta q_k \qquad ...(9)$$

Differentiating Eq. (3) partially with respect to $\dot{q}_k$,

$$\frac{\partial \dot{\mathbf{r}}_i}{\partial \dot{q}_k} = \frac{\partial \mathbf{r}_i}{\partial q_k} \qquad ...(10)$$

Differentiating Eq. (3) partially with respect to q_k,

$$\frac{\partial \dot{\mathbf{r}}_1}{\partial q_k} = \frac{\partial^2 \mathbf{r}_i}{\partial q_k \partial q_1} \dot{q}_1 + \frac{\partial^2 \mathbf{r}_i}{\partial q_k \partial q_2} \dot{q}_2 + ... + \frac{\partial^2 \mathbf{r}_i}{\partial q_k \partial q_k} \dot{q}_k + ... + \frac{\partial^2 \mathbf{r}_i}{\partial q_k \partial t} \qquad ...(11)$$

Also we have

$$\frac{d}{dt}\left(\frac{\partial \mathbf{r}_i}{\partial q_k} \right) = \frac{\partial}{\partial q_1}\left(\frac{\partial \mathbf{r}_i}{\partial q_k} \right) \dot{q}_1 + \frac{\partial}{\partial q_2}\left(\frac{\partial \mathbf{r}_i}{\partial q_k} \right) \dot{q}_2 + ... + \frac{\partial}{\partial q_k}\left(\frac{\partial \mathbf{r}_i}{\partial q_k} \right) \dot{q}_k$$

$$+ ... + \frac{\partial}{\partial q_f}\left(\frac{\partial \mathbf{r}_i}{\partial q_f} \right) \dot{q}_f + \frac{\partial}{\partial t}\left(\frac{\partial \mathbf{r}_i}{\partial q_k} \right)$$

$$= \frac{\partial^2 \mathbf{r}_i}{\partial q_1 \partial q_k} \dot{q}_1 + \frac{\partial^2 \mathbf{r}_i}{\partial q_2 \partial q_k} \dot{q}_2 + ... + \frac{\partial^2 r_i}{\partial q_k \partial q_k} \dot{q}_k + ... + \frac{\partial^2 \mathbf{r}_i}{\partial t \, \partial q_k} \qquad ...(12)$$

Comparing Eqs. (11) and (12).

$$\frac{d}{dt}\left(\frac{\partial \mathbf{r}_i}{\partial q_k} \right) = \frac{\partial \dot{\mathbf{r}}_i}{\partial q_k} \qquad ...(13)$$

From Eq. (10), $\frac{d}{dt}\left(\dot{\mathbf{r}}_i \cdot \frac{\partial \mathbf{r}_i}{\partial q_k}\right) = \frac{d}{dt}\left(\dot{\mathbf{r}}_i \cdot \frac{\partial \dot{\mathbf{r}}_i}{\partial \dot{q}_k}\right) = \frac{d}{dt}\frac{\partial}{\partial \dot{q}_k}\left(\frac{1}{2}\dot{\mathbf{r}}_i^2\right)$ ...(14)

Substituting (13) and (14) in Eq. (9),

$$\sum_i\sum_k m_i\left[\frac{d}{dt}\frac{\partial}{\partial \dot{q}_k}\left(\frac{1}{2}\dot{\mathbf{r}}_i^2\right) - \dot{\mathbf{r}}_i \cdot \frac{\partial \dot{\mathbf{r}}_i}{\partial q_k}\right]\delta q_k = \sum_i\sum_k \mathbf{F}_i \cdot \frac{\partial \mathbf{r}_i}{\partial q_k}\delta q_k$$

or $$\sum_i\sum_k\left[\frac{d}{dt}\frac{\partial}{\partial \dot{q}_k}\left(\frac{1}{2}m_i\dot{\mathbf{r}}_i^2\right) - \frac{\partial}{\partial q_k}\left(\frac{1}{2}m_i\dot{\mathbf{r}}_i^2\right)\right]\delta q_k = \sum_i\sum_k \mathbf{F}_i \cdot \frac{\partial \mathbf{r}_i}{\partial q_k}\delta q_k$$

or $$\sum_k\left[\frac{d}{dt}\frac{\partial}{\partial \dot{q}_k}\left(\sum\frac{1}{2}m_i\dot{r}_i^2\right) - \frac{\partial}{\partial q_k}\left(\sum\frac{1}{2}m_i\dot{r}_i^2\right)\right]\delta q_k = \sum_i\sum_k \mathbf{F}_i \cdot \frac{\partial r_i}{\partial q_k}\delta q_k \quad ...(15)$$

$$\sum\frac{1}{2}m_i\dot{r}_i^2 = T = \text{Total kinetic energy of the system of particles} \quad ...(16)$$

and $$\sum_i \mathbf{F}_i \cdot \frac{\partial \mathbf{r}_i}{\partial q_k} = \mathbf{Q}_k, \quad ...(17)$$

Here $\mathbf{Q}_k$'*s* are components of *generalised force*.

Eq. (15) become, $\sum_k\left\{\frac{d}{dt}\frac{\partial T}{\partial \dot{q}_k} - \frac{\partial T}{\partial q_k}\right\}\delta q_k = \sum \mathbf{Q}_k\delta q_k$ ...(18)

$$\therefore \quad \frac{d}{dt}\frac{\partial T}{\partial \dot{q}_k} - \frac{\partial T}{\partial q_k} = Q_k \quad ...(19)$$

This is the general form of *Lagrange's equation*. There are *f* such equations corresponding to *f* generalised co-ordinates.

When the system is wholly conservative,

$$\mathbf{F}_i = -\nabla V_i = -\frac{\partial V_i}{\partial \mathbf{r}_i}, \quad ...(20)$$

$$\mathbf{Q}_k = \sum_i \mathbf{F}_i \cdot \frac{\partial \mathbf{r}_i}{\partial q_k} = -\sum\frac{\partial V_i}{\partial \mathbf{r}_i}\cdot\frac{\partial \mathbf{r}_i}{\partial q_k} = -\sum_i\frac{\partial V_i}{\partial q_k} = -\frac{\partial}{\partial q_k}\left(\sum V_i\right) = -\frac{\partial V}{\partial q_k}$$

Here $V = \sum_i V_i$ = total potential energy of the system.

Putting this value of Q_k in Eq. (19), we get

$$\frac{d}{dt}\left(\frac{\partial T}{\partial \dot{q}_k}\right) - \frac{\partial T}{\partial q_k} = -\frac{\partial V}{\partial q_k}$$

or $$\frac{d}{dt}\left(\frac{\partial T}{\partial \dot{q}_k}\right) - \frac{\partial T}{\partial q_k} + \frac{\partial V}{\partial q_k} = 0$$

or $$\frac{d}{dt}\left(\frac{\partial T}{\partial \dot{q}_k}\right) - \frac{\partial}{\partial q_k}(T - V) = 0 \quad ...(21)$$

The potential energy V is the function of position co-ordinates q_k and not of the *generalised velocities* $\dot{q}_k$. Therefore, Eq. (21) may be written as

$$\frac{d}{dt}\frac{\partial}{\partial \dot{q}_k}(T - V) - \frac{\partial}{\partial q_k}(T - V) = 0 \quad ...(22)$$

But $L = T - V$, where L is known as *Lagrangian function.*

Eq. (22) becomes, $\frac{d}{dt}\left(\frac{\partial L}{\partial \dot{q}_k}\right) - \frac{\partial L}{\partial q_k} = 0$ ...(23)

This is Lagrange's equation for a *conservative system.*

23.9. LAGRANGE'S EQUATIONS FOR SYSTEMS CONTAINING DISSIPATIVE FORCES

Consider a system of particles containing dissipative or frictional forces.

The frictional force is proportional to the velocity of the particle, *i.e.,*

$$\mathbf{F}_i^{(d)} = -\lambda_i \dot{\mathbf{r}}_i, \quad ...(1)$$

Here $\mathbf{F}_i^{(d)}$ denotes the dissipative force, $\dot{\mathbf{r}}_i$ is the velocity of ith particle and λ_i is the corresponding constant of proportionality.

Forces of this type are derivable from *Rayleigh's dissipation function R* defined by

$$R = \frac{1}{2}\sum_i \lambda_i \dot{\mathbf{r}}_i^2 \quad ...(2)$$

Here $i = 1, 2,...n$ covers all the particles of the system.

$$\frac{\partial R}{\partial \dot{\mathbf{r}}_i} = \lambda_i \dot{\mathbf{r}}_i = -\mathbf{F}_i^{(d)} \quad \text{from Eq. (1)}$$

or

$$\mathbf{F}_i^{(d)} = -\frac{\partial R}{\partial \dot{\mathbf{r}}_i} \quad ...(3)$$

Here term $\frac{\partial R}{\partial \dot{\mathbf{r}}_i}$ represents the dissipative force.

The Lagrange's equation in terms of $\mathbf{r}$ is given by

$$\frac{d}{dt}\left(\frac{\partial L}{\partial \dot{\mathbf{r}}}\right) - \frac{\partial L}{\partial \mathbf{r}} = \mathbf{F}_i^{(d)} \quad ...(4)$$

Lagrange's equation in generalised coordinates q_k is

$$\frac{d}{dt}\left(\frac{\partial L}{\partial \dot{q}_k}\right) - \frac{\partial L}{\partial q_k} = \mathbf{Q}_k^{(d)}$$

where $\mathbf{Q}_k^{d}$ is component of generalised force.

In order to find the Lagrange's equation in generalised co-ordinates, we have to find the components of generalised force resulting from the dissipative force.

If $\mathbf{Q}_k^{(d)}$ is the component of generalised force along q_k, then

$$\mathbf{Q}_k^{(d)} = \sum_i \mathbf{F}_i^{(d)} \frac{\partial \mathbf{r}_i}{\partial q_k}$$

$$= -\sum \lambda_i \dot{\mathbf{r}}_i \frac{\partial \mathbf{r}_i}{\partial q_k} \quad \text{[from Eq.(1)]}$$

$$= -\sum \lambda_i \dot{\mathbf{r}}_i \frac{\partial \dot{\mathbf{r}}_i}{\partial \dot{q}_k} \quad \left[\because \frac{\partial \mathbf{r}_i}{\partial q_k} = \frac{\partial \dot{\mathbf{r}}_i}{\partial \dot{q}_k}\right]$$

or

$$\mathbf{Q}_k^{(d)} = -\sum_i \lambda_i \frac{\partial}{\partial \dot{q}_k}\left(\frac{1}{2}\dot{r}_i^2\right)$$

$$= \frac{\partial}{\partial \dot{q}_k} \sum_i \left(-\frac{1}{2} \lambda_i \dot{\mathbf{r}}_i^2 \right)$$

$$\therefore \qquad \mathbf{Q}_k^{(d)} = -\frac{\partial R}{\partial \dot{q}_k} \qquad ...(5)$$

Therefore the Lagrange's equation for a system containing dissipative force is given by

$$\frac{d}{dt}\left(\frac{\partial L}{\partial \dot{q}_k}\right) - \frac{\partial L}{\partial q_k} + \frac{\partial R}{\partial \dot{q}_k} = 0 \qquad ...(6)$$

Hence the term $\frac{\partial R}{\partial \dot{q}_k}$ takes into account the dissipative forces.

Thus, if dissipative forces are acting on the system, we must specify two scalar functions—the Lagrangian L and Rayleigh's dissipation function R- to derive the equations of motion.

23.10. APPLICATIONS OF LAGRANGE'S EQUATIONS

In order to use Lagrange's equations for the solution of a physical problem, one must use the following steps :

(*i*) Choose an appropriate coordinate system.

(*ii*) Write down the expressions for potential and kinetic energies.

(*iii*) Write down the equations of constraint, if any.

(*iv*) Choose the generalized coordinates.

(*v*) Set up the Lagrangian. $L = T - V$

(*vi*) Solve Lagrange's equations for each generalized coordinate using, if necessary, the equations of constraint.

(*a*) The Atwood's machine:

Let two small heavy particles of masses M_1 and M_2 be connected by a light inextensible rope of length l passing over a frictionless light pulley (Fig. 23.2).

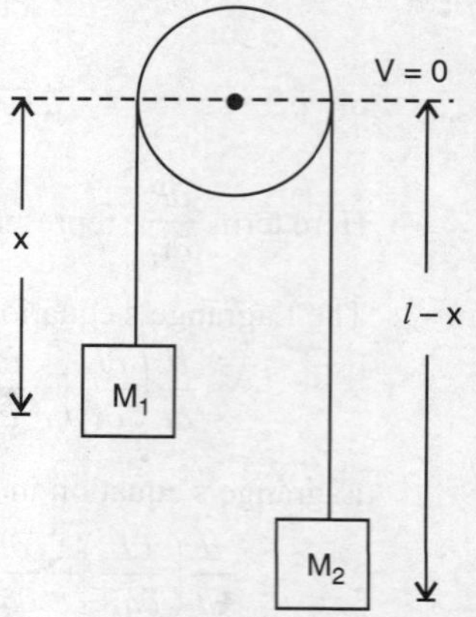

Fig. 23.2

It is found that the heavier particle descends while the lighter ascends, the system moving with a constant acceleration $\ddot{x}$.

The Atwood's machine is a conservative system with a holonomic constraint. There is only one independent coordinate x. The position of the other particle is determined by the constraint that the length of the rope between them is l.

The P.E. of the system = $V = -M_1gx - M_2g(l - x)$

The K.E. of the system = $T = \frac{1}{2}(M_1 + M_2)\dot{x}^2$

Hence, the Lagrangian function is given by

$$L = T - V = \frac{1}{2}(M_1 + M_2)\dot{x}^2 + M_1gx + M_2g(l - x)$$

The Lagrange's equation for a conservative system is

$$\frac{d}{dt}\left[\frac{\partial L}{\partial \dot{q}_i}\right] - \frac{\partial L}{\partial q_i} = 0$$

Since the system has only one degree of freedom, there is only one equation of motion, involving the derivatives.

$\therefore$ The equation of motion of the system is given by

$$\frac{d}{dt}\left(\frac{\partial L}{\partial \dot{x}}\right) - \frac{\partial L}{\partial x} = 0$$

From the expression for L we get, $\frac{\partial L}{\partial x} = (M_1 - M_2)g;$

$$\frac{\partial L}{\partial \dot{x}} = (M_1 + M_2)\dot{x}\,; \frac{d}{dt}\left(\frac{\partial L}{\partial \dot{x}}\right) = \frac{d}{dt}[(M_1 + M_2)\dot{x}] = (M_1 + M_2)\ddot{x}.$$

$\therefore$ We have, $(M_1 + M_2)\ \ddot{x} - (M_1 - M_2)g = 0$

or $$\ddot{x} = \frac{M_1 - M_2}{M_1 + M_2} g.$$

(b) A bead sliding on a Uniformly Rotating wire in a Force Free Space

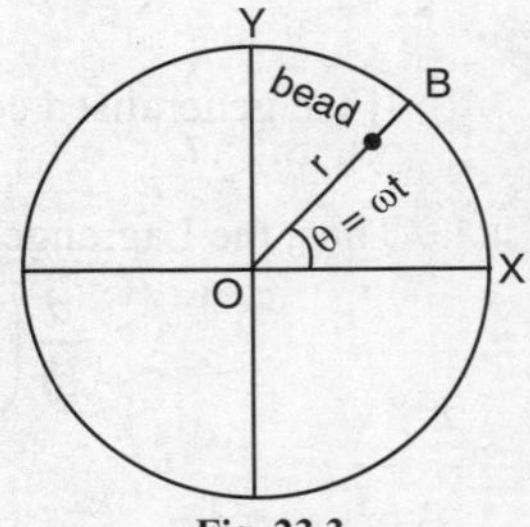

Fig. 23.3

Let OB be a straight frictionless wire fixed at a point O (Fig 23.3). Suppose the wire rotates about a perpendicular axis through O with constant angular velocity ω. Let r be the distance of the bead from point O of the wire at time t. In this example, constraint is time dependent given by the relation $\theta = \omega t$ where ω is the angular velocity of rotation. Then the rectangular coordinates of the bead are given by

$$x = r\cos\theta = r\cos\omega t$$
$$y = r\sin\theta = r\sin\omega t$$

Here it is assumed that at $t = 0$ the wire is along the X-axis.

Now, $$\dot{x} = \dot{r}\cos\theta - r\dot{\theta}\sin\theta$$
$$\dot{y} = \dot{r}\sin\theta + r\dot{\theta}\cos\theta$$

Kinetic energy T is given by

$$T = \frac{1}{2}m\left(\dot{x}^2 + \dot{y}^2\right) = \frac{1}{2}m\,[\dot{r}^2 + (r\,\dot{\theta})^2] = \frac{1}{2}m\,(\dot{r}^2 + r^2\,\omega^2)$$

The wire is rotating uniformly in a *force-free space.* Therefore the Lagrange's equation becomes

$$\frac{d}{dt}\left(\frac{\partial T}{\partial \dot{r}}\right) - \frac{\partial T}{\partial \mathbf{r}} = 0$$

From the expression for T we get, $\frac{\partial T}{\partial \dot{r}} = m\dot{r}$ and $\frac{\partial T}{\partial \mathbf{r}} = mr\omega^2$.

$\therefore$ Equation of motion is $m\,\ddot{r} - mr\omega^2 = 0$

or $$\ddot{r} = r\omega^2$$

This shows that the bead moves outward because of centripetal acceleration.

23.11. GENERALISED MOMENTUM

The generalised momentum conjugate to the generalised co-ordinate q_k is defined as the quantity $\frac{\partial L}{\partial \dot{q}_k}$. It is represented by p_k.

i.e., $$p_k = \frac{\partial L}{\partial \dot{q}_k}$$

23.12. CYCLIC OR IGNORABLE CO-ORDINATES

The Lagrangian L is written as function of q_i and $\dot{q}_i$. If the Lagrangian of a system does not contain a given coordinate, say q_k, then the coordinate is said to be *cyclic.* Since the coordinate q_k is absent in the expression for the Lagrangian L, the partial derivative of L with respect to q_k will vanish,

i.e., $\dfrac{\partial L}{\partial q_k} = 0.$

23.13. LAGRANGIAN FORMULATION OF CONSERVATION THEOREMS

***(a)* Conservation theorem for generalised momentum.**

According to Lagrange's equation,

$$\frac{d}{dt}\left(\frac{\partial L}{\partial \dot{q}_k}\right) - \frac{\partial L}{\partial q_k} = 0$$

If the generalised co-ordinate q_k is cyclic, then $\dfrac{\partial L}{\partial q_k} = 0.$

Thus, the Lagrange equation of motion reduces to

$$\frac{d}{dt}\left(\frac{\partial L}{\partial \dot{q}_k}\right) = 0$$

or $$\frac{dL}{d\dot{q}_k} = \text{constant}$$

i.e., $$p_k = \text{constant}$$

Thus, whenever a coordinate q_k does not appear explicitly in the Lagrangian function L, corresponding linear momentum p_k is a constant of the motion. Hence we can state as a general conservation theorem that *the generalised momentum conjugate to a cyclic coordinate is conserved.*

***(b)* Conservation theorem for energy.**

Consider a conservative system. The Lagrangian L does not depend upon time explicitly.

Then $$L = L\,(q_1, q_2, \ldots q_k, \ldots \dot{q}_1, \dot{q}_2, \ldots. \dot{q}_k, \ldots.)$$

Hence, the total time derivative of $L \equiv L\,(q_k, \dot{q}_k)$ is given by

$$\frac{dL}{dt} = \sum_k \frac{\partial L}{\partial q_k}\dot{q}_k + \sum_k \frac{\partial L}{\partial \dot{q}_k}\ddot{q}_k \qquad \text{...(1)}$$

Lagrange's Eq. is $\dfrac{d}{dt}\left(\dfrac{\partial L}{\partial \dot{q}_k}\right) - \dfrac{\partial L}{\partial q_k} = 0$ or $\dfrac{d}{dt}\left(\dfrac{\partial L}{\partial \dot{q}_k}\right) = \dfrac{\partial L}{\partial q_k}$

We can rewrite Eq. (1) as

$$\frac{dL}{dt} = \sum_k \frac{d}{dt}\left(\frac{\partial L}{\partial \dot{q}_k}\right)\dot{q}_k + \sum_k \frac{\partial L}{\partial \dot{q}_k}\ddot{q}_k = \sum_k \frac{d}{dt}\left(\dot{q}_k\,\frac{\partial L}{\partial \dot{q}_k}\right)$$

or $$\frac{d}{dt}\left(L - \sum \dot{q}_k\,\frac{\partial L}{\partial \dot{q}_k}\right) = 0$$

$\therefore$ $$L - \sum \dot{q}_k\,\frac{\partial L}{\partial \dot{q}_k} = \text{constant.} \qquad \text{...(2)}$$

Now $\frac{\partial L}{\partial \dot{q}_k} = \frac{\partial}{\partial \dot{q}_k}(T - V) = \frac{\partial T}{\partial \dot{q}_k} = \frac{\partial}{\partial \dot{q}_k}\left(\sum \frac{1}{2} m_k \dot{q}_k^2\right)$ $\qquad \left(\because T = \sum \frac{1}{2} m_k \dot{q}_k^2\right)$

$$\therefore \qquad \frac{\partial L}{\partial \dot{q}_k} = m_k \dot{q}_k.$$

Eq. (2) becomes $L - \sum m_k \dot{q}_k{}^2 = \text{constant}$

or $\qquad L - 2T = \text{constant}$ or $T - V - 2T = \text{constant}.$

or $\qquad -(T + V) = \text{constant}.$

$\therefore \qquad T + V = E = \text{total energy} = \text{constant}.$

Thus the energy conservation theorem states that, *if the Lagrangian function does not contain the time explicitly, the total energy of the conservative system is conserved.*

(c) Conservation Theorem for Linear Momentum.

Consider a conservative system so that the potential energy V is dependent on position only and the kinetic energy is independent of position, *i.e.*,

$$\frac{\partial V}{\partial \dot{q}_k} = 0 \text{ and } \frac{\partial T}{\partial q_k} = 0 \qquad \text{...(1)}$$

Then, Lagrange equation of motion for such a coordinate is,

$$\frac{d}{dt}\left(\frac{\partial L}{\partial \dot{q}_k}\right) - \frac{\partial L}{\partial q_k} = 0$$

i.e., $$\frac{d}{dt}\left(\frac{\partial L}{\partial \dot{q}_k}\right) + \frac{\partial V}{\partial q_k} = 0 \qquad \text{[using Eq. (1)]}$$

$$\therefore \qquad \dot{p}_k = \frac{d}{dt}\left(\frac{\partial L}{\partial \dot{q}_k}\right) = -\frac{\partial V}{\partial q_k} = \mathbf{Q}_k \qquad \text{...(2)}$$

Now, if we show that $\mathbf{Q}_k$ represents the component of total force along the direction of translation of the system and p_k is the component of total linear momentum along the same direction, then Eq. (2) will represent the equation of motion for total linear momentum.

Generalised force is given by

$$\mathbf{Q}_k = \sum_i \mathbf{F}_i \cdot \frac{\partial \mathbf{r}_i}{\partial q_k}$$

If $\mathbf{n}$ is the unit vector along the direction of translation, then

$$\delta \mathbf{r}_i = \mathbf{n}\delta q_k$$

or $$\frac{\partial \mathbf{r}_i}{\partial q_k} = \mathbf{n}$$

Thus $$\mathbf{Q}_k = \sum_i \mathbf{F}_i \cdot \mathbf{n} = \mathbf{n} \cdot \mathbf{F} \qquad \text{...(3)}$$

which represents the component of total force along the direction of $\mathbf{n}$.

The kinetic energy, $\quad T = \frac{1}{2}\sum_i m_i \dot{\mathbf{r}}_i^2$

So the generalised momentum is

$$p_k = \frac{\partial L}{\partial \dot{q}_k} = \frac{\partial T}{\partial \dot{q}_k} = \sum m_i \, \dot{\mathbf{r}}_i \, . \frac{\partial \dot{\mathbf{r}}_i}{\partial q_k}$$

$$= \sum m_i \mathbf{v}_i \, . \frac{\partial \mathbf{r}_i}{\partial q_k}$$

$$= \sum_i m_i \mathbf{v}_i \, . \, \mathbf{n} = \mathbf{n} \, . \sum_i m_i \mathbf{v}_i$$

which shows that p_k represents the component of total linear momentum along the direction of translation.

Now we can say that the equation $\dot{p}_k = \mathbf{Q}_k$

is the equation of motion for total linear momentum of the system.

If $\mathbf{Q}_k = 0, \; \dot{p}_k = 0$ *i.e.*, p_k = constant.

This gives the conservation theorem for linear momentum. It states that *if a given component of the total applied force vanishes, the corresponding component of the linear momentum is conserved.*

EXERCISE XXIII

1. State and prove the laws of conservation of linear momentum, angular momentum and energy for a system of interacting particles.
2. What are constraints ? How do they affect motion of a mechanical system? Distinguish clearly between
   (*a*) holonomic and non-holonomic constraints.
   (*b*) Scleronomic and rheonomic constraints.
3. What are Generalised coordinates ? What is the advantage of using them ?
4. What is meant by a configuration space ? How is this concept used to describe the motion of a system of particles ?
5. Explain the term "virtual displacement" and state the principle of "virtual work".
6. State and explain *D'*Alembert's principle.
7. Derive Lagrange's equations of motion from *D'*Alembert's principle for a holonomic conservative system.
8. Describe the use of Rayleigh's dissipation function.
9. State Lagrange's equations of motion in generalised coordinates. Apply them to the Atwood's machine to find the acceleration of the system.
10. Apply Lagrange's equation to determine the motion of a bead sliding on a uniformly rotating wire in a force-free space.
11. Define generalised momentum and a cyclic co-ordinate. Show that generalised momentum is conserved only when the corresponding generalised co-ordinate is cyclic or ignorable.
12. Prove that total energy of the system where Lagrangian does not depend upon time explicitly is conserved.